Two Manuals Combined into One Book



# Medical Evacuation

# And

# Rappelling

Department of Defense

Pentagon
Publishing

Field Manual
No. 4-02.2

Headquarters
Department of the Army
Washington, DC, 8 May 2007

# MEDICAL EVACUATION

## Contents



---

**Distribution Restriction:** Approved for public release; distribution is unlimited.

***This publication supersedes Chapters 1–7 and Appendixes A, B, D–F, K, L, and N of FM 8-10-6 dated 14 April 2000, and FM 8-10-26 dated 16 February 1999.**

# Figures

# Tables

# Preface

This field manual (FM) provides doctrine, as well as techniques and procedures for conducting medical evacuation and medical regulating operations. Medical evacuation encompasses both the evacuation of Soldiers from the point of injury (POI) or wounding to a medical treatment facility (MTF) staffed and equipped to provide essential care in theater and further evacuation from the theater to provide definitive, rehabilitative, and convalescent care in the continental United States (CONUS) and the movement of patients between MTFs or to staging facilities. Medical evacuation entails the provision of en route medical care; supports the joint health service support (JHSS) system; and links the continuum of care. In addition, it discusses the difference between medical evacuation and casualty evacuation (CASEVAC), as well as coordination requirements for and the use of nonmedical transportation assets to accomplish the CASEVAC mission. This publication is intended for use by medical commanders and their staffs, command surgeons, and nonmedical commanders involved in medical evacuation operations.

Users of this publication are encouraged to submit comments and recommendations to improve this publication. Comments should include the page, paragraph, and line(s) of the text where the change is recommended. The proponent for this publication is the United States (US) Army Medical Department (AMEDD) Center and School (USAMEDDC&S). Comments and recommendations should be forwarded, in letter format, directly to **Commander, USAMEDDC&S, ATTN: MCCS-FCD-L, 1400 East Grayson Street, Fort Sam Houston, Texas 78234-5052** or by using the e-mail address: medicaldoctrine@amedd.army.mil.

This publication applies to the Active Army, the Army National Guard (ARNG)/Army National Guard of the United States (ARNGUS), and the U.S. Army Reserve (USAR), unless otherwise stated.

Unless this publication states otherwise, masculine nouns and pronouns do not refer exclusively to men.

Use of trade or brand names in this publication is for illustrative purposes only and does not imply endorsement by the Department of Defense (DOD).

The staffing and organization structure presented in this publication reflects those established in base tables of organization and equipment (TOE) and are current as of the publication print date. However, such staffing is subject to change to comply with manpower requirements criteria outlined in Army Regulation (AR) 71-32. Those requirements criteria are also subject to change if the modification table of organization and equipment (MTOE) is significantly altered.

This publication implements the following North Atlantic Treaty Organization (NATO) International Standardization Agreements (STANAGs) and American, British, Canadian, and Australian (ABCA) Quadripartite Standardization Agreements (QSTAGs):

| TITLE | STANAG | QSTAG |
|---|---|---|
| Stretchers, Bearing Brackets, and Attachment Supports | 2040 | |
| Medical Employment of Air Transport in the Forward Area | 2087 | |
| Medical and Dental Supply Procedures | 2128 | |
| Minimum Labeling Requirements for Medical Materiel | | 436 |
| Documentation Relative to Medical Evacuation, Treatment, and Cause of Death of Patients | 2132 | 470 |
| Road Movements and Movement Control | 2454 | |

| TITLE | STANAG | QSTAG |
|---|---|---|
| Orders for the Camouflage of the Red Cross and Red Crescent on Land in Tactical Operations | 2931 | |
| Aeromedical Evacuation | 3204 | |

When amendment, revision, or cancellation of this publication is proposed which will affect or violate the international agreements concerned, the preparing agency will take appropriate reconciliatory action through international standardization channels. These agreements are available on request from the Standardization Documents Order Desk, 700 Robins Avenue, Building 4, Section D, Philadelphia, Pennsylvania 19111-5094.

The AMEDD is in a transitional phase with terminology. This publication uses the most current terminology; however, other FM 4-02-series and FM 8-series may use the older terminology. Changes in terminology are a result of adopting the terminology currently used in the joint and/or NATO and ABCA Armies publication arenas. Also, the following terms are synonymous and the current terms are listed first:

- Medical logistics (MEDLOG); health service logistics (HSL); and combat health logistics (CHL).
- Roles of care, echelons of care, and level of care.
- Combat and operational stress control (COSC) and combat stress control (CSC).
- Behavioral health (BH) and mental health (MH).
- Chemical, biological, radiological and nuclear (CBRN) and nuclear, biological, and chemical (NBC).
- Stability operations; stability, security, transition, and reconstruction (SSTR) operations; and stability operations and support operations.

# Introduction

The Army Health System (AHS) is a complex system of interrelated and interdependent systems which provides a continuum of medical treatment from the POI or wounding through successive roles of health care to definitive, rehabilitative, and convalescent care in the CONUS, as required. *Medical evacuation* is the system which provides the vital linkage between the roles of care necessary to sustain the patient during transport. This is accomplished by providing en route medical care and emergency medical intervention, if required, and to enhance the individual's prognosis and to reduce long-term disability.

Medical evacuation occurs at the tactical, operational, and strategic levels and requires the synchronization and integration of service component medical evacuation resources and procedures with the DOD worldwide evacuation system operated by the United States Transportation Command (USTRANSCOM).

Army medical evacuation is a multifaceted mission accomplished by a combination of dedicated ground and air evacuation platforms synchronized to provide direct support (DS), general support (GS), and area support within the joint operations area (JOA). At the tactical level, organic or DS medical evacuation resources locate, acquire, treat, and evacuate Soldiers from the POI or wounding to an appropriate MTF where they are stabilized, prioritized, and, if required, prepared for further evacuation to an MTF capable of providing required essential care within the JOA.

Although the most recognized mission of Army medical evacuation assets is the evacuation and provision of en route medical care to battlefield wounded, the essential and vital functions of medical evacuation resources encompass many additional missions and tasks that support the JHSS system. Medical evacuation resources are used to transfer patients between MTFs within the JOA and from MTFs to United States Air Force (USAF) mobile aeromedical staging facilities (MASFs) or aeromedical staging facilities (ASFs); emergency movement of Class VIII, blood and blood products, medical personnel and equipment; and serve as messengers in medical channels.

Medical regulating provides the interface with the DOD worldwide medical evacuation system by determining the patient's destination (the MTF best suited to provide the required care) and scheduling the means to transport the patient with the required en route medical care. Formal medical regulating begins at Role 3, however technological advances in information management (IM)/information technology (IT) are permitting this capability to be used at Role 1 and Role 2 MTFs in some situations.

# Chapter 1

# Overview of Army Health System Operations and Medical Evacuation

The AHS is the Army component of the military health system (MHS). Its capabilities are focused on delivering health care across the continuum of military operations—from the POI or wounding, through the JOA, to the CONUS support base. The two missions of the AHS are to provide health service support (HSS) (casualty care, medical evacuation, and medical logistics) and force health protection (FHP) (casualty prevention). The AHS is focused on promoting wellness, preventing casualties due to disease and nonbattle injuries (DNBIs), and providing timely and effective casualty care and management. Medical evacuation is the key factor in ensuring the continuity of care provided to our Soldiers by providing en route medical care during evacuation, facilitating the transfer of patients between MTFs to receive the appropriate specialty care, and ensuring that scarce medical resources (personnel, equipment, and supplies [to include blood]) can be rapidly transported to areas of critical need on the battlefield.

## SECTION I — ARMY HEALTH SYSTEM

1-1. The provision of AHS is governed by well established and time-tested principles and rules which ensure the care provided to our Soldiers is timely and effective. For an in-depth discussion of these principles, rules, and roles of medical care, refer to FM 4-02.

1-2. The AHS is comprised of 10 medical functions. They are—

- Medical command, control, communications, computers, and intelligence.
- Medical treatment.
- Medical evacuation.
- Hospitalization.
- Dental services.
- Preventive medicine (PVNTMED) services.
- Combat and operational stress control (COSC).
- Veterinary services.
- Medical logistics.
- Medical laboratory support.

1-3. Army health system resources are arrayed across the battlefield in successive levels of support. These successive levels have increased medical care capabilities at each higher level. Medical evacuation and the provision of en route medical care ensures an uninterrupted continuum of care is maintained while Soldiers are moved through the roles of medical care to the MTF best suited to treat the patient's specific injuries.

## PRINCIPLES OF ARMY HEALTH SYSTEM

1-4. The principles of AHS provide a framework in which medical planners can ensure that comprehensive plans are developed to support the tactical commander's operation plan (OPLAN). The principles of AHS are conformity, continuity, control, proximity, flexibility, and mobility. For this publication, the discussion of these principles has been focused toward the medical evacuation mission. For a general discussion of how these principles relate to the overall AHS mission, refer to FM 4-02.

## CONFORMITY

1-5. Participating in the development of the OPLAN or the operation order (OPORD) ensures that the medical planner conforms to strategic, operational, and tactical plans. This is the most fundamental element for effectively providing AHS and ensures medical influence over the execution of medical evacuation operations. Only by participating in the orders process and developing a medical evacuation plan, will the medical planner ensure that medical evacuation support is arrayed on the battlefield in the right place at the right time and synchronized across operational commands to maximize responsiveness and effectiveness.

### CONTINUITY

1-6. En route medical care provided during medical evacuation must be effective and continuous to prevent interruptions in the continuum of care. An interruption in medical treatment may result in an increase in morbidity, mortality, and disability. No patient is evacuated any farther than his physical condition or the military situation requires.

1-7. Medical evacuation resources provide the linkage between the roles of care within the JOA. They also provide interface with other deployed elements of the MHS operated by other services to enhance and facilitate the continuum of care from the POI to the CONUS support base.

### CONTROL

1-8. Medical planners must ensure medical control is exercised over the execution of ground medical evacuation operations and that medical influence (technical and operational supervision) is exercised over the execution of aeromedical evacuation (AE) operations. Furthermore, medical planners must ensure the medical evacuation system is responsive to changing requirements and tailored to effectively support the forces within an assigned area of operations (AO). Since medical evacuation resources are limited, it is essential that medical control and influence be retained at the highest level consistent with the tactical situation.

1-9. A thorough and comprehensive medical evacuation plan is essential to establishing and maintaining control of medical evacuation operations characterized by decentralized execution of the plan. The medical evacuation plan complies with the combatant command guidance and intent and maximizes the use of scarce medical evacuation resources. When directed by the combatant command, Army air and ground ambulances may support operations conducted by other services, allied and coalition partners, and the host nation (HN).

### PROXIMITY

1-10. The location of medical evacuation assets in support of combat operations is dictated by orders and the tactical situation (mission, enemy, terrain and weather, troops and support available, time available, and civil considerations [METT-TC]). Accurately determining time and distance factors and the availability of evacuation resources are critical to determining the disposition of evacuation assets. The speed with which medical evacuation is initiated is extremely important in reducing morbidity, mortality and disability. Medical evacuation time must be minimized by the effective disposition of resources, ensuring close proximity of both supported elements and MTF. Medical evacuation assets cannot be located so far forward that they interfere with the conduct of combat operations. Conversely, they must not be located so far to the rear that medical treatment is delayed due to lengthy evacuation routes.

1-11. Medical evacuation resources, both ground and air, are arrayed on the battlefield to best support both the tactical commander and the AHS. Depending upon the situation, evacuation resources may be placed in a DS role to support maneuvering forces or GS, during stability operations which are centrally located to accomplish an area support mission.

### FLEXIBILITY

1-12. Changes in tactical plans or operations may require redistribution or reallocation of medical evacuation resources. Therefore, the medical evacuation plan must be designed to ensure flexibility and agility as well as enhance the ability to rapidly task-organize and relocate medical evacuation assets to meet changing battlefield requirements. Medical planners must also ensure medical control and influence is exercised through the orders process and facilitates the synchronization of air and ground evacuation assets to rapidly clear the battlefield.

### MOBILITY

1-13. Medical evacuation assets must have the same mobility and survivability capability (such as armor protection) as the forces supported. This mobility and survivability ensures that medical evacuation resources can rapidly respond and that evacuation routes do not become too lengthy. Medical evacuation assets also enhance the mobility of forward deployed MTFs by rapidly evacuating their patients to the next role of medical care.

## BATTLEFIELD RULES

1-14. The AMEDD has developed a set of battlefield rules to aid in establishing priorities and to resolve competing priorities within AHS activities. These rules are intended to guide the medical planner to resolve system conflicts encountered in designing and coordinating AHS operations. Although medical personnel seek always to provide the full scope of AHS services and support in the best possible manner, during every combat operation there are inherent possibilities of conflicting support requirements. The planner or operator applies these rules to ensure that the conflicts are resolved appropriately. These battlefield rules are depicted in Table 1-1.

**Table 1-1. Army Medical Department Battlefield Rules**

| **AMEDD BATTLEFIELD RULES** |
|---|
| BE THERE (MAINTAIN A MEDICAL PRESENCE WITH THE SOLDIER) |
| MAINTAIN THE HEALTH OF THE COMMAND |
| SAVE LIVES |
| CLEAR THE BATTLEFIELD OF CASUALTIES |
| PROVIDE STATE-OF-THE-ART MEDICAL CARE |
| ENSURE EARLY RETURN TO DUTY OF THE SOLDIER |

### BE THERE

1-15. Ensure that medical evacuation assets are in close proximity to supported elements to enhance response time, increase Soldier confidence and be a combat multiplier. This is accomplished by complementing organic medical evacuation assets with medical evacuation assets placed in DS, GS, and area support roles.

### Maintain the Health of the Command

1-16. Ensure that medical evacuation assets are in close proximity to the MTF. The closer evacuation assets are to an MTF, the more rapidly care may be rendered.

### Save Lives

1-17. Ensure medical evacuation assets are the primary means of evacuating patients. Medical evacuation assets provide en route medical care that provides a continuum of care that is instrumental in preserving life and reducing long-term disability. The use of CASEVAC should only be used in extreme emergencies or when the medical evacuation system is overwhelmed. Refer to paragraphs 1-27 through 1-33 for a discussion on the differences between medical evacuation and CASEVAC operations.

### Provide State-of-the Art Medical Care

1-18. Ensure sufficient medical evacuation assets are available to facilitate the movement of patients between MTFs (intratheater evacuation) and the emergency movement of Class VIII, blood and blood products, and medical personnel and equipment. This capability permits the medical planner to maximize the use of scarce medical resources and exercise economy of force without sacrificing state-of-the-art medical care.

### Early Return To Duty of The Soldier

1-19. Ensure seamless integration with the intertheater medical evacuation system. This allows rapid evacuation and continuous medical treatment throughout the continuum of care resulting rapid return to duty (RTD) rates.

# SECTION II — MEDICAL EVACUATION

## PURPOSE

1-20. An efficient and effective medical evacuation system—

- Minimizes mortality by rapidly and efficiently moving the sick, injured, and wounded to an MTF.
- Serves as a force multiplier as it clears the battlefield enabling the tactical commander to continue his mission with all available combat assets.
- Builds the morale of Soldiers by demonstrating that care is quickly available if they are wounded.
- Provides en route medical care that is essential in improving the prognosis and reducing disability of the wounded, injured, or ill Soldiers.
- Provides medical economy of force.
- Provides connectivity of the AHS as appropriate to the MHS.

| Tactical | Operational | Strategic |
|---|---|---|
| Corps<br>Division<br>BCT<br>Emergency Class VIII Resupply<br>Emergency Movement of Medical Personnel/Equipment<br>Stability Operations | Coalition<br>Joint<br>Class VIII Resupply<br>Joint Blood Program<br>DOD Civilians/Contractors<br>Medical Personnel/Equipment<br>Military Working Dogs<br>Homeland Security Operations<br>Shore-to-Ship<br>Stability Operations | Host Nation<br>Nongovernment Organizations<br>Private Volunteer Organizations<br>DOD support to Stability Operations<br>International Disaster Relief |

**Figure 1-1. Army medical evacuation as it supports full spectrum operations**

1-21. The AHS is established in roles of increasing capability from the POI to definitive care. On the battlefield, casualties are evacuated rearward from one role of care to the next, the sequencing of this movement is dependent upon METT-TC factors. In a contiguous battlefield, well established lines of communication (LOC), large numbers of casualties, and wide array of MTFs can result in very deliberate evacuation from one sequential role to the next higher. However in many situations, such as noncontiguous battlefield, the array of medical resources across the battlefield, the availability of medical evacuation resources, and the number of patients being evacuated may facilitate procedures that permit bypassing roles of care in order to ensure the timely treatment and care of casualties. This evacuation plan will be established by the appropriate level of command in coordination with the command surgeon to ensure the best treatment is provided to all casualties.

# ATTRIBUTES

## DEDICATED RESOURCES

1-22. The Army medical evacuation system is comprised of dedicated air and ground evacuation platforms. These platforms have been designed, staffed, and equipped to provide en route medical care to patients being evacuated and are used exclusively to support the medical mission. The focus of the medical evacuation mission coupled with the dedicated platforms permit a rapid response to calls for support. The dedicated nature of this mission dictates that Army medical evacuation unit's posture themselves in a "ready alert" status, ready to rapidly respond to evacuation missions and not diverted to perform any other task. Medical evacuation resources are protected under the provisions of the Geneva Conventions from intentional attack by the enemy. (For a discussion of the Geneva Conventions in relation to air and ground evacuation operations, refer to Appendix A.)

## EN ROUTE MEDICAL CARE

1-23. En route care is provided on all Army medical evacuation platforms. This care is essential for minimizing mortality, enhancing survival rates, and reducing disability of wounded, injured, or ill Soldiers. Refer to paragraphs 1-27 through 1-33 for a discussion on the differences between medical evacuation and CASEVAC operations.

## CONNECTIVITY

1-24. The Army provides connectivity to the MHS. The synchronized employment of medical evacuation resources provides and maintains the seamless continuum of care from the POI through successive roles of essential care within the theater. In addition to evacuating patients and providing en route medical care, medical evacuation resources provide for the emergency movement of scarce medical resources such as critical Class VIII, blood, medical personnel, and medical equipment. Further, medical evacuation resources are used to transfer patients from one MTF to another within the theater, to facilitate specialty care as well as transferring patients from an MTF to a MASF to facilitate intertheater evacuation. Figure 1-1 depicts the connectivity provided by medical evacuation platforms between the roles of care within theater. In addition to providing connectivity within the AHS, Army medical evacuation resources provide medical evacuation support and interface with MTFs of the other services deployed in the theater. In joint and multinational operations, the geographic combatant commander (GCC) may direct Army assets to provide this support and connectivity to joint, allied, and coalition forces within the JOA.

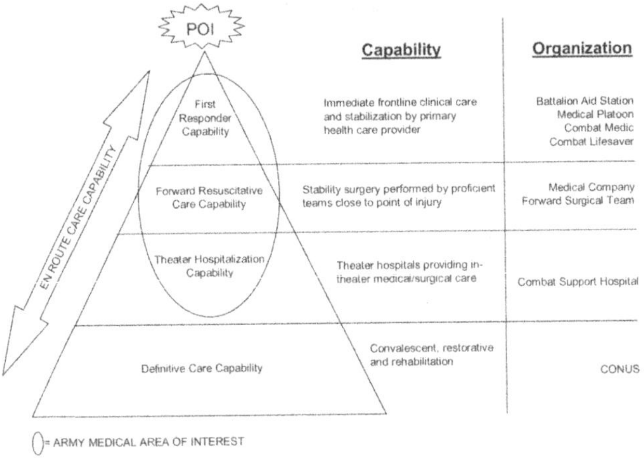


**Figure 1-2. Army air and ground evacuation platforms provide connectivity to assure a seamless continuum of medical care**

## Medical Economy of Force

1-25. The flexibility and versatility which medical evacuation resources afford the AHS to respond to urgent requirements throughout the battlefield is essential in ensuring the provision of seamless medical care. Medical economy of force is achieved for low density, high demand medical specialties (such as a neurosurgeon), medical supplies, and medical equipment (such as computer tomography scans or magnetic resonance imaging equipment) by having the capability to move the patient to the required care over long distances and by permitting the emergency of cross-leveling on medical supplies reducing the need for large Class VIII stockpiles. By providing medical economy of force, the deployed medical footprint is reduced without negatively impacting on the care provided to the Soldier.

## Force Multiplier

1-26. A highly effective, efficient, and responsive AHS increases Soldiers' confidence that if wounded in battle, they will receive timely and appropriate medical care to enhance their prognosis, speed their recovery, and reduce long-term disabilities. Medical evacuation resources locate and acquire wounded Soldiers and facilitate their entry into the AHS. These resources clear the battlefield of casualties which enables the tactical commander the ability to capitalize on and exploit battlefield opportunities with all available assets in the execution of his warfighting mission.

## SECTION III — MEDICAL EVACUATION VERSUS CASUALTY EVACUATION

# MEDICAL EVACUATION

1-27. Medical evacuation is performed by dedicated, standardized medical evacuation platforms, with medical professionals who provide the timely, efficient movement and en route care of the wounded, injured, or ill persons from the battlefield and/or other locations to MTFs. Medical evacuation is an AMEDD function that supports and is an integral part of the AHS. The provision of en route care on medically equipped vehicles or aircraft greatly enhances the patient's potential for recovery and may reduce long-term disability by maintaining the patient's medical condition in a more stable manner.

1-28. Medical evacuation ground/air ambulance platforms are defined as: Platforms especially for the medical evacuation mission with allocated medical equipment specifically designed for the purpose of enroute care and by trained medical personnel.

1-29. The gaining MTF in coordination with the losing MTF, is responsible for arranging for the evacuation of patients from the lower role of care. For example, Role 2 medical units are responsible for evacuating patients from Role 1 MTFs.

1-30. Medical evacuation begins when a request is initiated and continues until the patient is released from the AHS by medical authorities.

# CASUALTY EVACUATION

1-31. Casualty evacuation is a term used to refer to the movement of casualties aboard nonmedical vehicles or aircraft.

**WARNING**

Casualties transported in this manner may not receive proper en route medical care or be transported to the appropriate MTF to address the patient's medical condition. If the casualty's medical condition deteriorates during transport, or the casualty is not transported to the appropriate MTF, an adverse impact on his prognosis and long-term disability or death may result.

1-32. If dedicated medical evacuation platforms (ground and air) are available, casualties should be evacuated on these conveyances to ensure they receive proper en route medical care.

1-33. Since CASEVAC operations can reduce combat power and degrade the efficiency of the AHS, units should only use CASEVAC to move Soldiers with less severe injuries when medical evacuation assets are overwhelmed. Medical planners should ensure CASEVAC operations are addressed in the OPLAN/OPORD as a separate operation, as these operations require preplanning, coordination, synchronization, and rehearsals. The CASEVAC plan should ensure casualties with severe or life-threatening injuries are prioritized for evacuation and are evacuated on dedicated medical evacuation platforms.

1-34. When possible, nonmedical vehicles/aircraft transporting casualties should be augmented with a trauma specialist or combat lifesaver (CLS). (On nonmedical aircraft, sufficient space may not be available to permit a caregiver to accompany the casualties.) The type of en route monitoring and medical care/first aid provided is limited by the following factors:

- Skill level of the individual providing care. (The combat medic is military occupational specialty [MOS]-qualified [MOS 68W] to provide emergency medical treatment [EMT]; the CLS is trained to provide enhanced first aid.) The combat medic can provide emergency medical intervention, whereas the CLS can only monitor the casualty and ensure that the basic lifesaving first-aid tasks are accomplished.
- Medical equipment available.
- Number of casualties being transported.
- Accessibility of casualties—If nonstandard evacuation vehicle is loaded with the maximum number of casualties, the combat medic or CLS may not be able to attend to the casualties while the vehicle is moving. If the condition of a casualty deteriorates and emergency measures are required, the vehicle will have to be stopped to permit care to be given.

# SECTION IV — THEATER EVACUATION POLICY

## ESTABLISHING THE THEATER EVACUATION POLICY

1-35. The theater evacuation policy is established by the Secretary of Defense, with the advice of the Joint Chiefs of Staff and upon the recommendation of the theater commander. The policy establishes, in number of days, the maximum period of noneffectiveness (hospitalization and convalescence) that patients may be held within the theater for treatment. This policy does not mean that a patient is held in the theater for the entire period of noneffectiveness. A patient who is not expected to be ready for RTD within the number of days established in the theater evacuation policy is evacuated to the CONUS or other safe haven. This is done providing that the treating physician determines that such evacuation will not aggravate the patient's disabilities or medical condition. For example, a theater evacuation policy of seven days does not mean that a patient is held in the theater for six days and then evacuated. Instead, it means that a patient is evacuated as soon as possible after the determination is made that the Soldier cannot be returned to duty within seven days following admission to a Role 3 hospital.

### TEMPORARY REDUCTIONS

1-36. To the degree that unplanned for increases in patients occur (due perhaps to an epidemic or heavy combat casualties), a temporary reduction in the policy may be necessary. This reduction is used to adjust the volume of patients being held in the theater hospital system. A reduction in the evacuation policy increases the number of patients requiring evacuation out-of-theater and increases the requirement for evacuation assets. This action is necessary to relieve the congestion caused by the patient increases. A decrease in the theater evacuation policy increases the evacuation asset requirements.

### ESSENTIAL CARE IN THEATER AND LENGTH OF POLICY

1-37. Due to the reduction of the medical footprint within the theater, health care delivery to deployed forces is now provided under the concept of essential care in theater. This concept provides for essential and stabilizing care being rendered within the theater, with the patient being evacuated to CONUS or other safe haven for definitive, rehabilitative, and convalescent care. In current and ongoing operations, if the patient cannot be treated and returned to duty within seven days of admission to a combat support hospital (CSH), the patient is evacuated from the theater for continued care.

1-38. The time period established by the theater evacuation policy starts on the date the patient is admitted to the first hospital. The total time a patient is hospitalized in the theater (including transit time between MTFs) for a single, uninterrupted episode of illness or injury should not exceed the number of days stated in the theater evacuation policy. Though guided by the evacuation policy, the actual selection of a patient for evacuation is based on clinical judgment as to the patient's ability to tolerate and survive the movement to the next role of care.

### EXCEPTION TO POLICY

1-39. An exception to the theater evacuation policy may be required with respect to SOF personnel. This exception may be required to retain low density MOS skills within the theater. Retaining these personnel within the theater for an extended period of time is possible if the medical resources are available within the theater to treat their injuries and provide for convalescence and rehabilitation. If retention within the theater would result in a deterioration of their medical condition or would adversely impact on their prognosis for full recovery, they are evacuated from the theater for definitive care.

### EVACUATION POLICY CONSIDERATIONS

1-40. Paragraphs 1-41 through 1-45 discuss the various considerations of the theater evacuation policy.

#### Physicians and Dentists

1-41. To physicians and dentists engaged in direct patient treatment and decisions relating to patient disposition, it means that there is a maximum period which clinical staffs may complete the necessary treatment needed to return the patient to full duty within the theater. If the theater policy is seven days and full RTD can be predicted within that time, the patient is retained in the theater. If the patient cannot be returned to full duty within seven days, the patient is evacuated out-of-theater as early as clinically prudent.

#### Medical Planners and Staff

1-42. The medical staff computes mix, number, and distribution of hospital beds required in the theater.

1-43. The medical operator has a management tool which, when properly adjusted and used, provides the balance between patient care and tactical support requirements. The medical staff is able to tailor a medical package specifically designed to handle the patient workloads, with maximum benefit to the patients and with maximum economy of available resources.

#### Logisticians

1-44. The nonmedical logistician is able to estimate his total obligation to support this system.

#### United States Air Force Planner

1-45. The USAF planner can accurately plan the USAF AE requirements for both intra- and intertheater patient movements.

## FACTORS DETERMINING THE THEATER EVACUATION POLICY

1-46. To fully understand how the theater evacuation policy affects AHS operations, the medical operator should be aware of the factors that influence the establishment of this policy.

### NATURE OF THE OPERATION

1-47. A major factor in determining the theater evacuation policy is the nature of the tactical operations. For example: Will the operations be of short duration and with a low potential for conflict? Will there be operations of long duration with significant combat operations? Will chemical, biological, radiological, nuclear, and high yield explosives (CBRNE) be employed? Will only conventional weapons be used? Is a static combat situation expected? Is there a significant threat of terrorist activities? Are the majority of patients anticipated to be DNBI patients or those with combat-related trauma?

### NUMBER AND TYPE OF PATIENTS

1-48. Another factor affecting the policy is the number and types of patients anticipated and the rate of patient RTD. Admission rates vary widely in different geographical areas of the world and in different types of military operations.

### EVACUATION MEANS

1-49. The means (quantity and type of transportation) available for evacuation of patients from the theater to CONUS is an essential factor impacting on the evacuation policy.

### AVAILABILITY OF REPLACEMENTS

1-50. The capability of CONUS to furnish replacements to the theater is another consideration. For each patient who is evacuated from the theater to CONUS, a fully trained and equipped replacement must be provided. During a small-scale conflict overseas, the CONUS replacement capability is much greater than when compared to a large-scale conflict such as World War II.

### AVAILABILITY OF IN-THEATER RESOURCES

1-51. Limitations of all AHS resources (such as insufficient number and types of medical units in the division/corps to support the brigade combat teams [BCTs] and an insufficient amount of medical and nonmedical logistics) will have an impact on the theater evacuation policy. The more limitations (or shortages), the shorter the theater evacuation policy.

## IMPACT OF THE EVACUATION POLICY ON ARMY HEALTH SYSTEM REQUIREMENTS

### SHORTER EVACUATION POLICY

1-52. A shorter theater evacuation policy—

- Results in fewer hospital beds required in the theater and a greater number of beds required elsewhere.
- Creates a greater demand for intertheater USAF evacuation resources. (A shortened intratheater evacuation policy would likewise increase the number of airframes required in the theater.)
- Increases the requirements for replacements to meet the rapid personnel turnover which could be expected, especially in combat units. (The impact this would have on both intra- and intertheater transportation and other requirements must also be considered.)

### LONGER EVACUATION POLICY

1-53. A longer theater evacuation policy—

- Results in a greater accumulation of patients and a demand for a larger AHS infrastructure in the theater. It decreases bed requirements elsewhere.
- Increases the requirements for medical logistics (medical supplies, equipment, and equipment maintenance) and nonmedical logistics support.
- Increases the requirements for hospitals, engineer support, and all aspects of base development for deployed AHS force. (It demands the establishment of a larger number of hospitals within the theater and may require medical specialty augmentation.)
- Provides for a greater proportion of patients to RTD within the theater and, thus, reduces the loss of experienced manpower.
- A longer intratheater evacuation policy may decrease the demand on the intratheater evacuation assets and system.

1-54. The concept of essential care in theater does not support longer evacuation policies as the deployed hospitals are not designed to provide definitive, rehabilitative, and convalescent care/services. If the theater evacuation policy is extended in theaters operating under the essential care in theater concept, augmentation of medical specialty resources will be required.

## PATIENT STABILIZATION

1-55. The evacuation policy has no impact on the patient stabilization period for movement. This period is known as the *evacuation delay*. It is the period of time planned for between the time of patient reporting and the time of medical evacuation of the patient to the next role of care. Evacuation delays normally range from 24 to 72 hours and are designated by the Army service component command (ASCC) surgeon.

# Chapter 2
# Medical Evacuation Resources

This chapter discusses the mission, functions, and capabilities of medical evacuation units and elements, as specified in the unit's TOE. It also discusses the command and control (C2) headquarters to which they are assigned. The discussion of each organization also includes the basis of allocation (BOA), limitations, and dependencies.

## SECTION I — MANEUVER BATTALION MEDICAL PLATOON AMBULANCES

## MEDICAL PLATOON AMBULANCE SQUADS

2-1. The ground ambulance team is the basic medical evacuation element used within the BCTs and at the corps and theater AOs. These ground ambulance teams provide medical evacuation from the POI to supporting MTFs while ensuring the continuity of care en route. Ambulance squads consisting of two ambulance teams organic to the medical platoons of movement and maneuver and fires units and to the evacuation platoon of the brigade support medical companies (BSMCs).

2-2. The primary mission of the maneuver medical platoon ambulance squads is to provide ground ambulance evacuation support from supported infantry/armor companies or from POI back to a casualty collection point (CCP) or to the Role 1 MTF/battalion aid station (BAS). Maneuver medical platoon ambulance teams are assigned to heavy BCT (HBCT), infantry BCT, Stryker BCT (SBCT), reconnaissance squadrons, and reconnaissance and target acquisition (RSTA) squadrons. They also provide area support to other elements (which do not have organic medical evacuation resources) operating in their AO. Ground ambulance support provided is consistent with evacuation precedence, tactical standing operating procedures (TSOPs), and other operational considerations.

2-3. For definitive information on maneuver battalion medical platoon operations, see FM 4-02.4, FM 4-02.6, and FM 4-02.21.

## MANEUVER UNIT MEDICAL PLATOON

### ORGANIZATION

2-4. The medical platoon consists of a platoon headquarters section, a treatment squad, combat medic section, and an ambulance squad. The number of ambulance squads in an ambulance platoon varies and is based on the type of parent organization. The infantry, airborne, and air assault battalions' ambulance platoons have two ambulance squads equipped with high mobility multipurpose wheeled vehicle (HMMWV) ambulances. The mechanized infantry and armor combat maneuver battalions' ambulance platoons have four ambulance squads equipped with M-113 track ambulances. The SBCT infantry battalion's and RSTA squadron's ambulance platoons have two ambulance squads equipped with Stryker wheeled armored ambulances (referred to as the Stryker medical evacuation vehicle [MEV]).

#### Medical Platoon Headquarters

2-5. The medical platoon leader is a physician and also serves as the battalion surgeon. He is assisted by the medical operations officer (field medical assistant) in the operational, administrative, and logistical support aspects of the platoon.

### Ambulance Section

2-6. The ambulance section of the medical platoon is organized into ambulance squads and is supervised by the platoon sergeant. Each squad contains a noncommissioned officer (NCO) squad leader, three combat medics, and two ambulances.

### Ambulance Squad Leader

2-7. The ambulance squad leader ensures that ambulance team members perform operator maintenance on assigned vehicles. An ambulance team consists of one vehicle and two medical personnel (an aide/evacuation NCO and a combat medic).

### Ambulance Team

2-8. Ambulance team members are also responsible for performing operator maintenance on assigned ambulances. Specific duties of the ambulance team are to—

- Administer EMT as required.
- Initiate the Department of Defense (DD) Form 1380, US Field Medical Card (FMC), as required.
- Prepare patients for ground and air medical evacuation.
- Provide medical evacuation of wounded or injured Soldiers from the POI to supporting MTF or ambulance exchange point (AXP).
- Provide patient care en route.
- Provide emergency movement of medical personnel and emergency delivery of blood, medical supplies, and medical equipment.
- Assist in the care of COSR casualties.
- Maintain operational readiness of the ambulance.
- Resupply company and platoon trauma specialists, when required.
- Operate the vehicle and maintain contact with supported and supporting elements.
- Find and collect the wounded.
- Serve as messengers within medical channels.

### PRIMARY FUNCTIONS OF AMBULANCE TEAMS

2-9. The primary function of the ambulance team is to collect and treat the sick, injured, and wounded Soldiers on the battlefield and to provide medical evacuation support from the POI, CCP, or AXP to the supporting MTF.

## SECTION II — AMBULANCE/EVACUATION PLATOON—FORWARD SUPPORT, BRIGADE SUPPORT, OR AREA SUPPORT MEDICAL COMPANY

## MEDICAL COMPANIES

2-10. Medical companies provide Role 2 MTF and medical evacuation. Medical companies assigned to or in DS of a BCT are referred to as either a forward support medical company (FSMC) or a BSMC. Area support medical companies (ASMCs) are assigned to multifunctional medical battalions and operate Role 2 MTF and provide medical evacuation support for division, and theater units. In addition, all medical companies provide Role 1 area medical support to units without organic medical personnel and may augment/reinforce maneuver medical platoons/sections with treatment and medical evacuation support.

## Forward Support Medical Company

2-11. The FSMC, an Army of Excellence (AOE) designed unit is organized to provide triage and management of mass casualties, advanced trauma management (ATM), initial resuscitation and stabilization, care for patients with DNBI, and battle wounded and injured Soldiers. The FSMC also provides intervention for combat and operational stress disorders to include COSR and preparation of patients for further medical evacuation. The FSMC is organized into a company headquarters, a treatment platoon, an ambulance platoon, a PVNTMED section, and a BH section.

2-12. The ambulance platoon of FSMC-Light and Airborne units employ a platoon headquarters with four-wheeled ambulance squads (eight ambulances); the FSMC-Air Assault employs a platoon headquarters with three-wheeled ambulance squads (six ambulances); the FSMC-Heavy employs a platoon headquarters with three-tracked ambulance squads (six ambulances) and two-wheeled ambulance squads (four ambulances); and the FSMC-Airborne employs a platoon headquarters with four-wheeled ambulance squads (eight ambulances). A complete discussion on the organization, mission, and functions of the FSMC is provided in FM 4-02.6.

## Brigade Support Medical Company

2-13. The new BSMC is designed to support the modular BCT. The BOA for the BSMC is one per maneuver brigade supported. The mission of the BSMC is to provide HSS to all BCT units operating within the brigade AO. The BSMC locates and establishes its company headquarters and also operates a Role 2 MTF and may operate Role 1 MTFs on an area support basis for units that do not have organic medical assets. The company provides C2 for its organic and attached/operational control (OPCON) medical augmentation elements. The BSMC may be augmented with a forward surgical team (FST) providing the company surgical capability, see FM 4-02.25. The BSMC is organized into a company headquarters, a PVNTMED section, a BH section, a treatment platoon, and an evacuation platoon.

2-14. The evacuation platoon performs ground medical evacuation and en route patient care for the supported units. The evacuation platoon consists of a platoon headquarters, a GS evacuation section, and a DS evacuation section. The platoon employs five HMMWV evacuation squads (or ten evacuation teams). A complete discussion on the organization, mission, and functions of the BSMC is provided in FM 4-02.6 and Field Manual Interim (FMI) 4-90.1.

## Area Support Medical Company

2-15. The ASMC performs functions similar to those of the BCT medical companies. The ASMCs are employed primarily in support of division, corps, and echelons above corps (EAC) units. They are deployed to a geographical area to provide area HSS, or may be deployed to provide HSS for designated units. The ASMC also establishes its Role 2 MTF in a secure location centrally located for supported units. Medical treatment squads/teams of the ASMC may be deployed to establish Role 1 MTF and provide HSS support to concentrations of nondivisional units that do not have organic medical capabilities. The ASMC is organized into a company headquarters, a treatment and ambulance platoon, and a BH section.

2-16. The ambulance platoon performs ground medical evacuation and en route patient care for supported units. The ambulance platoon consists of a platoon headquarters, four ambulance squads (or eight ambulance teams), one HMMWV control vehicle, and eight HMMWV ambulances. A complete discussion on the organization, mission, and functions of the ASMC is provided in FM 4-02.24.

## Organizations

2-17. The BSMC/FSMC normally establishes its Role 2 MTF in the BSA, while the ASMC normally establishes its Role 2 MTF in echelons above brigade (EAB) and echelons above division sustainment areas and at theater level. The medical company provides DS/GS ground ambulance evacuation for supported units. It coordinates/ requests air ambulance support for air evacuation of patients from supported Role 1 MTFs/BASs, POI, CCP, and AXPs. The organizational structure, number and type of ambulances assigned to an ambulance/evacuation platoon may differ depending on their location on the battlefield and the types of units supported. For example, a BSMC in support of a HBCT has a mix of armored and wheeled evacuation vehicles while a BSMC in support of the SBCT has only wheeled evacuation vehicles.

2-18. The company headquarters section is organized into a command element; a support element; a unit supply element; a medical supply and medical maintenance element; and an operations and communications element.

## Treatment Platoon

2-19. The treatment platoon is composed of a platoon headquarters, treatment squads, and an area support section, which consists of an area support squad, an area support treatment squad, and an area patient holding squad.

2-20. The platoon headquarters is the C2 element of the platoon. The platoon headquarters:

- Determines and directs the disposition of patients and submits requests through the company command post (CP) for their evacuation of patients to supporting hospitals.
- Directs, coordinates, and supervises platoon operations.
- Directs the activities of the Role 2 MTF, monitors Class VIII supplies, blood usage and inventory levels, and keeps the commander informed.
- Manages platoon operations, operations security (OPSEC), communications, medical administration, organizational training, supply transportation, patient accountability, statistical reporting functions, and blood situation reporting.

2-21. The treatment squad element can contain up to four treatment squads, depending on the type company assigned. The treatment squads have the capabilities to:

- Provide emergency and routine sick call treatment services to Soldiers assigned to supported units.
- Operate as separate treatment teams (Teams A and B) for limited periods of time.
- Operate up to four treatment stations.
- Reinforce or reconstitute similar treatment squads and can operate for up to 48 hours while separated from their parent unit.

2-22. For medical companies deployed in the BSA, these squads can further be used to—

- Provide augmentation to maneuver battalion medical platoons.
- Facilitate the movement of the treatment platoon MTF from one location to another or echelons forward to establish an MTF at a new location. A treatment squad can be echeloned forward to establish an MTF at a new location. The echeloning of elements allows the old treatment site to remain operational until the new site is established.
- Augment an AXP, CCP, or positioned at designated points in the ambulance shuttle system, when required.

2-23. The area support section of the treatment platoon is composed of an area treatment squad, an area support squad, and a patient-holding squad. These squads form the medical company Role 2 MTF. The area support treatment squad provides trauma care and routine sick call services to personnel assigned to units located in the supported area of the medical company. The area support squad provides operational dental care that includes emergency and essential dental services, limited laboratory and radiological services, and blood support commensurate with Role 2 MTF. The patient-holding squad provides up to 40 patient beds (40 patient beds for heavy divisions and 20 patient beds for light divisions) for patients requiring minimal treatment. Patients held in the patient-holding area are those patients who are expected to be RTD within 72 hours from the time they are held for treatment.

2-24. The area treatment squad is identical in personnel and equipment as the treatment squads of the treatment section. It is the base medical treatment element of a Role 2 MTF. It provides sick call services and initial resuscitative treatment for supported units. For communications, the squad employs frequency modulated (FM) radios and is deployed in the medical radio and wires communications nets.

## Ambulance/Evacuation Platoon

2-25. The ambulance/evacuation platoon provides C2 for ambulance platoon operations. The ambulance platoon headquarters element maintains communications to direct ground ambulance evacuation of patients. It provides ground ambulance evacuation support for supported maneuver battalions and for supported units operating in the sustainment areas. The ambulance headquarters element performs route reconnaissance and develops and issues graphic overlays to all its ambulance teams. It also coordinates and establishes AXPs for both air and ground ambulances, as required.

2-26. Ambulance squads provide ground ambulance evacuation of patients from supported BASs/unit aid stations back to the Role 2 MTF that is located in the brigade support area (BSA). An ambulance squad consists of two ambulance teams (two ambulances, wheel or tracked vehicles). Ambulance squad personnel—

- Perform EMT, evacuate patients, and provide for their continued care en route.
- Operate and maintain assigned communication and navigational equipment.
- Perform preventive maintenance checks and services (PMCS) on ambulances and associated equipment.
- Maintain supply levels for the ambulance medical equipment sets (MES).
- Ensure that appropriate property exchange of medical items (such as litters and blankets) is made at sending and receiving MTF.
- Prepare patients for ground and air medical evacuation.
- Provide medical evacuation of wounded or injured Soldiers from the POI to supporting MTF, CCP or AXP.
- Initiate the DD Form 1380, FMC, as required.
- Provide emergency movement of medical personnel and emergency delivery of blood, medical supplies, and medical equipment.
- Assist in the care of COSR casualties.
- Maintain operational readiness of the ambulance.
- Resupply company and platoon trauma specialists, when required.
- Operate the vehicle and maintain contact with supported and supporting elements.
- Find and collect the wounded.
- Serve as messengers within medical channels.

## Preventive Medicine Section

2-27. Preventive medicine sections are organic to the divisions and the BSMC of the BCT. The PVNTMED section is also found in the multifunctional medical battalion (MMB) headquarters. The PVNTMED section assigned to medical companies has a primary responsibility for supervising the unit's PVNTMED program as described in AR 40-5. The section—

- Ensures measures are implemented to protect personnel against food, water, and arthropodborne diseases, as well as environmental injuries.
- Provides advice and consultation in the area of health threat assessment, FHP, environmental sanitation, epidemiology, sanitary engineering, and pest management.
- Assists the higher headquarters in determining requirements for medical intelligence assessments, particularly with respect to CBRN and disease prevalence.
- Coordinates with supporting veterinary teams for conducting and implementing food safety and quality assurance surveillance and assisting in foodborne and zoonotic disease surveillance and control.

## Behavioral Health Section

2-28. Combat and operational stress is controlled through vigorous prevention, consultation, and restoration programs by the BSMC BH section. These programs are designed to maximize the RTD rate of

COSR Soldiers by providing rest/restoration within or near their unit areas. The BSMC BH section has a primary responsibility for assisting commanders in controlling combat stress and serves as a consultant to the commander, staff, and others involved with providing prevention and intervention services to unit Soldiers and their families. For additional information on the operations and functions of BSMCs, and ASMCs, refer to FMs 4-02.6 and 4-02.24.

# SECTION III — MEDICAL COMPANY (GROUND AMBULANCE)

## OPERATIONAL INFORMATION

2-29. The AHS is a continuum of increasing roles of care extending from the POI through the CONUS base. All sick, injured, and wounded patients must be evacuated from the battlefield in the shortest possible time to the MTFs that can provide the required treatment. Ground ambulances serve as one of the primary means of evacuating patients from the battlefield.

### MISSION

2-30. The mission of the medical company, ground ambulance, TOE 08453A00, is to provide ground evacuation within the theater.

### ASSIGNMENT

2-31. The medical company, ground ambulance is normally assigned or attached to MMB (TOE 08485G00) or a medical brigade (MEDBDE) for C2.

### EMPLOYMENT

2-32. The medical company, ground ambulance, provides DS to BCTs and is employed in the division, corps, and EAC to provide area support. It is tactically located where it can best control its assets and execute its patient evacuation mission.

### BASIS OF ALLOCATION

2-33. The BOA within the combat zone is one per division supported and one per two divisions supported within the theater.

### CAPABILITIES

2-34. At Level I, the unit is capable of providing:

- A single-lift capability for evacuation of 96 litter patients or 192 ambulatory patients.
- Medical evacuation from BCT/division medical companies and ASMCs to supporting hospitals.
- Reinforcement of BCT/division medical company evacuation assets.
- Reinforcement of covering force and unassigned area operations.
- Movement of patients between hospitals and aeromedical staging squadrons/MASFs, railheads, or seaports in the division, corps, and EAC.
- Area evacuation support beyond the capabilities of the ASMC.
- Emergency movement of medical personnel and supplies.
- Preparation of patients for ground and air medical evacuation.
- Medical evacuation of wounded or injured Soldiers from the POI to supporting MTF, CCP, or AXP.
- Field feeding and vehicle refueling support when collocated with the MMB.

### DEPENDENCY

2-35. This unit is dependent upon the appropriate elements of the division, corps, EAC, or ASCC for—

- Religious, financial management, legal, personnel, mortuary affairs, and administrative services.
- Viable communications systems for C2 and adequate road networks.
- Laundry, shower, and clothing repair.
- Generator, communication equipment, and communication security (COMSEC) equipment maintenance.
- Health service support to include hospitalization.

### ORGANIZATION AND FUNCTIONS

2-36. This company is organized into a company headquarters section, two ambulance platoon headquarters, and two evacuation sections (Figure 2-1).

2-37. The company headquarters provides C2, communications, administration, field feeding, and logistical support (to include maintenance) for the subordinate ambulance platoons.

2-38. The ambulance platoon headquarters provides C2 for the subordinate ambulance squads.

2-39. The evacuation section operates ambulances and provides en route medical care for patients. Each evacuation section consists of 12 ambulances, each with a two-man crew. The members of the squad maintain the level of expendable Class VIII supplies in the ambulance MES by reconstituting supplies from medical companies or hospitals when they pick up or drop off patients. They are also responsible for performing operator maintenance on assigned vehicles.

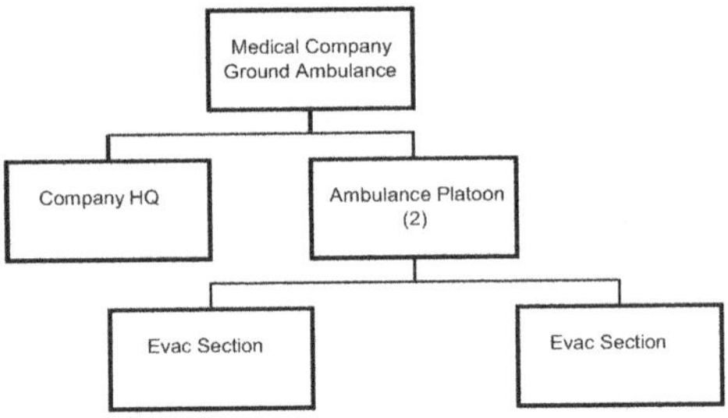


**Figure 2-1. Medical company, ground ambulance**

## SECTION IV — MULTIFUNCTIONAL MEDICAL BATTALION

# MODULAR FORCE

2-40. A modular force is a key aspect in the transformation of the Army. The modular force model is based on brigade-sized elements that are more responsive than division-sized elements and can perform joint and expeditionary-type missions. Using modular units, Army planners can tailor force structure, reduce strategic lift requirements, and create flexible forces with specialized capabilities based on everchanging mission requirements. To facilitate the synchronization and provision of AHS services and support, the AMEDD C2 organizations have been redesigned into a modular design. The MMB will replace the functional battalions (area support, evacuation, and medical logistics).

2-41. The new modular design of medical units gives the medical command (MEDCOM) the ability to—

- Assist in the employment of the optimal mix of medical capabilities within an expeditionary to campaign framework.
- Ensure seamless, state-of-the-art care, regardless of delivery location.
- Provide tested and proven systems to the battlefield and ensure the provision of the right care at the right place and time.
- Promote scalability through easily tailored, capabilities-based packages that result in improved tactical mobility, reduced footprint, and increased modularity for flexible task organization.
- Provide and enable the joint force commander (JFC) to choose among augmentation packages, thus enabling rapid synchronization of desired medical capabilities.

# MULTIFUNCTIONAL MEDICAL BATTALION

## MISSION

2-42. The mission of the MMB, TOE 08486G000, is to provide a scalable, flexible and modular C2, administrative assistance, logistical support, and technical supervision capability for assigned and attached medical organizations (companies, detachments, and teams) task-organized for support of deployed forces.

## ASSIGNMENT

2-43. This unit is assigned to a MEDCOM or a MEDBDE.

## EMPLOYMENT

2-44. The MMB is normally employed in a MEDCOM or MEDBDE AO.

## BASIS OF ALLOCATION

2-45. The BOA is one per three to six subordinate company-size units plus the blood detachment.

## CAPABILITIES

2-46. At Level I, this unit provides—

- Command and control, staff planning, supervision of operations, medical and general logistics support as required, and administration of the assigned and attached units conducting AHS operations in its supported AO.
- Task organization of medical assets to meet the projected patient workload.
- Advice to senior commanders in the AO on the medical aspects of their operations.
- Coordination of medical regulating and patient movement within the AO.
- Monitoring, planning, and coordinating ground and air evacuation within the battalion AO. Coordinating air evacuation support requirements and synchronizating the air evacuation plan into the overall medical evacuation plan with the supporting aviation unit.
- Consultation and technical advice on PVNTMED (medical entomology; medical, and occupational and environmental health [OEH] surveillance; and sanitary engineering), pharmacy procedures, COSC and BH, medical records administration, veterinary services, nursing practices and procedures, dental services, medical laboratory procedures, and automated medical information systems to supported units.
- Guidance for facility site selection and area preparation, as required.
- Monitoring and supervising medical logistics operations, to include Class VIII supply/resupply, medical equipment maintenance and repair support, optical fabrication and repair support, and blood management.
- Planning and coordination for Role 1 and Role 2 AHS operations, to include staff advice to units without organic medical assets within the AO.

- Unit level maintenance for wheeled vehicles and power generation equipment and wheeled vehicle recovery operations support to assigned or attached units.
- Organizational communications equipment maintenance support for the battalion.
- Food service support for staff and other assigned/attached medical units.
- Consolidated property book maintenance.
- Religious support for MMB staff, assigned/attached medical units, and casualties in MMB subordinate unit MTFs.

### DEPENDENCY

2-47. This unit is dependent on the—

- Corps for legal, administrative, financial management, human resources, and transportation services, area damage control, CBRN decontamination assistance, mortuary affairs, and laundry and bath services.
- Quartermaster supply company or equivalent for Class I rations.
- Engineer company or equivalent for site selection, waste disposal, and minor construction.
- Movement control battalion or equivalent for supplemental transportation requirements.

### ORGANIZATION AND FUNCTIONS

2-48. Figure 2-2 depicts the organization of the MMB. Specific information concerning the early entry and campaign elements is provided below.

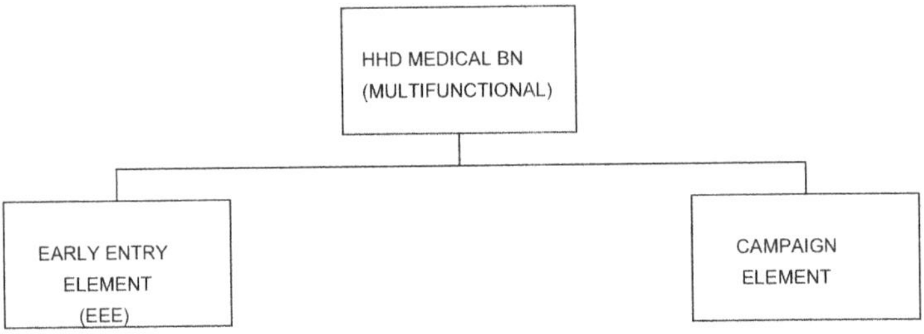


**Figure 2-2. Multifunctional medical battalion**

## EARLY ENTRY ELEMENT, MULTIFUNCTIONAL MEDICAL BATTALION

2-49. Deployed units will begin operations immediately upon arrival in the theater avoiding the delays experienced in past conflicts. There is a need to deploy an EEE medical force to assess, plan, and coordinate AHS operations in the AO and provide medical surveillance, PVNTMED, and area medical support early on. One of the keys to providing early medical care is the development of modular medical elements. These modular AHS elements will perform specific medical functions and will allow medical planners to bring necessary medical units into theater as required by the battlefield mission.

### MISSION

2-50. The mission of the EEE of the MMB, TOE 08486GA00, is to provide a scalable, flexible and modular C2, administrative assistance, logistical support, and technical supervision capability for assigned and attached medical organizations (companies, detachments, and teams) task-organized for support of deployed forces.

## Assignment

2-51. This EEE is assigned to an MMB.

## Employment

2-52. Normally employed in a MEDCOM/MEDBDE AO. It can also be deployed to provide C2 to expeditionary forces in early entry operations and facilitates the reception, staging, onward movement, and integration (RSOI) of medical forces.

## Basis of Allocation

2-53. The BOA is one per MMB.

## Capabilities

2-54. At Level I, this unit provides—

- Command and control, staff planning, supervision of operations, medical and general logistics support as required, and administration of the assigned and attached units conducting AHS operations in its supported AO.
- Task organization of medical assets.
- Advice to senior commanders in the AO on the medical aspects of their operations.
- Coordination of medical regulating and patient movement within the AO.
- Monitoring, planning, and coordinating ground and air evacuation within the battalion AO. Coordinating air evacuation support requirements with the supporting aviation unit, and synchronizing the air evacuation plan into the overall medical evacuation plan.
- Consultation and technical advice on PVNTMED (medical entomology, medical and OEH surveillance, and sanitary engineering), COSC and BH, medical records administration, veterinary services, nursing practices and procedures, dental services, and automated medical information systems to supported units.
- Guidance for facility site selection and area preparation, as required.
- Monitoring and supervising medical logistics operations, to include Class VIII supply/resupply.
- Consolidated property book maintenance.
- Unit-level maintenance for wheeled vehicles and power generation equipment and wheeled vehicle recovery operations support to assigned or attached units.
- Organizational communications equipment maintenance support for the battalion.
- Food service support for staff and assigned/attached medical units.

## Dependency

2-55. This unit is dependent on the—

- Corps for legal, administrative, financial management, human resources, religious support, transportation services, area damage control, CBRN decontamination assistance, mortuary affairs, and laundry and bath services.
- Quartermaster supply company or equivalent for Class I rations.
- Engineer company or equivalent for site selection, waste disposal, and minor construction.
- Movement control battalion or equivalent for supplemental transportation requirements.

## ORGANIZATION AND FUNCTIONS

2-56. The battalion command section provides C2 of assigned and attached medical companies, detachments, and teams. (See Figure 2-3.)

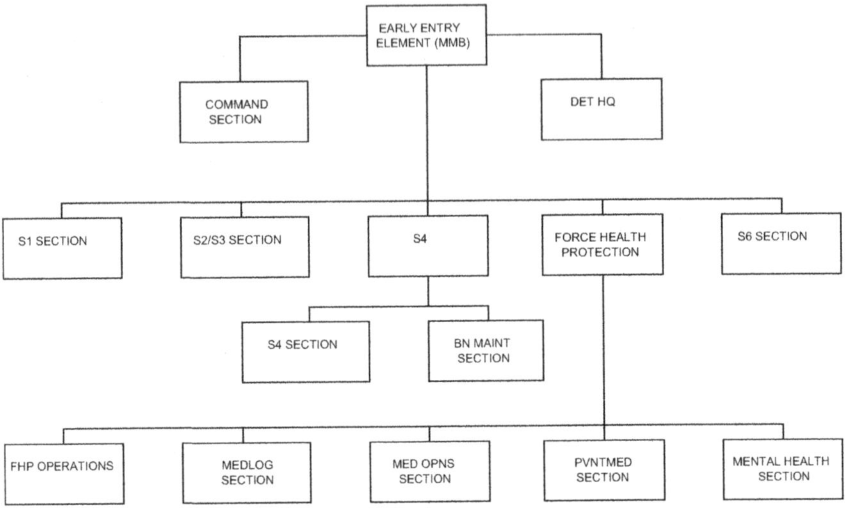


**Figure 2-3. Early entry element, multifunctional medical battalion**

2-57. The Adjutant (US Army) (S1) operates according to METT-TC and is the principle coordinating staff officer responsible for the delivery of human resources support. The S1 is responsible for the execution of all of the human resources (HR) core competencies to include: personnel readiness management (PRM); personnel accountability and strength reporting (PASR); personnel information management (PIM); reception, replacement, RTD, rest and recuperation (R&R), redeployment (R5) operations; casualty operations; essential personnel services (EPS); and postal operations. Morale, welfare, and recreation (MWR) operations, HR planning; and staff operations. The S1 also coordinates with elements of supporting agencies for finance, legal, religious, and administrative services.

2-58. The Intelligence Officer (US Army)/Operations and Training Officer (US Army) (S2/S3) section is responsible for security, plans and operations, deployment, relocation, and redeployment of the battalion and its assigned and attached units. The battalion's primary net control station is in this paragraph.

2-59. The Supply Officer (US Army) (S4) section is responsible for coordination, control, and management of logistics for assigned and attached units. The battalion maintenance section is under the staff supervision of the battalion S4. This section provides unit level maintenance for wheeled vehicles assigned to the HHD and assigned or attached units without unit level maintenance capability.

2-60. The FHP operations section is responsible for the planning, coordination, and execution of the AHS mission within the battalion's AO. This section supervises the operations of the medical logistics, operations, PVNTMED, and BH sections as well as coordinates medical regulation and patient movement within the AO. The medical operations section is responsible for—

- Management of area medical support (Roles 1 and 2).
- Planning, coordination, and execution of the Class VIII mission within the battalion's AO. This includes blood and medical maintenance management.

- Situational awareness for locations of ground and air ambulance assets.
- Planning and coordination for evacuation of patients using ground and air ambulance assets.
- Planning and coordination between air evacuation elements and the maneuver brigade S3. Information provided to the maneuver brigade S3 should include, but not be limited to the location of—
  - Medical treatment facilities, CCPs, and AXPs.
  - Forward air refueling equipment.
  - Location of supported units and liaison requirements. Air evacuation corridors, ground evacuation routes, and recommendations on usage.
- The PVNTMED section is responsible for the planning, coordination, and execution of the PVNTMED mission within the battalion's AO. This includes management of PVNTMED and veterinary assets.
- The BH section is responsible for the planning, coordination, and execution of the COSC mission within the battalion's AO. This section also collects and records social and psychological data.

2-61. The S6 section is responsible for all aspects of information management, automation, and communications-electronic (CE) support to assigned and attached units. The battalion's alternate net control station is in this paragraph.

2-62. The detachment headquarters provides for billeting, field feeding, discipline, security, training, and administration for personnel assigned to the EEE.

2-63. The battalion maintenance section is under the staff supervision of the battalion S4. This section provides unit level maintenance for wheeled vehicles assigned to the HHD and assigned or attached units without unit level maintenance capability.

# CAMPAIGN SUPPORT ELEMENT, MULTIFUNCTIONAL MEDICAL BATTALION

## MISSION

2-64. The mission of the campaign support element, MMB, TOE 08486GB00, is to complete the staffing of the headquarters of the MMB to enhance the delivery of campaign quality health care to deployed forces. To provide a scalable, flexible, and modular C2, administrative assistance, logistical support, and technical supervision capability for assigned and attached medical organizations (companies, detachments, and teams) task-organized for support of deployed forces.

## ASSIGNMENT

2-65. This unit is assigned to an MMB.

## EMPLOYMENT

2-66. Normally employed with the MEDBDE or a MEDCOM. If the EEE is deployed first, the campaign element falls in on this deployed module to provide a more robust C2 element and additional operational and planning capabilities, increased medical logistical staff, and a more diverse and robust clinical staff.

## BASIS OF ALLOCATION

2-67. The BOA is one per MMB.

## CAPABILITIES

2-68. At Role 1, this unit provides—

- Augmentation of C2 in personnel, logistics, HSS, FHP, medical operations, PVNTMED, BH, automation, maintenance, religious support, and food service to the EEE.
- Unit maintenance on all organic equipment except COMSEC equipment.

## DEPENDENCY

2-69. This unit is dependent on the HHD, MMB for legal, administrative, financial management, human resources, food service, transportation services, area damage control, CBRN decontamination assistance, mortuary affairs, and laundry and bath services.

## ORGANIZATION AND FUNCTIONS

2-70. The command section provides C2 for assigned and attached medical companies, detachments, and teams. (See Figure 2-4.)

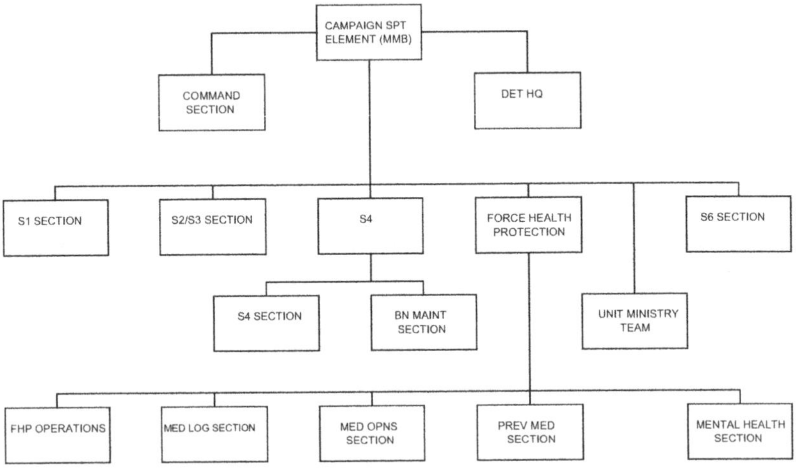


**Figure 2-4. Campaign support element, multifunctional medical battalion**

2-71. The S1 operates according to METT-TC and is the principle coordinating staff officer responsible for the delivery of HR support. The S1 is responsible for the execution of all of the HR core competencies to include: PRM, PASR, PIM, R5 operations, and casualty operations, EPS, postal operations, MWR operations, and HR planning and staff operations. The S1 also coordinates with elements of supporting agencies for finance, legal, religious, and administrative services.

2-72. The S2/S3 section coordinates with the EEE S2/S3 section of the MMB in the plans and operations support to subordinate organizational elements.

2-73. The S4 section is responsible for coordination, control, and management of logistics for assigned and attached units.

2-74. The FHP operations section plans and coordinates the HSS/FHP mission within the battalion's AO when collocated with the EEE. This section also provides advice on the operation of medical logistics, operations, PVNTMED, BH, medical regulating, and patient movement to supported organizations.

2-75. The medical logistics section plans and organizes Class VIII supply support (including blood and medical equipment management) in coordination with the medical logistics section of the EEE.

2-76. The medical operations section is responsible for—

- Management of area medical support (Roles 1 and 2).
- Situational awareness for locations of ground and air ambulance assets.
- Planning and coordination for evacuation of patients using ground and air ambulance assets.
- Planning and coordination between air evacuation elements and the maneuver brigade S3. Information provided to the maneuver brigade S3 should include, but not be limited to the location of—
  - Medical treatment facilities, CCPs, and AXPs.
  - Forward air refueling equipment.
  - Supported units and liaison requirements.
  - Evacuation corridors and evacuation routes and recommendations on usage.

2-77. The PVNTMED section plans and coordinates PVNTMED/veterinary support to subordinate organizations.

2-78. The BH section plans and coordinates BH support to subordinate organizations. This section also collects and records social and psychological data when collocated with the EEE.

2-79. The S6 section provides advice and support to subordinate organizations for CE support and information management.

2-80. The unit ministry team provides and coordinates religious support for all MMB staff, assigned/attached medical units and casualties in MMB subordinate unit MTFs.

2-81. The detachment headquarters provides coordination with the EEE in support of the field feeding and administration mission of the unit.

2-82. The battalion maintenance section provides unit level maintenance management support to assigned and attached units in coordination with the battalion S4 and the battalion maintenance section of the EEE.

## SECTION V — GENERAL SUPPORT AVIATION BATTALION, COMBAT AVIATION BRIGADE

# GENERAL SUPPORT AVIATION BATTALION

### MISSION

2-83. The mission of the general support aviation battalion (GSAB) (Figure 2-5) is to provide the combat aviation brigade with aerial sustainment and maneuver support.

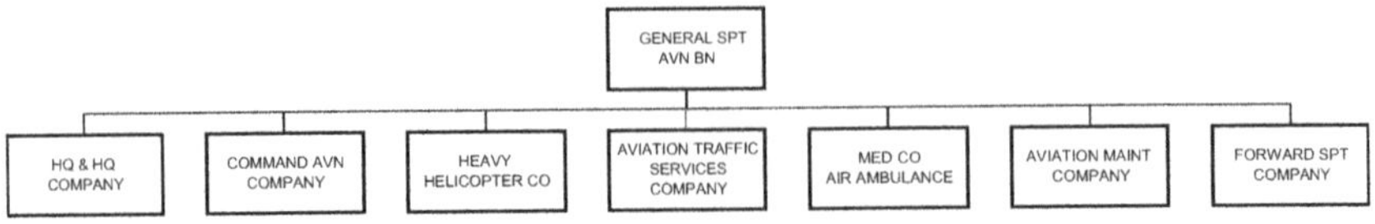


**Figure 2-5. General support aviation battalion**

### ASSIGNMENT

2-84. Organic to the division combat aviation brigade, theater assault aviation brigade, theater general support aviation brigade, and the Army National Guard (ARNG) aviation expeditionary brigade (TOE 01300G400).

### BASIS OF ALLOCATION

2-85. The BOA is one per division combat aviation brigade, one per ARNG aviation expeditionary brigade, one per theater assault aviation brigade, and three per theater GSAB (TOE 01300G400).

### CAPABILITIES

2-86. This organization provides the following when all of its subordinate units are organized at Level I as shown in their respective TOEs:

- Battle command on the move.
  - Employment of Army airspace command and control (A2C2) system platforms in support of the division command group, maneuver, and aviation commanders' C2 requirements.
  - Radio relay and transport of liaison officers as required.
- Air assault.
  - Airlift for air assault operations and air movement of tactical forces (other than assault), field artillery or other fire support assets, tactical air defense (AD) systems, engineer equipment and personnel, and other resources.
  - Casualty evacuation when medical aircraft are not readily available.
  - Evacuation of downed aircraft and personnel.
- Sustainment.
  - Division sustainment with heavy-lift helicopter company.
  - General aviation support to the aviation brigade.
  - Support aircraft to the aviation support battalion.
  - Personnel recovery (PR).
  - Downed aircraft recovery team operations.
  - Casualty evacuation operations.
  - Air movement of general supplies, equipment, and personnel, including logistics over-the-shore (LOTS) when applicable.
  - Aerial noncombatant evacuation operations when directed by applicable commander/authority.
  - Air movement of nuclear and other special munitions.
  - Self-deployment of organic CH-47 helicopters to a theater of operations when the aircraft have been configured with extended range fuel systems.

    - Retail level aircraft refueling when aircraft are configured with extended range fuel system (commonly referred to as "Fat Cow").
- Air traffic services support.
- Aeromedical evacuation.
    - Patient evacuation.
    - Patient movement between MTFs (patient transfers).
    - Class VIII resupply.
    - Joint blood program support.
    - Medical C2.
    - Movement of medical personnel and medical equipment.
    - Air crash rescue support.

2-87. Due to the CH-47's capabilities, heavy helicopter units can perform two unique missions:

- High-altitude operations.
- Oversized, heavy, and special munitions movement.

2-88. This GSAB is dependent upon the following organizations:

- Appropriate elements of the division aviation brigade for legal, HSS, FHP, financial management, and personnel and administrative services.
- United States Air Force weather team in the headquarters and headquarters company of the combat aviation brigade (TOE 01302G200) for air weather service support.
- Aviation support company (TOEs 01918G200 or 01919G000 [ARNG]) for intermediate level field maintenance support, DS maintenance for assigned aviation ground power units (AGPU), line item number (LIN) P44627, and support for aviation and common Class IX repair parts.

# MEDICAL COMPANY, AIR AMBULANCE (HH-60), GENERAL SUPPORT AVIATION BATTALION, COMBAT AVIATION BRIGADE

## OPERATIONAL INFORMATION

2-89. The medical company, air ambulance (Figure 2-6), provides aeromedical evacuation for all categories of patients consistent with evacuation precedence and other operational considerations within the division. In today's ongoing efforts in building a modular force, medical company, air ambulance will fall under the GSAB which will provide aircraft maintenance and logistics support, aviation communications, and real-time operational picture associated with today's combat environment.

## MISSION

2-90. The mission of the medical company, air ambulance (HH-60) (TOE 08443G00) is to provide air ambulance evacuation support within the division and theater.

## ASSIGNMENT

2-91. The medical company, air ambulance, (HH-60) is organic to the GSAB (TOEs 01125G000, 01225G100, and 01305G200) for C2.

## EMPLOYMENT

2-92. This unit is employed in the division and corps area of responsibility. It is tactically located where it can best control its assets and execute its patient evacuation mission.

## Basis Of Allocation

2-93. The BOA is one per GSAB and two per theater of operations (TOEs 01125G100, 01225G100, and 01305G100).

## Capabilities

2-94. At Level I, this unit provides—

- Twelve helicopter ambulances to evacuate critically wounded or other patients consistent with evacuation priorities and operational considerations, from points as far forward as possible, to division, corps, and EAC hospitals. Single patient lift capability for the company is 72 litter patients or 84 ambulatory patients, or some combination thereof.
- Four forward support medical evacuation teams (three helicopters each) that can be deployed as a team or an independent aircraft.
- Air crash rescue support.
- Expeditious delivery of whole blood, biological, and medical supplies to meet critical requirements.
- Rapid movement of medical personnel and accompanying equipment/supplies to meet the requirements for mass casualty (MASCAL), reinforcement/reconstitution, or emergency situations.
- Movement of patients between hospitals, ASFs, hospital ships, casualty receiving and treatment ships, seaports and railheads in the theater of operations.
- Military working dog evacuation and aeromedical evacuation support to a combat search and rescue task force.

## Dependency

2-95. This unit is dependent upon—

- Appropriate elements of the division or corps for religious, legal, financial management, personnel and administrative services, and CE and COMSEC maintenance.
- The headquarters and headquarters company of the GSAB for religious, legal, HSS, FHP, personnel and administrative service, and unit CBRN support.
- The forward support company of the GSAB for Class III, automotive and generator maintenance, and field feeding.
- The aviation support company of the GSAB for unit level field maintenance of organic aircraft, including unit level supply support for aircraft Class IX.
- United States Air Force weather team in the headquarters and headquarters company of the division aviation brigade for air weather service support.

## Organization and Functions

2-96. The company headquarters provides C2, logistical support, and administration for the company.

2-97. The four forward support MEDEVAC teams (FSMTs) provide a task-organized means for medical evacuation at the DS and GS level. They also, provide emergency movement of medical personnel and emergency delivery of whole blood, biological, and medical supplies and equipment.

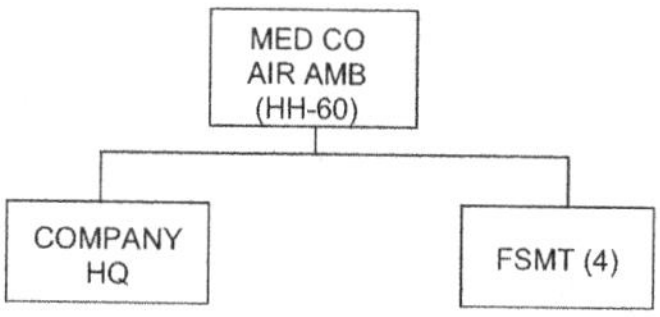


**Figure 2-6. Medical company, air ambulance (HH-60)**

# AVIATION MAINTENANCE COMPANY, AVIATION UNIT MAINTENANCE, GENERAL SUPPORT AVIATION BATTALION, COMBAT AVIATION BRIGADE

## MISSION

2-98. The mission of the aviation maintenance company is to perform aviation unit maintenance for the GSAB.

## ASSIGNMENT

2-99. This unit is organic to a GSAB (TOE 01305G100).

## EMPLOYMENT

2-100. The aviation maintenance company is employed as part of combined forces and provides unit level field maintenance support to the GSAB. The aviation maintenance company is located and its operations are normally conducted immediately behind the main battle area (MBA) in the EAB. The aviation maintenance company operations in forward areas are usually limited to daily inspections, minor adjustments, emergency and contact maintenance, and aircraft recovery. Contact teams and slice elements could be utilized as necessary.

## BASIS OF ALLOCATION

2-101. The BOA is one per GSAB (TOE 01305G100).

## CAPABILITIES

2-102. The aviation maintenance company provides unit level field maintenance support for 8 UH-60, 12 HH-60, and 12 CH-47 aircraft. Individuals of this organization can assist in the coordinated defense of the unit's area or installation. This unit performs unit level field maintenance on only the supported aircraft and some associated maintenance equipment.

## DEPENDENCY

2-103. This unit depends on—

- Appropriate elements of the division or corps for religious, legal, HSS, FHP, financial management, personnel and administrative services, and logistical support (including supplemental transportation).
- Headquarters and headquarters company of the GSAB (TOE 01306G100) for religious, legal, HSS, FHP, financial management, personnel, and administrative services, unit-level logistical support, and communications maintenance.

- Forward support company (TOE 63317G200) for Class III, automotive and generator maintenance, and field feeding service support.
- Aviation maintenance battalion (TOE 01918G200), for unit level field maintenance support, backup unit level field maintenance support, DS maintenance for assigned AGPU, and support for aviation and common Class IX repair parts.

### ORGANIZATION AND FUNCTIONS

2-104. This unit inspects aircraft systems, subsystems, components, and the maintenance performed by crew chiefs and repairers to determine airworthiness and compliance with applicable technical publications; performs operational checks and troubleshoots aircraft systems to isolate malfunctions; participates in maintenance test flights and maintains historical maintenance files.

2-105. This unit provides company maintenance operations. It maintains and regulates uniform flow of aircraft and associated components. Further, it maintains the company's aircraft logbooks, provides unit level Class IX support, and performs maintenance on aviation life support equipment (ALSE).

2-106. This unit provides unit level maintenance on UH/HH-60/CH-47 helicopters and associated equipment; services and lubricates helicopter systems and subsystems; and diagnoses and troubleshoots operational malfunctions.

## MEDICAL COMMAND

2-107. The MEDCOM serves as the senior medical command within the theater, and works directly for the theater commander. With augmentation, MEDCOM is joint capable. In the absence of a theater commander the MEDCOM works for the regional combatant commander during a major combat operation (MCO). It possesses a centralized capability to effectively and efficiently task-organize medical elements based on theater specific medical requirements. The MEDCOM is depicted in Figure 2-7.

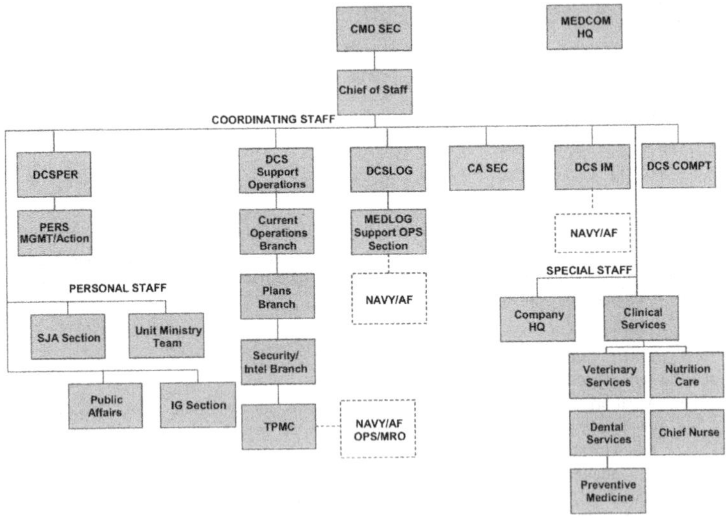


Figure 2-7. Medical command

2-108. The MEDCOM serves as the medical force provider (Figure 2-8) for the corps and EAC and provides the requisite medical C2 necessary to provide campaign-quality health care to deployed forces. The MEDCOM also focuses on theater medical operations plans and medical contingency plans.

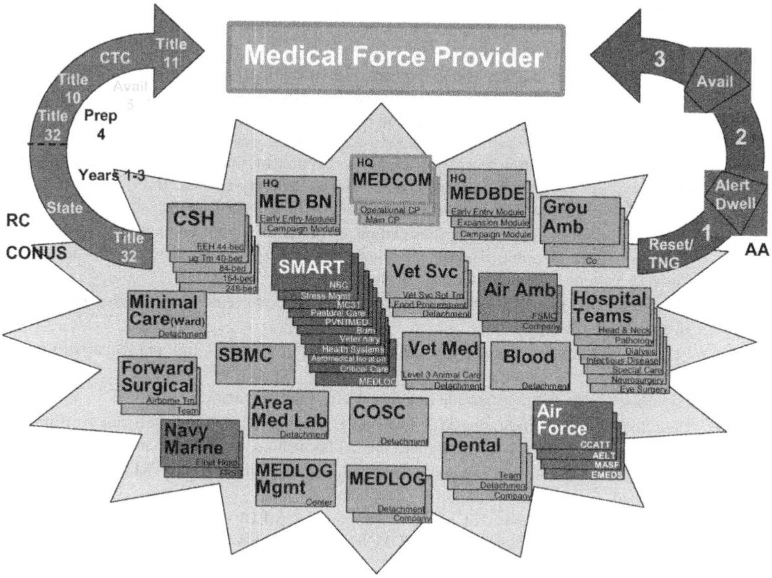


**Figure 2-8. Medical force pool**

2-109. The ability of the MEDCOM staff to develop accurate patient estimates based on real-time medical data ensures the right mix of medical specialties and resources are in the right place and at the right time to treat casualties on the battlefield. Care must be continuous and seamless from the POI or wounding through a phased system providing essential care in theater, with evacuation for definitive and rehabilitative care in CONUS.

2-110. The MEDCOM is a dedicated, regionally focused command with a basis of allocation of one per theater.

2-111. The MEDCOM is a versatile, modular medical C2 structure composed of a main command post (MCP) and an operational command post (OCP) which are 100 percent mobile. The MEDCOM is comprised of two standard requirements code (SRC) identified modules to provide a scalable medical C2 capability to the corps commander. Further, the MEDCOM is capable of—

- Providing subordinate medical organizations to operate under the subordinate MEDBDE and/or MMB and providing FSTs and/or other augmentation required by BCTs, and echelons above brigade (EAB) medical companies.
- Operating as the medical force provider for provisioning of required assets to the MEDBDEs.
- Advising the corps and other senior level commanders on the medical aspects of their operations.
- Providing staff planning, supervision of operations, and administration of assigned and attached medical units.
- Providing reach to the MHS.

- Providing medical Army support to other Services (ASOS)/Title 10 responsibilities of the ASCC commander.
- Coordinating with the USAF theater patient movement requirements center (TPMRC) for medical regulating and movement of patients from MEDCOM MTFs.
- Providing consultation services, technical advice, and clinical policy development in all aspects of medical and surgical services.
- Containing the functional staff to coordinate hospitalization, nursing, PVNTMED (entomology, epidemiology, OEH surveillance, potable water inspection, pest management, food facility inspection, and control of medical and nonmedical waste), medical evacuation, medical regulating, MEDLOG to include blood management, dental services, veterinary services (zoonotic disease control, investigation and inspection of subsistence, and animal medicine), pharmacy, optometry, nutrition care, and COSC, neuropsychiatric (NP), and BH services, medical laboratory services, and area medical support to supported units.
- Providing technical advice and consultation on medical automated information systems and programs such as the Theater Medical Information Program (TMIP) and Medical Communications for Combat Casualty Care (MC4).
- Functioning as the senior theater joint medical C2 element when augmented with joint augmentation package.
- Peacetime planning, training for and with the MEDBDEs, and participation in theater engagement plans.
- Providing advice and assistance in facility selection and preparation.
- Coordinating and orchestrating MEDLOG operations to include Class VIII supply, distribution, medical maintenance and repair support, optical fabrication, and blood management.
- Planning for the single integrated medical logistics manager (SIMLM), mission when designated.
- Planning and coordination of religious support for MEDCOM staff, assigned and attached medical units, and casualties in MEDCOM subordinate unit MTFs.

## Operational Command Post

2-112. The MEDCOM OCP is a deployable, versatile module, which provides medical C2, policy development, and technical guidance to subordinate medical units and provides interface and liaison with supported theater forces in EAC, corps, and subordinate BCTs. The MEDCOM OCP has an assigned SRC to facilitate the placement of personnel in the CP and to integrate the CP into the time-phased force deployment list (TPFDL). The MEDCOM OCP is depicted in Figure 2-9.

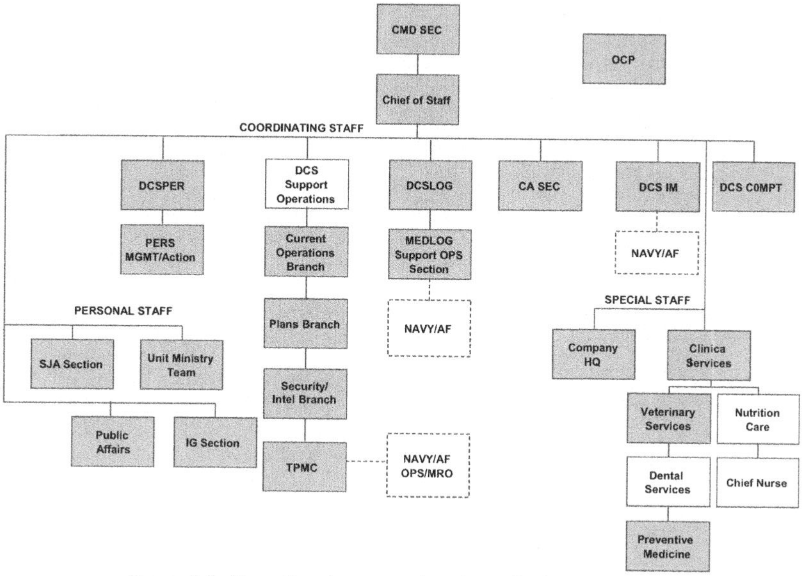


**Figure 2-9. Operational command post, medical command**

2-113. As the force continues to build within the theater, the MEDCOM's OCP can be incrementally expanded or augmented with additional planning and professional expertise. The deployment of personnel from the MCP model can then be built into the medical C2 required structure in order to provide campaign-quality health care to the deployed force.

## Main Command Post

2-114. The MEDCOM MCP is a deployable augmentation module, which completes the staffing of the MEDCOM headquarters to enhance the delivery of campaign-quality health care to deployed forces in EAC, corps, and supported BCTs. It facilitates medical C2, policy development, and technical guidance to subordinate MEDBDEs and provides interface and liaison with supported theater forces in EAC, corps, and subordinate BCTs. If the OCP is deployed and needs additional personnel or clinical skills, the MEDCOM MCP can deploy personnel to the deployed OCP. This will provide a more robust medical C2 element and additional operational and planning capabilities; increased medical logistics staff; automated medical information systems; a more diverse clinical staff; and more depth in the command judge advocate (CJA), inspector general (IG), civil affairs (CA), and public affairs sections. The MEDCOM MCP has an assigned SRC to facilitate the placement on and the integration into the TPFDL. The MCP is depicted in Figure 2-10.

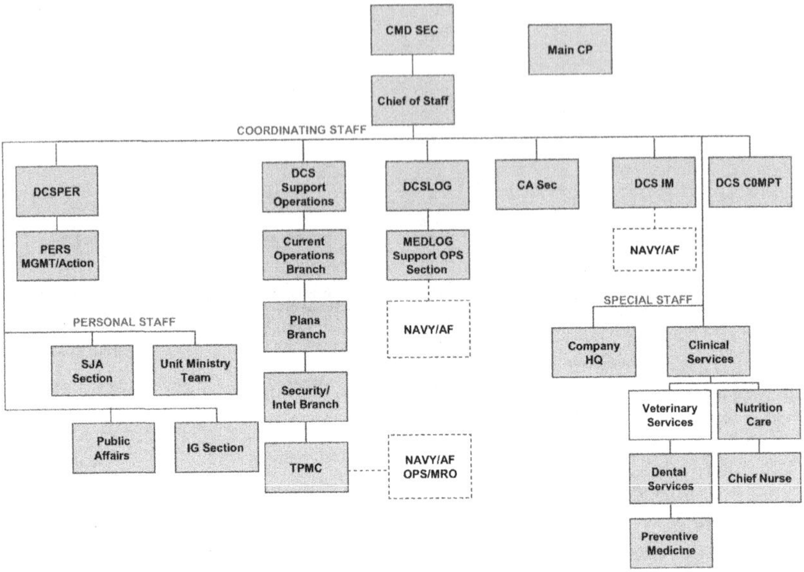


**Figure 2-10. Main command post, medical command**

# MEDICAL BRIGADE

2-115. The MEDBDE provides a scalable expeditionary medical C2 capability for assigned and attached medical functional organizations task-organized in support of the deployed division/corps forces. The MEDBDE brings all requisite medical C2 and planning capabilities to provide responsive and effective HSS/FHP throughout the corps AO. The MEDBDE is depicted in Figure 2-11.

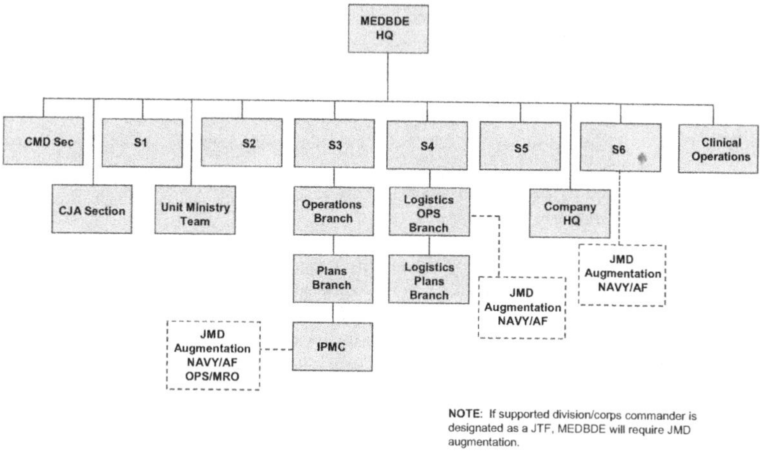


**Figure 2-11. Medical brigade**

2-116. The MEDBDE ensures the division/corps commander has the right mixture of medical professional (operational, technical, and clinical) expertise to synchronize the complex system of AHS functions required to maintain the health of his command by promoting fitness, preventing casualties from DNBI, and promptly treating and evacuating those injured on the battlefield.

2-117. The design and flexibility of the MEDBDE permits the division/corps commander the ability to meet expeditionary medical support requirements in support of early entry forces. As the supported corps commander's forces grow in both size and complexity, the MEDBDE can deploy additional modules that build upon one another to support full spectrum operations. The MEDBDE also provides the corps commander the appropriate medical C2 to continue to build his medical force capabilities through the integration of Army, joint, and multinational medical forces to ensure he can identify and counter any health threats in his supported areas. This will permit him to transition from expeditionary medical support operations to providing campaign-quality health care in his AO.

2-118. The MEDBDE has the capability to provide SRC elements for an early entry module (EEM), an expansion module, and the campaign module (CM), thus enabling its capability to be tailored to METT-TC factors of a specific division/corps operation.

2-119. The MEDBDE provides medical C2, administrative assistance, and technical supervision of assigned and attached medical units providing AHS in support of a division/corps. Further, it is capable of—

- Providing a rapidly responsive early entry medical C2 module that can quickly integrate into the early entry deployment sequence for crisis management.
- Providing full spectrum continuous medical C2 in support of all Army corps and joint (when augmented) forces.
- Providing operational medical augmentation to Role 2 BCT medical companies.
- Advising corps and BCT commanders on the medical aspects of their operations.
- Providing medical staff planning, operational and technical supervision, and administrative assistance for MMBs and CSHs task-organized under the MEDBDE.
- Coordinating with supporting TPMRC for medical regulating and medical evacuation from MMBs and hospitals to supporting EAC MTFs and CONUS.

- Providing medical consultation services and technical advice in—
  - Preventive medicine (medical and OEH surveillance, environmental health, sanitary engineering, and medical entomology).
  - Nursing services.
  - Dental services.
  - Behavioral health (to include COSC) and NP care.
  - Veterinary services (including food safety and inspection, animal medical care, and veterinary PVNTMED services).
  - Nutrition care and medical food service.
  - Medical laboratory support.
- Providing advice and recommendations for conduct of civil-military operations (CMO).
- Control and supervision of Class VIII supply and resupply movement to include blood management. When designated by the combatant command, serves as the SIMLM.
- Providing joint capable medical C2 capability when augmented with appropriate joint assets.
- Serves as the executive agent for veterinary services according to DODD 5200.31E.
- Coordinates ASOS for ship-to-shore/shore-to-ship medical evacuation mission.
- Plans and coordinates religious support for MEDBDE staff, assigned and attached medical units, and casualties in MEDBDE subordinate unit MTFs.

## Chapter 3

# Operational and Tactical Evacuation Planning

A comprehensive medical evacuation plan is essential to ensure effective, efficient, and responsive medical evacuation is provided to all wounded, injured, and ill Soldiers in the JOA. The Army medical evacuation plan flows from the GCC's guidance and intent and incorporates all missions and tasks directed by the GCC to be accomplished and is synchronized with supporting and supported units. In some scenarios, Army air and ground evacuation resources may be directed to provide support to sister Services, allied and coalition partners, and HN forces. Examples of a medical evacuation OPLAN and OPORD are contained in Appendix D.

## SECTION I — THEATER MEDICAL EVACUATION PLANNING RESPONSIBILITIES

### JOINT PLANNING

3-1. All medical evacuation planning within the theater reflects the GCC's guidance and intent. Planning at the joint level is governed by Joint Publications (JPs) 5-0, 5-00.1, and 5-00.2. Guidance on medical evacuation in joint operations is addressed in JP 4-02. When directed by the GCC, Army medical evacuation assets may be tasked to support other than Army forces engaged in the execution of the joint mission. These additional support missions will be clearly articulated in the GCC's OPLAN and OPORD. The ASCC surgeon, with the advice of the senior medical evacuation planner, will coordinate and synchronize these support operations with the combatant command surgeon, joint task force (JTF) surgeon, and the other Services and/or multinational partners as required to ensure that a comprehensive and effective, efficient, and responsive plan is developed and implemented.

### MEDICAL COMMAND AND CONTROL ORGANIZATIONS

3-2. Army evacuation assets support the full spectrum of all Army operations and are capable of supporting Army forces engaged in the execution of the joint mission, as directed. Major combat operations supported by an ASCC will have a deployed MEDCOM. The C2 coordination and orders flow for AHS is illustrated in Figure 3-1. Evacuation planning should occur at each major level of command in order to effectively execute command, control, and influence over evacuation planning.

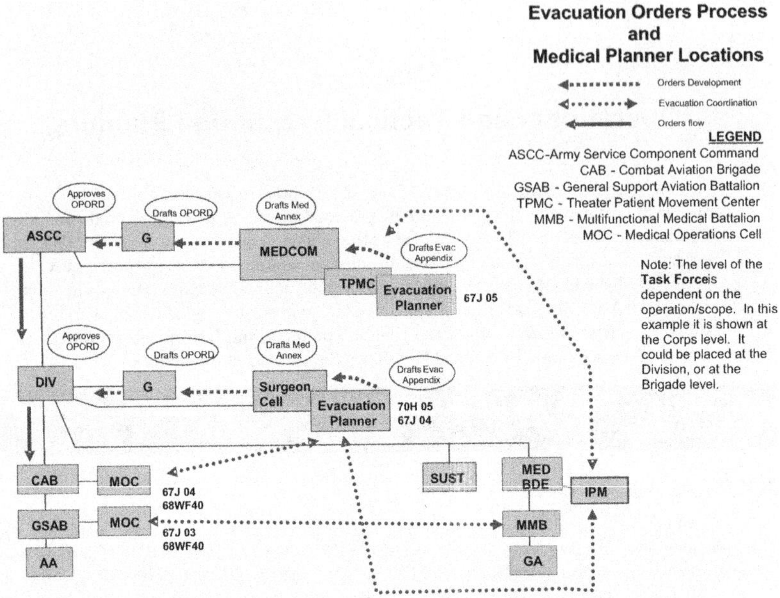


**Figure 3-1. Command and control organizations for coordination and orders flow**

3-3. Under the new Army modular design, some early entry operations may not require the deployment of a theater or a brigade C2 structure. In these cases, medical evacuation planning, coordination, and synchronization are initiated by the most senior medical command/staff level (such as the division surgeon, medical brigade, GSAB, and/or MMB). The senior leaders responsible for air and ground medical evacuation provide estimates and recommendations for the plan. This plan is normally provided as an appendix to the HSS annex and is disseminated down through the chain of command.

3-4. Dedicated Army rotary-wing aircraft (air ambulances) are under the C2 of the GSAB within the transformed Army aviation structure. Ground ambulances remain under the C2 of medical units/elements that are organic to movement and maneuver, fires, and sustainment units. The medical company, ground ambulance is under the C2 of the MMB.

3-5. To ensure an integrated medical evacuation system, synchronization of air with ground ambulance assets is essential. To affect this synchronization, the medical C2 headquarters must maintain influence over the execution of the medical evacuation plan. This influence is accomplished through the orders process.

## Theater Patient Movement Center

3-6. In the design of the modular MEDCOM, the medical reengineering initiative (MRI) MEDCOM's medical regulating office (MRO) staff was reevaluated, realigned, and augmented with medical evacuators to more effectively coordinate and synchronize the various ground and air evacuation aspects of the operation and to supervise and facilitate the medical regulating functions. This staff section was renamed as the theater patient movement center (TPMC). The addition of medical evacuators enables this section to develop, coordinate, and synchronize the medical evacuation portion of the MEDCOM AHS hour coordination and management responsibility for patient regulating and administration within the MEDCOM AO and directs patient evacuation between facilities within its AO as required to avoid patient overflow. This staff is responsible for preparing patient statistical reports for the command. This section also coordinates with the supporting TPMRC for the transfer of patients out of the MEDCOM/theater. The TPMRC regulates the flow of patients out of the theater to an MTF capable of providing the required specialty care and arranges for the transport of these patients by USAF aeromedical evacuation assets.

### INTRATHEATER PATIENT MOVEMENT CENTER

3-7. As with the modular MEDCOM, the MRO of the modular MEDBDE was also redesigned in a similar fashion to enhance the capabilities for medical evacuation planning within the MEDBDE AO. This staff section was renamed the intratheater patient movement center (IPMC). This office is responsible for coordinating and synchronizing medical evacuation operations with the TPMC and the division surgeon cell. It regulates patients within its subordinate hospitals and arranges for transportation to transfer patients between facilities to relieve surgical backlogs, to obtain specialty care, and to ensure bed availability prior to start of operations, as appropriate. It also coordinates with the MEDCOM TPMC and the TPMRC for the evacuation of patients out of theater. This office maintains a 24-hour coordination and management responsibility for patient regulating and administration with the MEDBDE AO. It also prepares patient statistical reports for the MEDBDE commander. In the absence of the TPMC, IPMC assumes duties of the TPMC.

# SECTION II — PLANNING PROCESS

## EVACUATION PLANS AND ORDERS

### CHARACTERISTICS

3-8. Operation plans and orders are the means by which commanders express their visualization, commander's intent, and decisions. They focus on results the commander expects to achieve. These OPLANs and OPORDs provide the basis for ensuring that medical evacuation operations are comprehensively planned, synchronized, responsive, and ensure a seamless continuum of care.

3-9. Evacuation plans and orders help form the basis commanders use to synchronize military operations. They encourage initiative by providing the *what* and *why* of a mission, and leave the *how to accomplish the mission* to subordinates. They give subordinates the operational and tactical freedom to accomplish the mission by providing the minimum restrictions and details necessary for synchronization and coordination. As medical evacuation operations are complex and may cross command and service boundaries, it is essential that the medical evacuation annex to the medical OPLAN/OPORD clearly delineates responsibilities, communications interoperability and procedures, and coordination requirements.

### PLANS AND ORDERS

3-10. Publishing the commander's OPLAN/OPORD permits subordinate commanders to prepare supporting plans and orders. They can discern and implement instructions derived from a higher commander's plan and order. Appendix D provides an example of operational planning and a sample OPORD. Additionally, the commander's OPLAN/OPORD—

- Focuses subordinates' activities.
- Provides tasks and activities, constraints, and coordinating instructions necessary for mission accomplishment.
- Encourages agility, speed, and initiative during execution.
- Conveys instructions in a standard, recognizable, clear, and simple format.

3-11. The amount of detail provided in an evacuation plan or order depends on several factors, to include the experience and competence of subordinate commanders, cohesion and tactical experience of subordinate units, and the complexity of the operation.

3-12. Commanders balance these factors with their guidance and commander's intent, and determine the type of plan or order to issue. To maintain clarity and simplicity, plans and orders include annexes only when necessary and only when they pertain to the entire command. Annexes contain the details of support and synchronization necessary to accomplish the mission.

3-13. For a complete discussion of the military decision-making process and development of OPLANs/OPORDs refer to FM 5-0 and FMI 5-0.1.

## MEDICAL EVACUATION PLANNING

3-14. Medical evacuation planning which supports the HSS/FHP plan is an ongoing process and is accomplished by medical operations element at battalion, brigade, division, corps, and theater levels. The HSS/FHP operations planning process must take under consideration all issues that could affect or influence HSS/FHP support for tactical operations. These issues should be considered in the initial developmental stages of theater, corps, division, or brigade plan as the supporting HSS/FHP annex is developed.

3-15. Planning is a dynamic and continuous process. The framework established in a robust initial plan can be built on in subsequent phases of an operation. The evacuation plan for a given AO will require adjustment and change during different phases of an operation. For example, transitioning from offensive operations to stability operations will generally dictate a significant change in evacuation coverage priorities.

3-16. Medical evacuation planning considerations, although essentially the same at all levels of command, may entail a broader scope, complexity, and detail at the higher levels of command due to the coordination and synchronization requirements which exist across command lines and service or multinational boundaries. Table 3-1 provides medical evacuation planning considerations at various levels of command based on METT-TC.

**Table 3-1. Medical evacuation planning considerations**

| | |
|---|---|
| ***Mission*** | 1. Define the protocol for a valid evacuation mission (mission statement). This protocol will serve as the authority to execute the mission by each medical evacuation unit/element/crew.<br>2. Use the HSS/FHP annex and mission analysis to:<br>• Task evacuation units where needed. This may involve tasking part or all of DS units to perform a GS or area support mission.<br>3. Based on the type of operation:<br>• Define trigger points for changes in the evacuation support plan.<br>• Develop a concept of operations for all of the corps evacuation resources.<br>4. Define task organization of units assigned to the corps (and when directed to the JTF) and support relationships, if possible. Lines of coordination and mutual support relationships should be clearly delineated.<br>5. Area support and GS coverage areas will be |

| | |
|---|---|
| | defined with flexibility built into the plan. Coverage areas along division boundaries, for example, should not restrict or hinder evacuation of neighboring units that are geographically close.<br>6. Define pilot certifications for shore-to-ship medical air evacuation.<br>7. Develop standing operating procedure (SOP) requirements for single versus dual ship medical air evacuation. Address theater-specific procedures for in-flight link-up between medical evacuation aircraft and nonmedical evacuation aircraft for the purpose of dual ship requirements and/or security of pick-up zone.<br>8. Delineate any nonstandard air evacuation missions and allocate resources accordingly, such as support to Army special operations, joint and coalition forces.<br>9. Coordinate with any units that will have shared responsibilities for evacuation in a specific area and/or a specific mission.<br>10. Evaluate medical evacuation specific A2C2 considerations. Specifically, coordinating the crossing of brigade/division/corps boundaries/ area of operations. Medical evacuation must be afforded special expedited procedures for medical evacuation missions. Additionally, METT-TC dependent air corridors may have to be developed for the medical evacuation mission. |
| ***Enemy*** | 1. Define the level or air-to-air threat and the threat faced by ground evacuation resources and the impact of this threat on the overall evacuation plan. This will be a major consideration in dictating which means of evacuation will be the primary means.<br>2. Based on the estimate of the number of enemy prisoners of war (EPW) or other categories of detainees involved in the operation—<br>• Define procedures for evacuating wounded, injured, or ill persons through medical channels to include responsibilities of the echelon commander for designating guards for those evacuated directly from the point of capture to an MTF.<br>• Allocate assets as necessary to provide this support and to support detainee collection points and holding areas. |
| ***Terrain and Weather*** | Define the types of terrain in the theater and use information as a major consideration in dictating the primary means of evacuation (ground or air in |

| | |
|---|---|
| | each sector) and any factors that may impair evacuation efforts. |
| ***Troops and Support Available*** | 1. Disseminate locations of facilities used in evacuation, MTFs, MASFs, and aerial ports of debarkation (APOD).<br>2. Disseminate the capabilities and availability of joint evacuation assets and how to request their assistance.<br>3. Collect contact information from all evacuation units, assemble and publish for theater use.<br>4. Standardize a mandatory evacuation mission data collection format and collection method and schedule. |
| ***Time Available*** | 1. Define the acceptable limits of evacuation time based on the distance from evacuation units.<br>2. Determine the proper allocation of resources to support the entire theater requirements.<br>3. Plan for trigger points (in time or distance) for changes in the evacuation plan to occur.<br>4. Define phases of the operation that reflect the commander's intent. Major changes in evacuation coverage usually occur when changes occur in the operational tempo (OPTEMPO).<br>5. Define briefing levels for launch authority. Consider developing a uniform theaterwide aviation risk assessment sheet. |
| ***Civil Considerations*** | 1. Based on the HSS/FHP estimate and plan, determine requirements for the evacuation of HN civilians and others.<br>2. Define and disseminate protocols and procedures for evacuating civilians.<br>3. Allocate resources to support this mission, when directed. |

3-17. For an effective HSS/FHP plan, medical staffs must coordinate and share operations information in order to synchronize support operations for mission accomplishment. See FMs 4-02, 4-02.1, 4-02.6, 4-02.12, 4-02.17, 4-02.25, 8-42 (4-02.42), and 8-55 (4-02.55), for doctrinal guidance on HSS/FHP operations. After the HSS/FHP plan is completed, it is incorporated into the C2 headquarters plan. After all annexes and appendixes of the plan are approved by the commander, it is incorporated into the OPORD.

## SURGEON'S RESPONSIBILITIES

3-18. The surgeon is responsible for supervision and development of AHS input for the OPORD. The HSS plan serves as the base document for this input. The HSS plan is revised or updated based on mission analysis or changes in HSS requirements. The surgeon is tasked by the Assistant Chief of Staff (Operations and Plans) (G3)/S3 for HSS/FHP input to the OPORD for support of operations. The G3/S3 indicates timeline requirements. The surgeon and his staff are involved in all stages of the planning process, to identify all HSS requirements. Information on development of running estimates and the OPLAN/OPORD are discussed below. The medical plan/operations cell develops a HSS/FHP plan based on guidance received from the commander and the surgeon and provides HSS/FHP operational planning updates to the surgeon. The plan is briefed to the commander for approval, as required and provided to the G3/S3 according to the format of the in FM 8-55 (4-02.55).

# MEDICAL EVACUATION TOOLS

3-19. It is essential that the evacuation plan for all combat operations be well planned, coordinated, and disseminated. In designing the medical evacuation plan, the medical planner should consider the following:

- Casualty collection points normally are predesignated along the axis of advance or evacuation routes.
- Forward of the BAS, the trauma specialist, CLS, and combat troops take casualties to the CCPs. These points facilitate acquisition by supporting ambulance teams and reduce evacuation time.
- When used by the BAS, CCPs help preserve BAS mobility, preclude carrying casualties forward, and reduce evacuation time to the sustainment area. When designating a CCP, the designating authority makes a decision whether or not to provide medical staff at the location. This decision is based upon the assessment of risk versus the availability of personnel.
- Normally, the role of care designating the point is responsible for staffing. Medical personnel may not be available to staff these points, and CLSs and ambulatory patients may be required to perform self-aid, buddy aid, or enhanced first aid. Casualty collecting points should be identified on operational overlays.

---

***Note***. A CCP staffed by a trauma treatment team is designated as a BAS (minus) rather than as a CCP.

---

3-20. Ambulance exchange points are a predetermined position where patients are exchanged from one evacuation platform to another.

3-21. These points are normally preplanned and are a part of the HSS/FHP annex to the OPLAN. In the forward area, the threat of enemy ground activities, large concentrations of lethal weapons systems, and effective use of antiaircraft weapons may dictate that the AXP be a predetermined rendezvous point for the rapid transfer of patients from one evacuation platform to another. The location of AXPs should be frequently changed to preclude attracting enemy fires.

3-22. Ambulance exchange points are established for many different reasons:

---

For example, the ambulance platoon of the heavy BSMC possesses a mixture of wheeled and tracked ambulances.

---

- The brigade tracked ambulances are provided so that they may keep up with maneuver elements. These vehicles carry the patients from the BAS to an AXP where the brigade wheeled ambulances take over for the relatively longer trip to the sustainment area.
- Ambulance exchange points are not limited to ground evacuation assets. Another example is a situation where the threat air defense artillery capability is such that air ambulances cannot fly as far forward as the BASs. However, an AXP could be established a few kilometers to the rear, still well forward of the BSA.
- The brigade combat team (BCT) tracked or wheeled ambulances could then transfer the patients to the air assets, thereby facilitating the rapid evacuation of patients and realizing a significant time savings.

3-23. By using AXPs evacuation assets are returned to their supporting positions faster. This facilitates evacuation as the returning crews are familiar with the road network and the supported unit's tactical situation. In the case of air ambulance (AA) assets, it is important because of the requirements for integration into the A2C2 system at each role and the enhancement to survivability provided by current threat and friendly air defense information.

3-24. The ambulance shuttle system is an effective and flexible method of employing ambulances during combat. It consists of one or more ambulance loading points, relay points, and when necessary, ambulance control points, all leveled forward from the principal group of ambulances, the company location, or basic relay points as tactically required. The ambulance—

- Loading point is a point in the shuttle system where one or more ambulances are stationed ready to receive patients for evacuation.
- Relay point is a point in the shuttle system where one or more empty ambulances are stationed. They are ready to advance to a loading point or to the next relay post to replace an ambulance that has moved from it. As a control measure, relay points are generally numbered from front to rear.
- Control points consists of a Soldier (from the ambulance company or platoon) stationed at a crossroad or road junction where ambulances may take one more directions to reach loading points. The Soldier, knowing from which location each loaded ambulance has come, directs empty ambulances returning from the sustainment area. The need for control points is dictated by the situation. Generally, they are more necessary in forward areas.

3-25. In the establishment of the ambulance shuttle system, once the relay points are designated, the required numbers of ambulances are stationed at each point. If the tactical situation permits, the ambulances may be delivered to the relay points by convoy.

3-26. Staffing of relay, loading, and ambulance control points. Important points may be manned to supervise the blanket, litter, and splint exchange and to ensure that messages and medical supplies to be forwarded are expedited.

3-27. Advantages of the ambulance shuttle system are that the system—

- Places ambulances at CCPs and BASs as needed.
- Permits a steady flow of patients through the system to MTFs.
- Avoids unnecessary massing of transport in forward areas.
- Minimizes the danger of damage to ambulances by the enemy.
- Permits the commander or platoon leader to control his elements and enables him to extend their activities without advancing the headquarters.
- Facilitates administration and maintenance.
- Maximizes the use of small battle command elements (sections or platoons) to operate the ambulance shuttle without employing the entire parent unit.
- Provides for flexible use of other ambulance assets for specific situations.

## OBSTACLES MARKING

3-28. Ambulance crews must know and recognize the standard land/gap marking patterns. Unit TSOPs on the types of materials used should be available to ambulance crews. Refer to FM 3-34.2 for additional information.

## Chapter 4

# Army Medical Evacuation

Medical evacuation operations are planned to provide comprehensive, responsive, flexible, and agile support to the tactical commander in conformity with the commander's intent and OPLANs. This chapter discusses the employment of medical evacuation resources and the coordination and synchronization required to effectively execute medical evacuation operations (to include the transfer of patients between MTFs and to staging facilities) by air and ground evacuation assets.

## SECTION I — MEDICAL EVACUATION SUPPORT

## EVACUATION PRECEDENCE

4-1. Casualties requiring evacuation are prioritized to ensure the most seriously injured or ill receive timely medical intervention consistent with their medical condition. As with medical treatment, medical urgency is the only factor used to determine the medical evacuation precedence. (Appendix A provides an in-depth discussion of the provisions of the Geneva Conventions.)

**This paragraph implements STANAG 3204.**

4-2. The determination to request medical evacuation and assignment of a precedence is made by the senior military person present or, if available, the senior medical person at the scene. This decision is based on the advice of the senior medical person at the scene (if available), the patient's condition, and the tactical situation. Assignment of a medical evacuation precedence is necessary. The precedence provides the supporting medical unit and controlling headquarters with information that is used in determining priorities for committing their evacuation assets. For this reason, correct assignment of precedence cannot be overemphasized; over classification may result in an increase in evacuation which could burden the HSS system.

4-3. The patient's medical condition is the overriding factor in determining the evacuation platform and destination facility. The AA operates wherever needed on the battlefield, dependent on risk and METT-TC factors. The crew of the AA, assisted by onboard patient monitoring and diagnostic equipment, is trained in aeromedical procedures to provide optimum en route patient care. It is the preferred method of evacuation for most categories of patients. Air ambulances are a low-density, high-demand resource and must be managed accordingly. To conserve these valuable resources, medical planners should plan to use AA to primarily move Priority I, URGENT and Priority IA, URGENT-SURG patients with the other categories on a space-available basis. Health service support planners must plan for a synchronized air and ground evacuation plan. Depending on the length of time required for an AA to be dispatched and arrive at the POI, it may be prudent to evacuate the casualty by ground evacuation assets to a BAS and/or Role 2 MTF for stabilization by a physician. Further evacuation could then be accomplished by AA.

4-4. Patients will be picked up as soon as possible, consistent with available resources and pending missions. Table 4-1 depicts the categories of evacuation precedence and the criteria used to determine the appropriate precedence.

**Table 4-1. Categories of evacuation precedence**

| | |
|---|---|
| ***Priority I—URGENT*** | Is assigned to emergency cases that should be evacuated as soon as possible and within a maximum of 2 hours in order to save life, limb, or eyesight, to prevent complications of serious illness, or to avoid permanent disability. |
| ***Priority IA—URGENT-SURG*** | Is assigned to patients who must receive far forward surgical intervention to save life and to stabilize them for further evacuation. |
| ***Priority II—PRIORITY*** | Is assigned to sick and wounded personnel requiring prompt medical care. This precedence is used when the individual should be evacuated within 4 hours or his medical condition could deteriorate to such a degree that he will become an URGENT precedence, or whose requirements for special treatment are not available locally, or who will suffer unnecessary pain or disability. |
| ***Priority III—ROUTINE*** | Is assigned to sick and wounded personnel requiring evacuation but whose condition is not expected to deteriorate significantly. The sick and wounded in this category should be evacuated within 24 hours. |
| ***Priority IV—CONVENIENCE*** | Is assigned to patients for whom evacuation by medical platform is a matter of medical convenience rather than necessity. |
| The NATO STANAG 3204 has deleted the category of Priority IV—CONVENIENCE; however, it will still be included in the US Army evacuation priorities as there is a requirement for it on the battlefield. | |

## SECTION II — MEDICAL EVACUATION REQUESTS

4-5. This section discusses requests for medical evacuation. Specific procedures, frequencies, and security requirements for transmittal of medical evacuation requests are delineated through the orders process and are made a part of the unit/command SOPs. Each sector based on the METT-TC may be designated with a different method of evacuation as the primary means to effect evacuation. In sectors which have a high ground-to-air or air-to-air threat may rely on ground evacuation assets to move the majority of patients. In other sectors where the ground threat is high and comprised of small arms, improvised explosive devices, and bombs, medical evacuation operations may be more efficiently and effectively executed by AAs. An additional consideration in planning medical evacuation operations is to determine whether armed escorts are required for either the ground or AA mission. Those missions that require armed escort must be thoroughly coordinated and synchronized between the medical assets and force protection assets that will accompany them.

4-6. Medical evacuation requests often are sent from the POI, through intermediaries, such as higher headquarters, who then transmit the request up to the nearest medical evacuation unit. The unit relaying the request must ensure that it relays the exact information originally received. The radio call sign and frequency relayed (Line 2 of the request) should be that of the requesting unit and not that of the relaying unit. However, the intermediaries contact information can be given as additional information if a callback is necessary to clarify details of the mission. Figure 4-1 depicts the communication flow for evacuation requests.

4-7. The necessity for secure communication in transmitting a 9-line request is METT-TC dependent. A technically capable enemy may be able to intercept a nonsecure 9-line request and strike at the pickup zone causing additional harm to the patients and the medical evacuation crew. In combat and when directed by higher headquarters, always use secure communications when requesting medical evacuation.

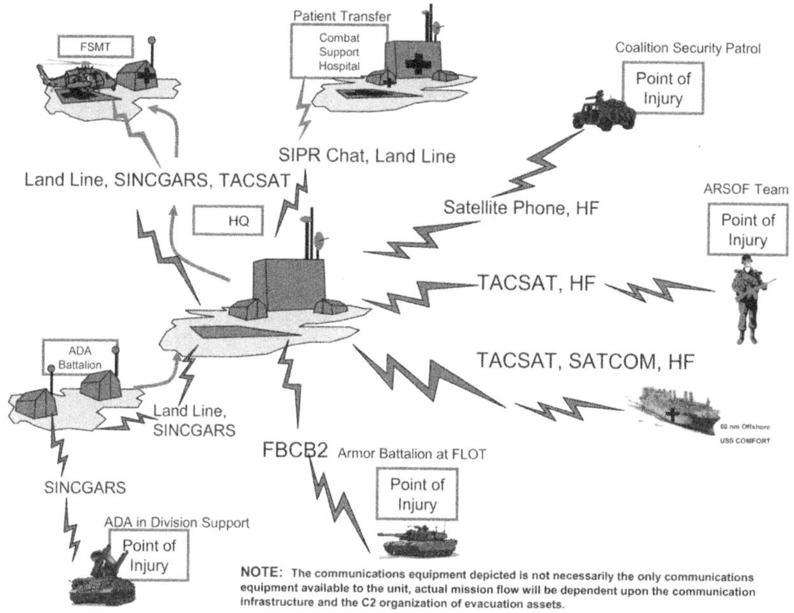


**Figure 4-1. Flow of communication for evacuation request**

# GROUND EVACUATION REQUEST

4-8. Figure 4-2 provides an example of a request from the POI for ground evacuation. The patient is a ROUTINE precedence patient requiring evacuation from the POI to the supporting BAS (Role 1 MTF) with follow-on evacuation to an AXP for further evacuation to the rear. Should the patient require further treatment, either an air or ground evacuation could be used.

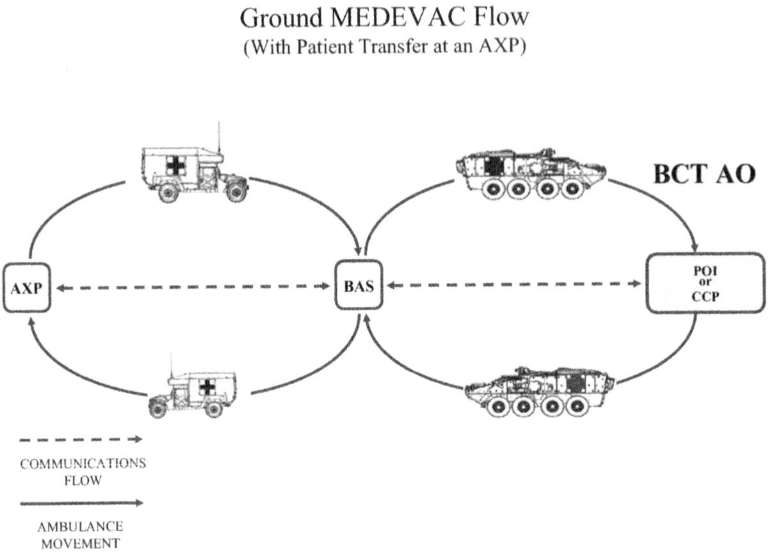


**Figure 4-2. Ground evacuation request in a maneuver unit**

4-9. Ground medical evacuation assets also provide DS and GS to supported units in the EAC. Figure 4-3 and Figure 4-4 depict the processes of providing DS to an ASMC and the GS support role of transferring a patient from a corps hospital to a MASF for further evacuation out of the theater.

4-10. In the corps area, the ASMC provides Roles 1 and 2 medical support on an area basis. The ASMC has organic evacuation assets and receives augmentation when required from the medical company ground ambulance assigned to the MMB.

4-11. Medical evacuation resources are also used to transfer a patient between hospitals within the theater and from a hospital to MASF for further evacuation out of theater. In Figure 4-3, this mission is being accomplished by a ground ambulance. However, depending upon the distance from the hospital to the MASF and/or APOD, this mission can also be accomplished by AA assets.

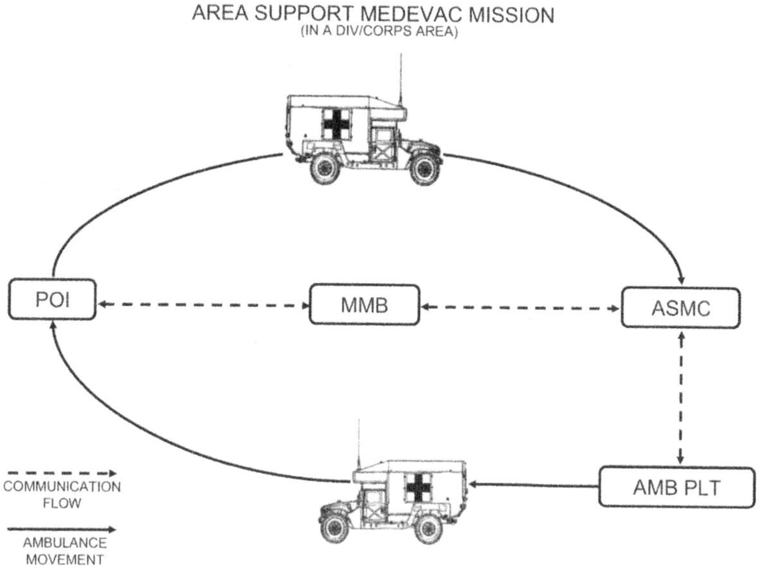


**Figure 4-3. Ground evacuation request in the corps**

# AIR EVACUATION REQUEST

4-12. Air ambulances may fly as far forward as possible on the battlefield. Although evacuation by AA is the preferred means for all casualties, when high evacuation workloads exist, evacuation by AA should be the primary means used for URGENT and URGENT-SURG patients. Figure 4-5 depicts the zones of evacuation. The GSAB/CAB will place AA assets where they can best support the tactical commander's plan and facilitate the timely and responsive evacuation. To coordinate and synchronize the ground and air medical evacuation plans, a staff element was designed within the GSAB headquarters. This cell consists of the medical planners on the GSAB and CAB staff. The medical operations cell (MOC) will:

- Establish flight procedures specific to air evacuation missions within the CAB. This may include special routes or corridors as well as procedures for escort aircraft linkup.
- Ensure lines of communication (LOC) to supported units and higher levels of medical command are available. Further the MOC ensures supported units understand medical evacuation procedures and capability.
- Establish medical evacuation briefing and launch procedures. Ensure there is 24-hour access to those able to launch high and very high-risk missions.
- Maintain awareness of tactical and medical situation. Coordinate with MROs at higher levels to efficiently conduct the general support mission and to work in concert with adjacent units.
- Assist the AA CO and GSAB/CAB staff conduct medical evacuation operations.

- Coordinate missions with supported command surgeons. To ensure situational awareness and coordination of medical evacuation efforts, this staff keeps command surgeons informed on air evacuation missions being executed in their AO.
- Consults and coordinates with supported command surgeons when AAs cannot be launched to execute a requested mission. Ensures that the appropriate command surgeon is notified to ensure mission can be accomplished by ground evacuation assets.

4-13. Figure 4-4 depicts a request for air medical evacuation. The scenario shows the request being sent to the FSMT, the AA company, and the GSAB. The routing of the request would be dependant on how our forces are arrayed. In a similar situation where ground evacuation may not be readily available or the tactical situation precludes its use the requesting unit could transmit their medical evacuation request directly to a supporting FSMT as a primary means for a medical evacuation request or the AA company as a secondary option with a third being sent directly to the GSAB.

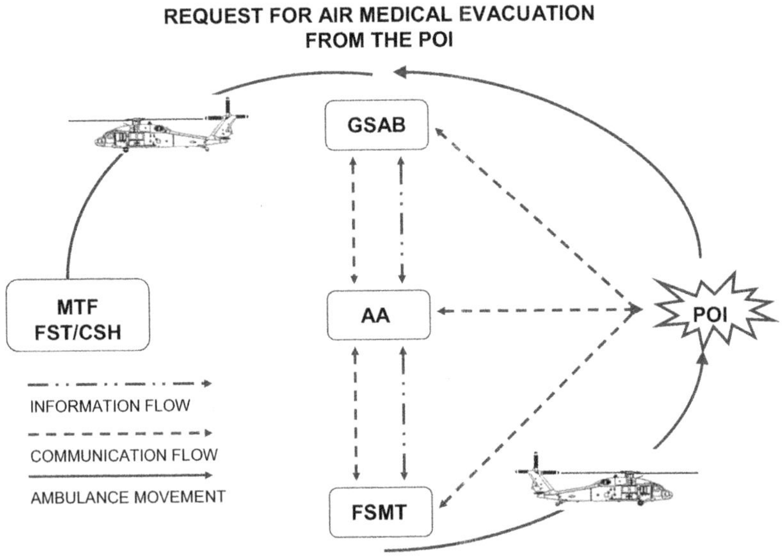


**Figure 4-4. Air evacuation request in the corps**

4-14. Figure 4-5 illustrates evacuation zones during a noncontiguous conflict. Here, no forward line of own troops (FLOT) is clearly visible.

- Forward support medical evacuation teams are placed where they are most needed. This can be with troops most often engaged in combat, high population density areas, area of famine or disease with high civilian casualties, refugee areas, geographically centralized location, and such.

- Area or general support FSMTs accomplish the patient transfer mission that develops between FSMTs and MTFs, between MTFs, and between MTFs and intertheater movement locations. These FSMT must also be responsible for the DS mission in their immediate vicinity.

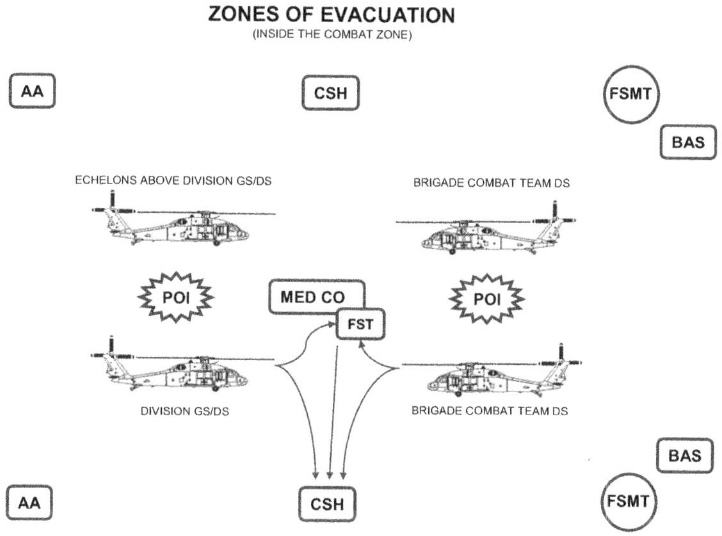


**Figure 4-5. Air ambulance zones of evacuation**

4-15. Figure 4-6 illustrates a comparison in aviation mission planning/execution cycle for medical evacuation and typical aviation.

- In contract to a typical aviation mission cycle, continuous aeromedical evacuation coverage results in extended operational duty days that often exceed 24 hours in length in order to provide seamless coverage to our forces. Medical evacuation units must plan and develop a detailed battle rhythm that addresses composite risk management, mission execution processes, and tactics, techniques, and procedures unique to 24-hour continuous evacuation operations.
- A medical evacuation crew cycle begins much like any other mission planning however they will not know the location or the time of execution for their mission. Since during these cycles it is critical the crew responds as rapidly as safely possible when the last information is received, it is imperative that crews are afforded the ability to manage rest while ensuring the procedures are in place to maintain battlefield situational awareness and facilitate a rapid, safe execution when a 9-line mission is received. This includes but is not limited to crew rest cycles, briefing procedures, and aviation and logistical support to remote/split-based crews to ensure the safe and timely execution of missions.

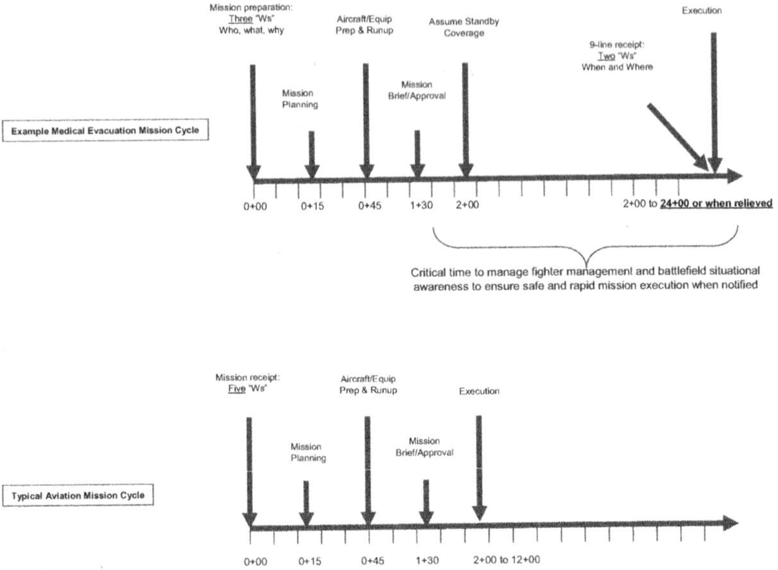


**Figure 4-6. Aviation mission planning/execution cycle**

# JOINT INTERCONNECTIVITY

4-16. When directed by the GCC, Army evacuation assets may be used to evacuate patients from the other services or may be placed in direct support of other service, allied, or coalition units participating in the operation. Prior to the initiation of the operation, communications, and procedural issues should be synchronized to ensure the seamless execution of the operation. Interoperability issues can result in slower response times and may adversely effect medical evacuation operations. Figure 4-7 depicts the communications and patient flow in a joint scenario.

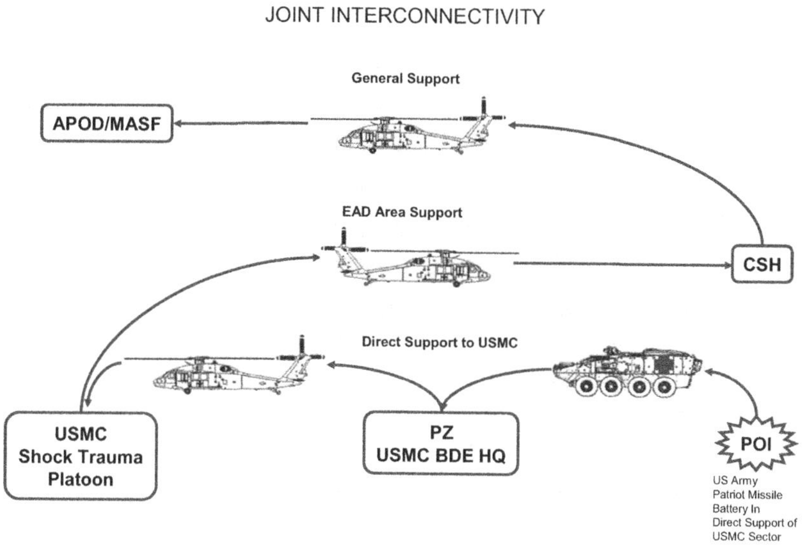


**Figure 4-7. Air and ground ambulance evacuation in a joint environment**

## SECTION III — MEDICAL EVACUATION UNITS, ELEMENTS, AND PLATFORMS CONSIDERATIONS

# MEDICAL EVACUATION SUPPORT PROTOCOL

4-17. Medical evacuation is performed by the higher role of medical care supporting forward and evacuating from the lower role. Medical evacuation support for theater, corps, and division is provided by organic ambulance teams, squads, and platoons. In addition, it is provided by the MMB medical company, ground ambulance medical company (GAMC), ASMC, and the GSAB medical company, AA.

4-18. The MMB maximizes the effective employment of ground ambulances resources whether in support of a division, corps, or at theater level. It coordinates through the combat aviation brigade's GSAB for effective use of AA support of its ASMCs. The medical brigade, MMBs, GSAB, sustainment brigade, BCTs and division surgeon section (DSS) coordinate and synchronize medical evacuation operations for the division AOs. The DSS coordinates with the MEDBDE for the evacuation of patients from division MTFs to supporting CSHs. The MEDBDE coordinates and arranges the required intratheater medical evacuation of patients to and from supporting CSHs. The MEDCOM coordinates for evacuation of patients out of the theater. To ensure that patients are evacuated to the appropriate MTFs, MROs are organic to the MEDCOM, and MEDBDE.

### Medical Evacuation Platforms Factors

4-19. Evacuation platforms must be capable of keeping pace with the troops supported. The patient's medical condition is the overriding factor in determining the evacuation platform and destination MTF. The AA operates wherever needed on the battlefield, dependent on risk and METT-TC factors. Use of hardened armored MEV may be the vehicle of choice for some missions for short evacuation to an MTF or to a secure AXP for transfer to an air or wheeled ground ambulance. The criteria for an efficient and effective evacuation system were identified in Chapter 1.

4-20. The crew of the AA, assisted by onboard patient monitoring and diagnostic equipment, is trained in aeromedical procedures to provide optimum en route patient care. It is the platform of choice for most categories of patients. However, insufficient numbers of AAs are available to evacuate all patients expected in a battle zone. To conserve these valuable resources, medical planners should plan to use AAs to primarily move Priority I, URGENT, Priority IA, URGENT-SURG, and Priority II patients with other categories on a space-available basis.

## SECTION IV — MEDICAL EVACUATION AT UNIT LEVEL

## CONSIDERATIONS FOR MEDICAL EVACUATION MISSIONS

4-21. As METT-TC factors affect all medical evacuation missions and the employment of ambulance assets. The medical commander must consider the basic tenets that influence the employment of medical evacuation assets. These factors include the patient's medical condition and the:

- Tactical commander's plan for employment of combat forces.
- Enemy's most likely course of action.
- Anticipated patient load.
- Expected areas of patient density.
- Availability of medical evacuation resources to include ground and air crews.
- Availability, location, and type of supporting MTFs.
- Protection afforded medical personnel, patients, and medical units, vehicles, and aircraft under the provisions of the Geneva Conventions.
- Army airspace command and control plan.
- Engineer obstacle plans.
- Fire support plan (to ensure medical evacuation assets are not dispatched onto routes and at the times affected by the fire support mission).
- Road network/dedicated medical evacuation routes (contaminated and clean).
- Weather conditions.

### Patient Acquisition

4-22. Units with organic medical evacuation assets have the primary responsibility for patient acquisition. Methods of employment and evacuation techniques differ depending upon the nature of the operation.

4-23. Units without organic ambulance assets are provided medical evacuation support on an area support basis. Units must develop techniques which facilitate the effective employment of their combat medics/company/platoon medics, enhance the ability to acquire patients in forward areas, and rapidly request medical evacuation support. The techniques developed should be included in the unit TSOP. As a minimum, the TSOP should include the—

- Vehicle assignment for the organic medical personnel.
- Vehicles designated to be used for casualty transport and/or patient evacuation.
- Procedures for requesting medical evacuation support (during routine operations or during mass casualty situations).
- Role of the first sergeant (1SG), platoon sergeants, and CLSs in medical evacuation.

## MANEUVER BATTALION MEDICAL PLATOONS

4-24. The medical platoon leader (a physician) along with the medical platoon field medical assistant should be included in all battalion tactical planning. This officer—

- Maintains knowledge of the concept of operations, commander's intent, and the anticipated medical requirements.
- Develops the HSS/FHP plan (FM 8-55 [FM 4-02.55] and FM 4-02.4) and provides HSS/FHP overlays with preplanned evacuation routes and CCPs, to ambulance squads or teams for inclusion in the battalion OPLAN.
- Ensures that the platoon sergeant provides strip maps or other navigational tools to the ambulance drivers, if needed.
- Requests augmentation support from the supporting medical company in advance of the operation, if required. When elements of other maneuver battalions are attached as part of a task force, then ensures that adequate medical elements are included in the support package.
- Ensures that orientation and support are provided for the medical personnel. This precludes taxing the medical elements of the receiving unit. These responsibilities are normally delegated to the medical operations officer (field medical assistant).

4-25. The ambulance/evacuation section squad leader/NCO ensures that the ambulance teams have a working knowledge of the terrain features in the AO. Whenever possible, he familiarizes with primary and secondary medical evacuation routes through route reconnaissance conducted by the squad leaders. This platoon sergeant and squad leader manages the employment of the ambulance teams and monitors the communications net to remain abreast of the tactical situation.

4-26. The following factors should be considered when selecting ambulance routes:

- Tactical mission.
- Coordinating evacuation plans and operations with the unit movement officer.
- Security of routes and security escort.
- Availability of routes.
- Physical characteristics of roads and cross-country routes (to include natural obstacles).
- Requirements to traverse roads in urban areas and potential obstructions from rubble and debris.
- Traffic density.
- Time and distance factors.
- Proximity of possible routes to areas that may be subject to enemy fire.
- Lines of patient drift.
- Cover, concealment, and available defilade for moving and stationary vehicles.
- Engineer obstacle plans.
- Fire support plan (to ensure medical evacuation assets are not dispatched onto routes and at the times affected by the fire support mission).

4-27. Depending upon the combat situation, the modes of evacuation may include walking soldiers who are wounded, manual and litter carries, nonmedical transportation assets, or dedicated medical evacuation platforms. Evacuation in the battalion area normally depends on the organic ambulances assigned. Evacuation by AA is dependent upon the patient's medical condition, availability of air assets, tactical situation, air superiority, and weather conditions.

4-28. The ambulance team is normally deployed to each maneuver company trains and one remains with the Role 1 MTF/BAS in the combat trains. Ambulances operate as far forward as the tactical situation permits. An ambulance team operating in a maneuver company AO, is normally under the tactical control of the maneuver company executive officer or 1SG. This ambulance team, however, remains under the OPCON of the medical platoon.

4-29. The medical operations officer ensures that the ambulances are located close to the anticipated patient workload. An ambulance team consists of one ambulance and two combat medics (on track vehicles, a third emergency care specialist is required to provide en route medical care). One or two of these teams serve in DS of a maneuver company. To become familiar with the specific terrain and battlefield situation, the team maintains contact with the supported maneuver company during most combat operations. Maneuver medical platoon ambulances not in DS are positioned strategically throughout the battalion area or are sited at the BAS to—

- Evacuate patients from the company aid posts, CCPs to the BAS.
- Reinforce the forward positioned ambulance teams.
- Support the combat forces held in reserve and/or scout, mortar and forward observer platoons.

4-30. Another employment option is to forward site the additional ambulance teams at company CCPs, as well as at the BASs. Many times the ambulance team finds battlefield casualties who have not been seen by a trauma specialist. In these cases, the team members dismount and then find, treat, and evacuate the patients. Ambulance teams not specifically dedicated to support combat elements can be used as messengers in medical channels and to provide transport of emergency medical personnel, equipment, and supplies.

4-31. During static situations where the maneuver company is not in enemy contact or is in reserve, the ambulance team returns to the BAS to serve as reinforcement to other elements in contact. However, during movement to contact, the ambulance team immediately deploys with its supported unit. In moving patients back to the CCP, the team may be assisted by nonmedical personnel. Specific duties of the ambulance team are to—

- Maintain contact with supported elements.
- Find and collect the wounded.
- Administer EMT.
- Initiate or complete the FMC.
- Evacuate patients to the BAS.
- Direct or guide ambulatory patients to the BAS.
- Resupply company/platoon medics.
- Serve as messengers in medical channels.
- Route recons from Role 1 MTF to Role 2 MTF or the higher role of medical care.

## Casualty Collection Points

4-32. During the offense, CCPs may be used to avoid hampering the movement of the maneuver elements. In fast-moving situations, preplanned CCPs are included in the HSS plan and activated based on the crossing of phase lines, upon the occurrence of predetermined events, or on the execution of other control measures. It may be necessary to set up multiple CCPs for each phase of an operation. Rotating the use of these points precludes the enemy from using them to pinpoint maneuver elements or from attracting enemy fires. When the situation permits, medical evacuation of patients from CCPs, or AXPs may be accomplished by AAs.

4-33. Ambulance teams move using available terrain features for cover and concealment. They avoid prominent terrain features and likely targets. When stationary, the ambulance crew should conceal the vehicle as much as possible. When a casualty occurs in a tank, Bradley infantry fighting vehicle (BIFV), a Stryker wheeled armored vehicles, the ambulance team moves as close to the armored vehicle as possible. Assisted by the armored crew, if possible, the casualty is extracted from the vehicle and then administered EMT. The ambulance team moves the patient to the Role 1 MTF/BAS or to a CCP point to await further evacuation.

4-34. The senior company medic/trauma specialist normally remains with the company combat trains, but may be used anywhere in the company area, even assisting the ambulance teams in some situations. He may be used to direct ambulance teams to locations where vehicle crews need assistance or where injured or wounded crew members have been left. In some situations, crew members may have to rely on self-aid or buddy aid until the CLS or the platoon medic/trauma specialist arrives.

### Area Medical Evacuation Support

4-35. Medical evacuation on an area basis is required at all roles in the AHS system. The division units without organic medical evacuation resources, such as combat engineers, will require evacuation support on an area basis. To ensure that these elements receive adequate support, the medical planner must include their requirements into the OPLAN. Prior coordination is essential to ensure that the locations of CCPs, AXPs, and BASs are disseminated to these elements and that any unique support requirements are included in the OPLAN. Further, the triggers for relocating these assets and their next locations should also be included.

## MEDICAL EVACUATION SUPPORT FROM ROLE 1 AND ROLE 2 MEDICAL TREATMENT FACILITIES

4-36. Evacuation from Role 1 MTFs operated by treatment squad or team is normally provided by the BSMC/FSMC ambulance platoon and the GSAB AA company. Further, these ambulance assets provide medical evacuation support on an area basis to other units in the brigade area, as required.

### Ground Ambulances

4-37. The ambulance teams from the ambulance platoon are normally collocated with the BSMC/FSMC treatment platoon for mutual support. They establish contact and locate one ambulance team with the medical platoon of each maneuver battalion. The remaining ambulances are used for BCT operations and area support. The ambulances are positioned at AXPs or CCPs, or are field-sited with Role 1 MTFs operated by the BSMC/FSMC treatment teams.

4-38. A FSMT from the supporting GSAB medical company, AAs provides DS AA support for the BSMC/FSMC Role 2 MTF. The teams normally remain under the operational control of the GSAB. The DS relationship provides authority to the BSMC/FSMC in coordination with the brigade support battalion/forward support battalion (BSB/FSB) distribution support section, brigade surgeons section to direct the integrated air and ground medical evacuation support for the BCT. The section leader of the FSMT should be included in the brigade tactical planning process.

### Air Ambulances

4-39. The AA team evacuates Priority I, URGENT patients from as far forward as possible to the BSMC/FSMC. Further, when a FST is colocated with a BSMC/FSMC, AAs evacuate Priority IA, URGENT-SURG to this facility if the patient can not be flown to a Role 3 MTF condition due to distance or patient condition. External lift capabilities of aeromedical evacuation helicopters add an important dimension to its role on the battlefield in moving medical supplies and equipment.

4-40. The GSAB AAs may operate from the sustainment area and BSAs providing around-the-clock, immediate response, evacuation aircraft, as well as provide the BSMC/FSMC commander flexibility and agility in the movement of treatment teams and equipment to the forward battle area. It also provides the capability to rapidly resupply Class VIII supplies to combat units.

4-41. The medical elements within the BSB/FSB will request AA support through the brigade aviation element (BAE). The FSMT leader will provide coordination between the BSB/FSB, BAE, or the BCT S3.

4-42. In the BSMC/FSMC, the executive officer (XO) is the principal assistant to the commander for the tactical employment of the company assets. The XO should be included in all brigade tactical planning. Officer preparation will be required to reinforce or reconstitute forward HSS elements and to request augmentation through the brigade surgeon section (BSS), if required. The FSMT leader keeps the BSMC/FSMC XO apprised of the operational capability. This enables the XO to effect timely reinforcement or augmentation. The BSMC/FSMC XO must be familiar with the specific terrain and battlefield situation. Further, the XO should have a thorough understanding of the division and brigade commanders' ground tactical plan.

### Brigade Medical Evacuation Plan

4-43. The BSS is responsible for developing the BCT medical evacuation plan. The BSMC/FSMC commander is responsible for the execution of brigade medical evacuation plan, to include the use of both ground and air assets. The BSS medical planner should include the medical company commander, XO, brigade S1, brigade XO, medical platoon leaders, FSMT leader, the BSB/FSB medical plans and operations officer, and the BAE company commander (if available) in the planning process.

### Ground Ambulance Evacuation Brigade Role 2 Medical Treatment Facilities

4-44. Evacuation from the BSMC/FSMC is provided by ground ambulances from the GAMC or from AAs from the GSAB AA company. The GAMC provides support to the BSMCs and medical elements in the division AO. They usually evacuate only those patients who will not RTD within 72 hours.

### Ground Ambulance Support for Echelons Above Brigade/Division Units

4-45. The ASMC ambulance platoon normally collocates with the treatment platoon for mutual support and area taskings. It performs ground evacuation and en route patient care for supported units in the sustainment area. The ASMC may also evacuate patients from the supported Role 2 MTFs in the sustainment area to the ASMC Role 2 MTF. The ambulance platoon is mobile in operations as its assets may be totally deployed at one time. The platoon normally forward stations a portion of its teams in support of those units in the sustainment area.

4-46. The remaining teams are used for GS support to division sustainment, movement and maneuver and fires units, and for reinforcing support or ambulance shuttles. Platoons or squads from the GAMC will be in DS, or OPCON to, and collocated with supported division units.

## SECTION V — EXCHANGE OF PROPERTY

## PROPERTY EXCHANGE AND PATIENT MOVEMENT ITEMS

4-47. United States Army medical evacuation operations require that whenever a patient is evacuated from one MTF to another, or is transferred from one ambulance to another, medical items of equipment (casualty evacuation bags [cold weather-type bags], blankets, litters, and splints) remain with the patient. To prevent rapid and unnecessary depletion of supplies and equipment, the receiving Army element exchanges like property with the transferring element. This reciprocal procedure will be practiced to the fullest extent possible through all phases of evacuation from the most forward element through the most rearward hospital.

4-48. With USAF aeromedical evacuation operations, a major factor in the evacuation of patients is that specific medical equipment and durable supplies designated as patient movement items (PMI) must be available to support the patient during the evacuation. Examples of PMI include—

- Ventilators.
- Litters.
- Patient monitors.
- Pulse oximeters.

These items will be available for exchange at the supporting ASFs and MASFs. Refer to JP 4-02 and FM 4-02.1 for additional information on PMI.

**This subparagraph implements STANAG 2128 and QSTAG 436.**

4-49. Medical property of allied nations (NATO and ABCA armies) accompanying patients of allied nations will be returned to the parent nation at once, if possible. If it is not possible, like items will be exchanged as in the paragraph above.

4-50. When medical property of coalition forces without ratified standardization agreements accompanies patients of coalition forces, it is returned as soon as practical. Commanders should consult with their staff judge advocate early in the planning process to ensure appropriate policy and procedures are developed and disseminated.

## SECTION VI — MEDICAL EVACUATION SUPPORT FOR THE OFFENSE, THE DEFENSE, AND STABILITY OPERATIONS

### MEDICAL EVACUATION SUPPORT FOR THE OFFENSE

4-51. The offense is the decisive form of war, the commander's only means of attaining a positive goal or of completely destroying an enemy force (FM 3-0). The offense is characterized by rapid movement, deep penetrations, aggressive action, and the ability to sustain momentum regardless of counterfires and countermeasures.

4-52. When considering the evacuation plans to support an offensive action, the medical planner must consider many factors (FM 8-55 [FM 4-02.55]). The forms of maneuver, as well as the enemy's capabilities, influence the character of the patient workload and its time and space distribution. The analysis of this workload determines the allocation of AHS resources and the location or relocation of MTFs. Evacuation support of offensive operations must be responsive to several essential characteristics. As operations achieve success, the areas of casualty density move away from the supporting facilities. This causes the routes of medical evacuation to lengthen. Heaviest patient workloads occur during disruption of enemy main defenses, at terrain or tactical barriers, during the assault on final objectives, and during enemy counterattacks. The accurate prediction of these workload points by the medical planner is essential if medical evacuation operations are to be successful.

4-53. In traditional combat operations, the major casualty area of the division BCTs is normally the zone of the main attack. As the main attack accomplishes the primary task of the division, it receives first priority in the allocation of combat power. The allocation of combat forces dictates roughly the areas which are likely to have the greatest casualty density. As a general rule, all BCT MTFs are located initially as far forward as combat operations permit. This allows the maximum use of these facilities before lengthening evacuation lines force their displacement forward. In operations that feature deep battles with CBRN weapons targeted at supporting logistical bases, MASCAL operations may be conducted in sustainment area.

4-54. As advancing combat formations extend control of the battle area forward, supporting medical elements overtake patients. This facilitates the acquisition of the battle wounded and reduces the vital time elapsed between wounding and treatment. In offensive operations, two basic problems confront the supporting evacuation units.

- First, contact with the supported unit must be maintained. Responsibility for the contact follows the normal HSS pattern of rear to front. The contact is maintained by forward deployed air and ground evacuation resources.
- Secondly, the mobility of the MTFs supporting the combat formations must be maintained. Periodically, BCT medical companies and collocated FSTs, are cleared of their patients so that they may move forward. In some scenarios, CSHs may also have to evacuate patients to reduce patient overflow, surgical backlogs, or to increase bed availability.

4-55. This requirement for prompt evacuation of patients from forward MTFs requires available ambulances to be leveled well forward from the outset. The requirement for periodic movement of large numbers of patients from divisional and corps facilities further stresses the evacuation system.

## TYPES OF OFFENSIVE OPERATIONS

### Movement to Contact/Attack

4-56. Medical evacuation support in movement to contact/attack is key to the tactical plan. Prior deployment of evacuation resources with parent and supported units permits uninterrupted and effective evacuation support.

### Exploitation and Pursuit

4-57. Evacuation support of exploitation and pursuit operations resembles those discussed for the envelopment (paragraph 4-59). Since exploitation and pursuit operations can rarely be planned in detail, evacuation operations must adhere to TSOPs and innovative C2. These actions are often characterized by—

- Decentralized operations.
- Unsecured ground evacuation routes.
- Exceptionally long distances for evacuation.
- Increased reliance on convoys and AAs.

## MEDICAL EVACUATION SUPPORT FOR DIFFERENT FORMS OF OFFENSIVE MANEUVER

### Penetration and Attack

4-58. In this tactic, the attack passes through the enemy's principal defensive position, ruptures it, and neutralizes or destroys the enemy forces. Of all forms of offensive maneuver, the penetration of main enemy defenses normally produces the heaviest medical evacuation workload. Patient acquisition starts slowly, but becomes more rapid as the attack progresses. The evacuation routes lengthen as the operation progresses. The penetration maneuver is often preceded by heavy preparatory fires which may evoke heavy return fire. These enemy fires may modify the decision to place evacuation assets as far forward as possible. Patient evacuation may be slow and difficult due to damage to roads or the inaccessibility of patients. Evacuation support problems multiply when some combat units remain near the point of original penetration. This is done to hold or widen the gap in enemy defenses while the bulk of division forces exploit or pursue the enemy. Treatment elements are placed near each shoulder of the penetration; ground evacuation cannot take place across an avenue of heavy combat traffic. Besides the heavy traffic, the area of the penetration is normally a target for both conventional and nonconventional CBRN weapons.

### Envelopment

4-59. In the envelopment, the main or enveloping attack passes around or over the enemy's principal defensive positions. The purpose is to seize objectives which cut the enemy's escape routes and subject him to destruction in place from flank to rear. Since the envelopment maneuver involves no direct breach of the enemy's principal defensive positions, the medical evacuation system may not be confronted with a heavy workload in the opening phase—

- Ambulances are echeloned well forward in all levels of AHS to quickly evacuate the patients generated by suddenly occurring contact.
- Medical treatment facilities moving with their respective formations overtake patients during evacuation and reduce delays in treatment. After triage and treatment, the patients are evacuated to division/corps-level facilities by accompanying division/corps ground ambulance assets or AAs.

- When the isolated nature of the envelopment maneuver precludes prompt evacuation, the patients are carried forward with the treatment element. Again, nonmedical vehicles may be pressed into emergency use for this purpose. When possible, nonmedical vehicles should be staffed by a combat medic or CLS.
- When patients must be carried forward with the enveloping forces, AHS commanders use halts at assembly areas and phase lines to arrange combat protection for ground ambulance convoys through unsecured areas.
- Further, the commander may take advantage of friendly fires and suppression of enemy air defenses to call for prearranged AA support missions, or emergency use of medium-lift helicopter backhaul capabilities.

## Infiltration

4-60. Infiltration is a form of maneuver used during offensive operations.

- The division can attack after infiltration or use it as a means of obtaining intelligence and harassing the enemy.
- Though it is not restricted to small units or dismounted actions, the division employs these techniques with a portion of its units, in conjunction with offensive operations conducted by the remainder of its units.

### *Army Health System of Infiltration/Maneuver*

4-61. The AHS of infiltration/maneuver is restricted by the amount of medical equipment, supplies, and transportation assets that can be introduced into the attack area. No deployment of division-level medical units without their organic transportation should be attempted.

4-62. Elements of unit-level AHS should be accompanied by their organic vehicles, and ambulances should receive priority for deployment. It may be necessary to man-carry enough BAS equipment into the attack area to provide EMT and ATM; however, this results in degrading mobility. When the element is committed without its ambulances, patients are evacuated to the BAS by litter bearer teams. This requires reinforcement of the medical platoon by division or corps medical personnel or improvisation of litter teams using combat troops (if available and approved by the tactical commander). Patient evacuation from the BAS and medical resupply of the force may be provided by litter bearers, depending upon distances and degree of secrecy required.

4-63. When airborne and air assault forces are used, infiltrating elements may land at various points within the enemy's rear area and proceed on foot to designated attack positions. As in surface movement, the amount of medical equipment taken may be limited. In airborne operations, the evacuation of patients will be by litter bearers or frontline ambulances to CCPs or the BAS and then by division-level ambulances to the Role 2 MTF. In air assault operations, the evacuation is by litter bearers to CCPs or to the BAS and then by AAs to a Role 2 MTF. Once the combat element begins the assault on the objective, secrecy is no longer important and its isolated location requires HSS characteristic to airborne and air assault operations until ground linkup.

## Turning Movement

4-64. The turning movement is a variant to the envelopment in which the attacker attempts to avoid the defense entirely; rather, the attacker seeks to secure key terrain deep in the enemy's rear and along his LOC. Faced with a major threat to his rear, the enemy is thus "turned" out of his defensive positions and forced to attack rearward at a disadvantage.

> *Note.* General MacArthur's invasion at Inchon during the Korean War is an example of a classic turning movement. Casualties were initially light as the main defenses were avoided; however, as the invasion developed, resistance stiffened and higher casualty rates were experienced. Further, as fighting occurred in a populated area (Seoul), significant civilian casualties resulted. The lack of Korean health care providers caused many of these civilians to seek medical aid from US field medical units.

### *Medical Evacuation for Turning Movement*

4-65. Medical evacuation support to the turning movement is provided basically in the same manner as to the envelopment. As the operation is conducted in the enemy's rear area, LOC and evacuation routes may be unsecured resulting in delays in resupply and evacuation. In the Inchon example, a hospital ship was located off the coast to accept patients evacuated from the fighting. However, due to the precarious tides, evacuation and resupply were often delayed for hours and sometimes days since the harbor could not be navigated by small vessels. It was not until Kimpo Airfield fell that timely evacuation could occur. The deployed HSS units must be able to quickly clear the battlefield of patients, evacuate them from the forward areas, and sustain the patients in rear areas until evacuation routes are established.

# MEDICAL EVACUATION SUPPORT FOR THE DEFENSE

## TYPES OF DEFENSE

4-66. There are three forms of the defense:

- **Area defense** concentrates on denying enemy access to designated terrain for a specific period of time, rather than on the outright destruction of the enemy.
- **Mobile defense** focuses on denying the enemy force by allowing him to advance to a point where he is exposed to a decisive counterattack by the striking force. The primary defeat mechanism, the counterattack, is supplemented by the fires of the fixing force.
- **Retrograde** is an organized movement to the rear and away from the enemy. The enemy may force these operations or a commander may execute them voluntarily. Within the retrograde operation there are three forms: delay, withdrawal, and retirement.

4-67. Support is generally more difficult to provide in the defense. The patient load reflects lower casualty rates, but forward area patient acquisition is complicated by enemy actions and the maneuver of combat forces.

4-68. Medical personnel generally are permitted much less time to reach the patient, complete vital EMT, and remove him from the battle site. Increased casualties among exposed medical personnel further reduce the medical treatment and evacuation capabilities. Heaviest patient workloads, including those produced by enemy artillery and CBRN weapons, may be expected during the preparation or initial phase of the enemy attack and in the counterattack phase.

4-69. The enemy attack may disrupt ground and air routes and delay evacuation of patients to and from treatment elements. The depth and dispersion of the defense create significant time and distance problems for evacuation assets. Combat elements may be forced to withdraw while carrying their remaining patients to the rear. The enemy exercises the initiative early in the operation which may preclude accurate prediction of initial areas of casualty density. This makes the effective integration of air assets into the evacuation plan essential. The use of AAs must be coordinated and synchronized with the supporting GSAB to ensure the synchronized execution of evacuation operations occurs.

4-70. The support requirements for retrogrades may vary widely depending upon the tactical plan, the enemy reaction, and the METT-TC factors. Firm rules that apply equally to all types of retrograde operations are not feasible, but considerations include:

- Requirement for maximum security and secrecy in movement.

- Influence of refugee movement that may impede medical evacuation missions conducted in friendly territory.
- Integration of evacuation routes and obstacle plans should be accomplished.
- Difficulties in controlling and coordinating movements of the force which may produce lucrative targets for the enemy.
- Movements at night or during periods of limited visibility.
- Time and means available to remove patients from the battlefield. In stable situations and in the advance, time is important only as it affects the physical well-being of the wounded. In retrograde operations, time is more important. As available time decreases, medical managers at all levels closely evaluate the capability to collect, treat, and evacuate all patients.
- Medical evacuation routes will also be required for the movement of troops and materiel. This causes patient evacuation in retrograde movements to be more difficult than in any other type of operation.
  - Command, control, and communications may be disrupted by the enemy.
  - Successful medical evacuation requires including ambulances on the priority list for movement.
  - Providing for the transportation of the slightly wounded in cargo vehicles and providing guidance to subordinate commanders defining their responsibilities in collecting and evacuating patients.
  - Special emphasis must be placed on the triage of patients and consideration given to the type of transportation assets available for evacuation.
- When the patient load exceeds the means to move them, the tactical commander must make the decision as to whether patients are to be left behind. The medical staff officer keeps the tactical commander informed in order that he may make a timely decision. Medical personnel and supplies must be left with patients who cannot be evacuated (refer to FM 4-02 for additional information).

## AIR AMBULANCE EVACUATION SUPPORT TO COVERING FORCES IN DEFENSIVE OPERATIONS

4-71. The division covering force uses combat power to screen the FLOT and force the early deployment of the enemy's main force. Organic AHS elements of the covering forces establish aid stations of minimal size or operate in the split-team mode (FM 4-02.4). Ambulances are deployed well forward to promptly clear patients from combat units. Medical care is limited to EMT and ATM, followed by rapid evacuation whenever possible.

4-72. The covering force surgeon maintains communications with attached aviation elements and uses these assets (augmented by medical personnel to provide en route medical care when feasible) to provide backhaul of casualties, rather than use ground ambulances. The early evacuation of patients from BASs ensures their mobility for rearward displacement. The BCT or armor cavalry regiment (ACR) medical company/troop is equipped to receive patients generated in the covering force area. Depending upon the fluidity of the defensive situation, the FSMT (if available) may have to rapidly evacuate patient's further rearward to a BSMC/FSMC and collocated FST or a CSH.

4-73. The wide dispersion of units and the manner in which they withdraw make patient acquisition difficult. When covering forces withdraw, patients are transported to the rear by the medical element that acquired them. Seriously wounded or injured patients are given priority for evacuation by air. There are usually few COSRs during covering force operations, although delayed symptoms may occur once the element is withdrawn to a safe area. Ground ambulances augmented by nonmedical transportation assets evacuate the remaining patients.

# MEDICAL EVACUATION SUPPORT FOR STABILITY OPERATIONS

4-74. Stability operations promote and protect US national interests by influencing the threat, political, and information dimensions of the operational environment. They include developmental, cooperative activities during peacetime and coercive actions in response to crisis.

4-75. Some of the purposes for which the US Forces are employed to conduct stability operations are as follows:

- Civil control.
- Civil security.
- Provide essential services.
- Support economic and infrastructure development.
- Governance.

4-76. Stability operations may complement the main effort or they may themselves constitute the main effort. Table 4-2 identifies the types of stability operations and the types of missions the BCT may conduct.

**Table 4-2. Types of stability operations**

| *Types Of Operations* | *Types Of Missions* |
|---|---|
| Civil Control | Establish police forces training. |
| | Integrate trained police into operations. |
| | Counter organized crime. |
| | Establish judicial system. |
| | Transition to indigenous police forces. |
| Civil Security | Initiate security forces training. |
| | Establish National Security Forces. |
| | Integrate security forces into operations. |
| | Isolate population from insurgency. |
| | Defeat insurgency. |
| | Transition to National Security Forces. |
| Essential Services | Sewage treatment plants operating. |
| | Establish trash disposal. |
| | Water treatment plants operating. |

**Table 4-2. Types of stability operations**

| | |
|---|---|
| | Restore electrical power.<br>Reopen hospitals and clinics. |
| | Reopen schools and universities. |
| Economic/Infrastructure Development | Implement Employment Programs. |
| | Secure vital natural resources. |
| | Repair/rebuild distribution infrastructure. |
| | Prioritize reconstruction projects. |
| | Essential banking services available. |
| | Implement public works projects. |
| Governance | Identify and recruit local leaders. |
| | Facilitate establishment of sector representation. |
| | Facilitate establishment of neighborhood councils. |
| | Facilitate establishment of district councils. |
| | Support and secure elections. |

4-77. Medical evacuation support to forces deployed in stability operations is dependent upon the specific type of operation, anticipated duration of the operation, number of forces deployed, theater evacuation policy, medical troop ceiling, and anticipated level of violence. In most situations, medical evacuation support follows the traditional support provided to combat forces. If there is a shortened theater evacuation policy, a limited medical troop ceiling, and limited hospitalization assets within the AO, organic and DS ambulance support is provided from the POI to the supporting Roles 1 or 2 MTF and, once the patient is stabilized for further evacuation, from the treatment element to an airfield for evacuation out of the theater.

4-78. Reconstruction operations use Army forces to assist civil authorities, foreign or domestic, as they prepare for or respond to crises and relieve suffering. In reconstruction operations, Army forces provide essential support, services, assets, or specialized resources to help civil authorities deal with situations beyond their capabilities. The purpose of reconstruction operations is to meet the immediate needs of designated groups for a limited time, until civil authorities can do without Army assistance.

4-79. Humanitarian and civic assistance operations can include a number of activities such as disaster relief, civil support, refugee assistance, the provision of medical care to isolated populations, and refeeding programs resulting from famines or natural disasters. Medical evacuation assets may be used to evacuate the injured from disaster sites, to provide the emergency transport of critically needed medical supplies and personnel to remote locations, or to perform emergency rescues during times of flooding, wildfires, or other natural disasters.

## SECTION VII — MEDICAL EVACUATION SUPPORT FOR ENABLING/SHAPING OPERATIONS

# PASSAGE OF LINES

4-80. Passage of lines presents a challenge for the medical planner. There will be a number of medical evacuation units using the same air and road networks. Coordination and synchronization are essential if confusion and over evacuation are to be avoided. The information required to operate in the BCT or division AO includes:

- Radio frequencies and call signs.
- Operations plans and TSOPs.
- Location of MTFs.
- Location of CCPs and AXPs.
- Main supply route, forward arming and refueling points (FARPs), and A2C2 data.

### SECURITY OPERATIONS

4-81. In security operations, the covering forces are dependent upon organic resources found in the maneuver battalion medical platoon for initial support. The level of command for the covering force (division or corps) determines the responsibility for the subsequent evacuation plan. In a corps covering force, for example, the corps AHS structure has the responsibility for establishing and operating the medical evacuation system to support the forward deployed corps forces. This is done to prevent the divisions following the covering forces from becoming overloaded with patients prior to the hand off and passage of lines. The use of CCPs, AXPs, and nonmedical transportation assets (CASEVAC) to move the wounded is essential. The covering force battle may be extremely violent. Patient loads will be high and the distance to MTFs may be much longer than usual. The effectiveness of the medical evacuation system depends upon the forward positioning of a number of ground ambulances and the effective integration of AAs into the evacuation plan.

### ADVANCE, FLANK, AND REAR GUARDS

4-82. In advance, flank, and rear guards, these forces normally receive medical evacuation support through the attachment of evacuation teams. The teams evacuate patients to predesignated CCPs along a main axis of advance or to the nearest treatment element providing area support. Employment of AAs provides a measure of agility and flexibility.

### RIVER CROSSING OPERATIONS

4-83. The river barrier itself exerts decisive influence on the use of division and BCT medical units. Attack across a river line creates an AHS delivery problem comparable to that of the amphibious assault. Army health system protection elements cross as soon as combat operations permit.

4-84. Early crossing of treatment elements reduces turnaround time for all crossing equipment that is used to load patients on the far shore. Maximum use of AA assets is made to prevent excessive patient buildup in far shore treatment facilities. Near shore MTFs are placed as far forward as assault operations and protective considerations permit to reduce ambulance shuttle distances from off-loading points. For detailed information on river crossing operations, refer to FM 90-13.

### RECONNAISSANCE OPERATIONS

4-85. The reconnaissance in force is an attack to discover and test the enemy's position and strength or to develop other intelligence. The division usually probes with multiple combat units of limited size, retaining sufficient reserves to quickly exploit known enemy weaknesses.

4-86. Ambulances are positioned well forward at both unit and division levels. Ambulances are moved at night to enhance secrecy. The echeloning of ambulances is an indication to the enemy that an attack is imminent due to the forward placement of HSS. Role 2 MTFs are not established until a significant patient workload develops.

4-87. Patients received at BASs of reconnoitering units are evacuated to Role 2 MTFs as early as practical or are carried forward with the force until a suitable opportunity for evacuation presents itself. Maximum possible use of AA assets is made to cover extended distances and to overcome potentially unsecured ground evacuation routes.

## RETROGRADE OPERATIONS

4-88. A retrograde operation is a movement to the rear or away from the enemy. This type of operation may be forced by enemy action or may be executed voluntarily.

## AEROMEDICAL EVACUATION SUPPORT IN RETROGRADE OPERATIONS

4-89. Aeromedical evacuation support in retrograde operations may vary widely depending upon the enemy reaction and the situation. Firm rules that apply equally to all types of retrograde operations are impossible to establish. Factors to consider in planning aeromedical evacuation support for retrograde operations include—

- Patient's condition/status.
- Mission, enemy, terrain and weather, troops and support available, time available and civil considerations.
- Requirement for maximum efforts in secrecy and employment of aircraft survivability equipment.
- Influence of refugee movement (which may impede military medical movements conducted in friendly territory).
- Integration of evacuation routes and other Army aviation routes.
- Difficulty in controlling and coordinating movements of the force which may produce lucrative targets for the enemy.

4-90. During retrograde operations, battalion, brigade, and division surgeons must evaluate their capability to collect, treat, and evacuate patients. Three factors influence the number of patients evacuated from the battlefield and the time and place of patient treatment:

- Time available prior to MTF displacement.
- Location of the enemy advance.
- Means of evacuation.

4-91. Due to congestion on main supply routes and general evacuation routes, the AAs may provide the best means of evacuation during retrograde operations. Seriously wounded patients should be evacuated by the fastest and most comfortable means and should receive medical care en route. Proper sorting (triage) and rapid evacuation of patients lessen the need for establishing a complete Role 2 MTF operation. Therefore, the FSMT (if available) leader should be prepared to surge his efforts in the retrograde. Combat and operational stress reaction casualty rates are usually low as compared to wounded in action (WIA) rates during controlled retrograde operations, but increase after safety is reached.

4-92. Commanders must decide to displace medical evacuation units before they hamper the maneuver forces conducting the retrograde movement. During mission planning, the medical evacuation units/elements integrate with the supported unit's displacement plan. The medical evacuation element leaders must coordinate with the supported unit commander and the BSMC for displacement decision points and locations.

## WITHDRAWAL OPERATIONS

4-93. A withdrawal operation is one in which a deployed force disengages from an enemy force. It may be forced by enemy pressure or conducted voluntarily.

4-94. Although the deployed force disengages from the enemy, contact is maintained by security elements while the main force moves to the sustainment area, forms march columns, and moves to a predesignated location. Aeromedical evacuation elements must be prepared to support the withdrawing force as well as any forces left in contact. Time available to acquire, treat, and evacuate patients from the line of contact is limited.

## AEROMEDICAL EVACUATION SUPPORT IN WITHDRAWAL OPERATIONS

4-95. The AA medical company element supporting the withdrawing force must ensure it maintains sufficient station capability at its present location to support the patients while taking action to move to a new location.

4-96. In addition to the evacuation support provided to the force left in contact, there is a requirement to support the force moving to the sustainment area. If the main force infiltrates to the sustainment area, patients are carried by their parent units to the assembly area or MTFs in the sustainment areas. Medical assets are positioned in the assembly area to receive patients.

4-97. The division establishes Role 2 MTFs at a minimal size and well to the rear of the first line of alternate positions. If combat and environmental conditions indicate a light patient workload and the road network permits rapid ambulance movement, the division may use another Role 2 MTF to provide GS to all withdrawing brigades.

4-98. Role 2 MTFs conduct leapfrog rearward, occupying successive positions placed along the withdrawal route to minimize the requirement for multiple displacements by any one MTF. This avoids unnecessary interference with combat operations while providing continuous medical support. With the exception of an ambulance squad to support the covering force the division normally does not employ ambulances further forward than the BAS. AA elements will normally evacuate to sustainment area and corps positions to avoid impeding the withdrawal of brigades.

4-99. Preparation for the withdrawal operation includes distribution of extra medical consumable supplies and nonexpendable exchange items to each medical element. This is done to overcome the potential effects of isolated MTFs or the possible intermittent operation of the medical evacuation system. The AA medical company may receive the mission to assist with this distribution.

4-100. If withdrawal is made under enemy pressure, the provision of aeromedical evacuation is modified. Since the time available is critical, aeromedical resources cannot remain in the forward areas. The assets are usually withdrawn as a unit. Patients occurring during the withdrawal are carried to the sustainment areas by the parent unit, normally using nonmedical vehicles.

## DELAYING OPERATIONS

4-101. Delaying operations occur when forces are insufficient to attack or to defend and when the defensive plan calls for drawing the attacker into an unfavorable situation. The delay maneuver may include friendly maneuver to successive positions across a broad front.

4-102. Delaying brigades split their combat power, moving their less mobile forces directly to the next defensive position while the elements remaining in contact fight to the sustainment area.

4-103. Aeromedical support for a delaying operation is conducted like withdrawal operations.

## RETIREMENT OPERATIONS

4-104. A retirement operation is a rearward movement of a force not in contact with the enemy. It is conducted according to the force's OPLAN and without pressure by enemy forces.

4-105. Because the division is no longer in contact with the enemy, it can march (in multiple columns) directly to the rear. The AHS requirements for this type of operation are similar to those in a movement to contact. The patient workload is light. The FSMT (if available) leader will coordinate with the BSMC commander, and the AA medical company CP will coordinate with the DSS for guidance in the positioning of company assets to support this operation.

## SECTION VIII — SPECIAL FORCES OPERATIONS

4-106. Army special operations forces (ARSOF) have limited organic AHS resources. They are, therefore, dependent upon the theater AHS for the majority of their health care requirements. To ensure that ARSOF receive comprehensive and timely AHS and logistics, medical planners must consider each medical function and determine the support organic and theater AHS assets will provide. The significant challenge for ARSOF medical planners is to ensure the efficient delivery and use of the conventional Class VIII (medical materiel) system for support of ARSOF missions.

4-107. Army special operations forces consist of special forces (SF), rangers, special operations aviation (SOA), psychological operations (PSYOP), and CA. The SOA, CA, and PSYOP have limited medical capabilities and are very dependent upon the theater HSS assets and the organizations they are supporting. Special forces and rangers often operate far removed from conventional AHS and must be more self-reliant and sustaining than conventional forces. Accordingly, SF and ranger medical personnel receive enhanced medical training above that provided for a trauma specialist. The Army educates the SOF medic to be an independent care practitioner qualified to provide ATM to combat casualties. When deployed on independent operations, the two SF medics are the sole source of medical care for their operational detachment, the indigenous forces, and their families that the detachment supports. They can train the indigenous populace in basic medical skills and establish an austere AHS system. Nonmedical ARSOF personnel receive medical training at the combat lifesaver level. Special operation combat medics (SOCMs) are trained in emergency medical treatment and in combat trauma management, but they do not possess the invasive surgical skills of the SF medical sergeant. The SOCMs provide medical care in combat environments where resuscitative surgery is more rapidly accessible than in the small unit missions of SF. Physicians and physician assistants assigned to units with SOCMs can provide ATM. Army health system planning for missions where the risk of penetrating trauma is high and further complicated by time and extended distance requires extensive planning and coordination. The capability for medical personnel and units that can perform resuscitative surgery and invasive stabilization must be provided regardless of the ARSOF unit involved.

4-108. Although the ARSOF health care provider receives enhanced medical training exceeding the level and scope found in conventional forces, he depends heavily on the conventional AHS system to conserve the combat strength of the ARSOF (particularly in the area of medical evacuation where the ARSOF does not have a dedicated system).

4-109. Ideally, medical evacuation for ARSOF personnel should follow the doctrinal flow sequence. The ARSOF medical planner must be innovative and follow the tenets of immediate far forward stabilization. He directs evacuation to the appropriate MTF when the condition of the patient warrants it, with whatever means of transportation are available.

4-110. Medical evacuation of ARSOF casualties is an operational matter. That is, it must reflect the commander's concept of the operation. It can only succeed when the medical planner integrates the medical evacuation plan with the tactical plan and logistics airflow.

4-111. Conventional medical evacuation assets are not normally used when operations occur in hostile and denied areas. Use the conventional medical evacuation system once the casualty is extracted from the hostile or denied area. For additional information on AHS to ARSOF, refer to FM 3-05, FM 4-02.6. FM 8-42 (FM 4-02.42), and FM 4-02.43.

## SECTION IX — URBAN OPERATIONS

# TERRAIN AND ENVIRONMENT

4-112. Throughout history, battles have been fought on urbanized terrain. Some recent examples urban operations (UO) include Hue, Beirut, Panama City, and Baghdad. Military operations on urbanized terrain are those military actions planned and conducted on a terrain where man-made structures impact on the tactical options available to the commander. This terrain is characterized by a three-dimensional battlefield, having considerable rubble, ready-made fortified fighting positions, and an isolating effect on all movement and maneuvers and fires, and sustainment units. In this environment, the requirement for a sound and understandable evacuation plan cannot be overstated. Of concern to medical and tactical planners is the need to plan, train, prepare, and equip for evacuation from under, above, and at ground level.

### TERRAIN CONSIDERATIONS

4-113. Medical evacuation in the urban environment is a labor-intensive effort. Due to rubble, debris, barricades, and destroyed roadways, much of the evacuation effort will be by manual litter teams. When this occurs, establish a litter or ambulance shuttle system. The shuttle system reduces the distance that litter teams have to carry the wounded or injured Soldiers. This enhances the litter teams' effectiveness by providing brief respites, reducing fatigue. Further, the litter teams stay in the forward areas. They are familiar with the geography of the area and what areas have or have not been searched for casualties. In moving patients by litter, you should:

- Use covered evacuation routes such as subways, whenever possible.
- Use easily identifiable points for navigation CCPs.
- Rest frequently by alternating litter teams.

### EVACUATION PLANNING CONSIDERATIONS FOR URBAN OPERATIONS

4-114. Conducting medical evacuation operations in the urban environment challenges the medical planner. He must ensure that the HSS/FHP plan includes special or unique materiel requirements or improvised use of standard equipment. The plan must be sufficiently flexible to support unanticipated situations.

4-115. Special equipment requirements include, but are not be limited to:

- Axes, crowbars, and other tools used to break through barriers.
- Special harnesses; portable block and tackle equipment; grappling hooks; collapsible litters; lightweight collapsible ladders; heavy gloves; and casualty blankets with shielding. This equipment, using pulleys, is for lowering casualties from buildings or moving them from one building to another at some distance above the ground.
- Equipment for the safe and quick retrieval from craters, basements, sewers, and subways. Casualties may have to be extracted from beneath rubble and debris.

4-116. Air ambulances equipped with a rescue hoist may be able to evacuate casualties from the roofs of buildings or may be able to insert needed medical personnel and supplies.

### COMMUNICATIONS IN THE URBAN ENVIRONMENT

4-117. The urban environment degrades communications. Line of sight radios will be ineffective and individual Soldiers will not have access to radio equipment. The medical evacuation teams will have difficulty in locating injured or wounded Soldiers due to their isolation within buildings, or by being hidden in rubble and debris. Injured Soldiers can display alternate forms of communications, such as markers, panels, or field expedients (field/uniform jackets or T-shirts) indicating where they may be found.

## Establishing Casualty Collection Points

4-118. Preplan and establish CCPs at relatively secure areas accessible to both ground and AAs. The location of these points are depicted on the HSS/FHP overlay to the OPLAN. Casualty collection points should—

- Offer cover from enemy fires.
- Be as far forward as the tactical situation permits.
- Be identifiable by an unmistakable feature (natural or man-made).
- Allow rapid turnaround of ambulances.

4-119. The tactical commander must approve route markings to the MTF and if to display of the Geneva Red Cross at the facility. (Camouflaging or not displaying the Geneva Red Cross can forfeit the protections, for both medical personnel and their patients, afforded under the Geneva Conventions. Refer to Appendix A and FM 4-02 for additional information.) The location of the MTF must be as accessible as possible, but well separated from fuel and ammunition depots, motor pools, reserve forces, or other lucrative enemy targets, as well as civilian hazards such as gas stations or chemical factories.

## Ground Evacuation For Urban Operations

4-120. When using ground evacuation assets in support of UO, the medical planner must be aware that built-up areas may have significant obstacles to vehicular movement. Factors requiring consideration include the following:

- Rubble and other battle damage complicate and canalize transportation operations within the urban terrain.
- Bypassed pockets of resistance and ambushes pose a constant threat along evacuation routes.
- Land navigation with most tactical maps proves to be difficult. Using commercial city maps when available can aid in establishing evacuation routes.
- Ambulance teams must dismount from the ambulance, search for, and rescue casualties.
- Movement of patients becomes a personnel intensive effort. There are insufficient medical personnel to search for, collect, and treat the wounded. Medical units may require assistance in the form of litter bearers and search teams from supported units.
- Refugees may hamper movement into and around urban areas.
- Dislocated civilians, EPW, and detainees receive medical treatment according to the command policy and the Geneva Conventions.

## Air Evacuation for Urban Operations

4-121. When using AA assets in support of UO, the medical planner must consider enemy air defense capabilities and terrain features, both natural and man-made, within and adjacent to the built-up areas. Air ambulances are the preferred means of evacuation in UO. Considerations in the use of AAs include the following:

- Movement is highly restricted and is canalized over secured areas, down wide roads, and open areas.
- Telephone and electrical wire and communication antennas hinder aircraft movement.
- Secure landing zones (LZs) must be available.
- Landing zones may include buildings with helipads on their roofs or sturdy buildings, such as parking garages.
- Snipers with air defense capabilities may occupy upper stories of the urban area's taller buildings.

### SPECIALTY TRAINING

4-122. Medical personnel require special training in the tactics, techniques, and procedures required to operate in an UO environment. If they are to survive in this environment, they must know how to:

- Cross open areas safely.
- Avoid barricades and mines.
- Enter and depart safely from buildings.
- Recognize situations where booby traps or ambushes are likely and would be advantageous to the enemy. Detailed information on the conduct of combat operations in the urban environment is contained in FM 3-06 and FM 3-06.11.

---

***Note.*** Medical personnel do not engage in offensive-type actions. They must rely on the supported unit to provide covering fires and to clear rooms and buildings prior to entry.

---

4-123. Units can modify and apply many of the extraction and patient evacuation techniques used in a mountainous terrain for evacuation in UO.

## SECTION X — HOMELAND SECURITY

4-124. To preserve the freedoms guaranteed by the Constitution of the US, the Nation must have a homeland that is secure from threats and violence, including terrorism. Homeland security (HS) is the Nation's first priority, and it requires a national effort; in which the DOD has a key role in that effort. The National Strategy for Homeland Security (NSHS) complements the National Security Strategy of the US by providing a comprehensive framework for organizing the efforts of federal, state, local, and private organizations whose primary functions are often unrelated to national security. Critical to understanding the overall relationship is an understanding of the distinction between the role that DOD plays with respect to securing the Nation and HS, and the policy in the NSHS, which has the Department of Homeland Security (DHS) as the lead. Homeland security at the national level has a specific focus on terrorist threats. The DOD focus in supporting HS is broader. Military application of the NSHS calls for preparation, detection, deterrence, prevention, defending, and responding to threats and aggression aimed at the homeland. The DOD also provides military assistance to civil authorities (MACA), including consequence management activities. The Armed Forces of the US support the NSHS through two distinct but interrelated mission areas–homeland defense (HD) and civil support. The United States Northern Command (USNORTHCOM) is the unified command that has the DOD responsibility for HD and civil support.

## HOMELAND DEFENSE

4-125. Homeland defense is the protection of US territory, sovereignty, domestic population, and critical infrastructure against external threats and aggression. The purpose of HD is to protect against and mitigate the impact of incursions or attacks on sovereign territory, the domestic population, and defense critical infrastructure. For HD missions, as directed by the President of the US and/or the Secretary of Defense, DOD serves as the lead federal agency.

### HOMELAND DEFENSE CHARACTERISTICS

4-126. The DOD is the lead, supported by other agencies, in defending against traditional external threats/aggression (such as air and missile attack). However, against internal asymmetric, nontraditional threats (such as terrorism), DOD may be in support of DHS. When ordered to conduct HD operations within US territory, DOD will coordinate closely with other federal agencies or departments.

### TYPES OF HOMELAND DEFENSE OPERATIONS

4-127. The types of homeland defense operations that the future brigade combat teams is expected to perform include—

- Offensive operations.
- Defensive operations.
- Humanitarian and civic assistance.
- Support to counterdrug operations.
- Show of force.
- Domestic support operations.
- Security operations (screen, guard, area security, and local security).
- Reconnaissance operations (zone, route, and area reconnaissance).

## CIVIL SUPPORT

4-128. Civil support missions consist of military support to US civil authorities for domestic emergencies, and for designated law enforcement within the scope of restrictions required by the Posse Comitatus Act and other support approved by the Secretary of Defense (see FM 3-07).

4-129. Employment of military forces within the US, its territories, and possessions, under the auspices of civil support, typically falls under the broad mission of MACA. The MACA missions consist of three mission subsets. These mission subsets consist of military support to civil authorities, military support to civilian law enforcement agencies, and military assistance for civil disturbances.

### MILITARY SUPPORT TO CIVIL AUTHORITIES

4-130. Military support to civil authorities (MSCA) is the most widely recognized form of DOD civil support because it usually consists of support for high-profile emergencies such as natural or man-made disasters that often invoke Presidential or state emergency/disaster declarations:

- **Natural Disasters.** In the event of a natural disaster or emergency, there will be a concerted US Government effort to support the affected areas. The DOD may be asked to provide assistance to DHS/Federal Emergency Management Agency (FEMA) in an attempt to save lives, protect property, and lessen the threat of a catastrophe in the US. Examples of natural disasters include, but are not limited to severe weather, wildland fire fighting, and animal disease eradication. When natural disasters occur and military assistance is anticipated, DHS/FEMA will request a defense coordinating officer (DCO) who serves as the single DOD point of contact within the disaster area. The DCO will be under the OPCON of the designated supported combatant command. Units supporting the event are normally OPCON to the supported combatant command.
- **Special Events.** The DOD provides support to a special event, a planned program of athletic competition or related activities involving participants from the US and/or other nations. Historic examples of these events are the Olympic Games and the Pan American Games.
- **Man-made Disasters.** Manmade disasters can be accidental or intentional. An example of an accidental event is an oil spill. The implications of a deliberate or unintentional large release of a CBRNE are severe. A catastrophic CBRNE event or attack may occur with little or no warning, resulting (either immediately or over time) in mass casualties and producing a mass exodus of evacuees. The response capabilities and resources of the local jurisdiction (to include mutual aid from surrounding jurisdictions and response support from the state) may be insufficient and quickly, if not immediately, overwhelmed.

### Domestic Support Operations

4-131. Domestic disasters occur due to various causes and many may require the assistance of DOD and more specifically, the Army medical evacuation assets to fill response shortfalls. In the post-9/11 environment, terrorism or other intentional acts are recognized as a more likely cause of contingencies within the US than many had acknowledged since the civil defense era of the forties and fifties. This threat increases the likelihood that a large-scale disaster will occur that requires the augmentation of civil authorities' resources by DOD. The US DHS and USNORTHCOM have been established to coordinate the preparedness for and response to such events for civil authorities and DOD respectively.

4-132. Domestic support operations are typically conducted to prevent loss of life and property. Certain operations may be conducted in the form of immediate and automatic response by US military commanders as defined in DOD Directive 3025.15, to save lives, prevent human suffering, or to mitigate great property damage under imminently serious conditions. Other requests for support for civil disasters are made by civil authorities or a DCO (in a Stafford Act declared emergency) through the Joint Directorate of Military Support per the directive and work down through command channels from the Secretary of Defense.

4-133. Homeland Security Presidential Directive 5 recognized that there currently exists an artificial separation of local, state, and federal response to an incident and the role of DOD is not well streamlined or defined. This recognition led the White House and the DHS to initiate the transition from the Federal Response Plan to the National Response Plan, which serves as the doctrine for how the US responds to disasters no matter the cause.

4-134. Medical evacuation support plans to augment civil authorities are necessary for a variety of contingencies. Domestic support planning must address a range of problems such as—

- Preidentifying medical evacuation capabilities, units and personnel available to support various contingencies large enough to require civil support.
- Command and control relationships between civil authorities and DOD forces especially when DOD units are divided.
- Support for deployed DOD forces when no DOD logistics operations are deployed, including medical support.
- Cost capture and reimbursement from civil authorities to DOD in non-Stafford Act emergencies.

## SECTION XI — OTHER TYPES OF MEDICAL EVACUATION SUPPORT MISSIONS

## EVACUATION OF MILITARY WORKING DOGS

4-135. Military working dogs (MWDs) when injured or ill may be evacuated on any transportation means available. The using unit is responsible for the evacuation of the animal. Use of dedicated medical evacuation assets (air or ground ambulances) is authorized based on mission priority and availability. When possible, the handler should accompany the animal during the evacuation. Using units should include the location of veterinary treatment facility/support units on operational overlays. See FM 4-02.18 for information on veterinary support for MWDs.

## PERSONNEL RECOVERY OPERATIONS

4-136. Air ambulances do participate in PR operations in a support role and are not protected from attack while engaged in a PR mission over contested or denied territory. If the involvement in these operations consists solely of evacuating wounded crewmembers from a crash site in friendly territory, AAs retain the protection accorded to them under the provisions of the Geneva Conventions. Air ambulances flying in contested or denied areas are not protected from attack unless flying at times and routes agreed to by the enemy and may be summoned to land. Air ambulances must obey a summons to land. Personnel retain their protections under the Conventions. However, if AAs participate in the actual search and rescue phases of the operation, they are not solely engaged in the provision of AHS and are, therefore, not afforded the protections. Further, because the mission does not fall within the protected activities, the AAs participating in the PR operation must remove the Geneva Conventions emblem from the aircraft.

# SHORE-TO-SHIP EVACUATION OPERATIONS

4-137. It is imperative that Army AA units be able to interface on the first day of battle with US Navy (USN) air-capable ships. Lessons learned from past operations, have shown that US Army helicopters should be able to operate to and from USN air-capable ships. An interservice agreement between the Army and the Navy allows for deck-landing qualification of Army pilots (refer to FM 1-564 for additional information). It is important that units having contingency missions requiring Navy support establish training requirements to obtain naval-operations orientation, water egress training, water survival, and deck-landing qualification. This enhances the successful accomplishment of the aeromedical evacuation mission to naval vessels.

## COMMUNICATIONS

4-138. In past joint operations, communications have been burdensome for both Army and Navy elements. It is essential to establish commonality of communication requirements during training exercises and to establish communication equipment and frequencies for medical evacuation to Navy vessels. This will provide smooth integration of Army helicopters into the Navy airspace management system during actual operations.

## NAVIGATION

4-139. As Navy vessels may operate relatively long distances from the ground combat operations, Army AA units need to be proficient in over water navigation. The use of navigational aids (NAVAIDS) from the Navy element in support of the operation is the first priority for over water navigation. Basic dead-reckoning remains a secondary measure.

# MEDICAL EVACUATION OF ENEMY PRISONERS OF WAR AND DETAINEES

4-140. Sick, injured, and wounded EPW and detainees are treated and evacuated in military police (MP) channels when possible. They must be physically segregated from US, allied, and coalition patients. Guards for these prisoners are provided according to the BCT, division or corps TSOP and are from other than medical resources. The echelon commander is normally responsible for this support. The US provides the same standard of medical care for wounded, sick, and injured EPW and detainees as that given to US, allied, and coalition Soldiers. Wounded, sick, or injured EPW and detainees in the CZ may be treated and returned to MP channels for evacuation; or the EPW and detainees may be stabilized and moved through medical channels to theater hospitals for treatment. Enemy prisoners of war and detainees are not evacuated from the theater for medical treatment.

4-141. When EPW are evacuated through medical channels, medical personnel—

- Report this action through medical channels to the next higher headquarters.
- Request disposition instructions from the MEDBDE IPMC.
- The MEDBDE IPMC is responsible for—
  - Coordinating the transportation means.
  - Identifying the MTF to which the EPW/detainees will be taken.
  - Coordinating, in conjunction with the hospital commander, with the Detainee Reporting System to account for EPW within medical channels.

# SECTION XII — MEDICAL EVACUATION REQUEST

4-142. Procedures for requesting medical evacuation support must be institutionalized down to the unit level. Procedural guidance and standardization of request procedures are provided below. The same format used to request aeromedical evacuation is also used for requesting ground evacuation.

**Table 4-3. Procedures for information collection and medical evacuation request preparation**

| *Line* | *Item* | *Explanation* | *Where/How Obtained* | *Who Normally Provides* | *Reason* |
|---|---|---|---|---|---|
| 1 | Location of pickup site | Encrypt the grid coordinates of the pickup site. When using the DRYAD Numeral Cipher, the same "SET" line will be used to encrypt the grid zone letters and the coordinates. To preclude misunderstanding, a statement is made that the grid zone letters are included in the message (unless unit SOP specifies it's use at all times) | From map | Unit leader(s) | Required so evacuation vehicle knows where to pickup patient. Also, so that the unit coordinating the evacuation mission can plan the route for the evacuation vehicle (if the evacuation vehicle must pick up from more than one location). |
| 2 | Radio frequency, call sign and suffix | Encrypt the frequency of the radio at the pickup site, not a relay frequency. The call sign (and suffix if used) of person to be contacted at the pickup site may be transmitted in the clear. | From SOI | RTO | Required so that evacuation vehicle can contact requesting unit while enroute (obtain additional information or change in situation or directions). |
| 3 | Number of patients by precedence | Report only applicable information and encrypt the brevity codes.<br>A—URGENT<br>B—URGENT-SURG<br>C—PRIORITY<br>D—ROUTINE<br>E—CONVENIENCE<br>If two or more categories must be reported in the same request, insert the word "BREAK" between each category. | From evaluation of patient(s) | Medic or senior person present | Required by unit controlling vehicles to assist in prioritizing missions. |

| | | | | | |
|---|---|---|---|---|---|
| 4 | Special equipment required | Encrypt the application brevity codes.<br>A—None<br>B—Hoist<br>C—Extraction equipment<br>D—Ventilator | From evaluation of patient/situation | Medic or senior person present | Required so that the equipment can be placed on board the evacuation vehicle prior to the start of the mission. |
| 5 | Number of patients by type | Report only applicable information and encrypt the brevity code. If requesting medical evacuation for both types, insert the word "BREAK" between the litter entry and ambulatory entry.<br>L+# of patients–Litter<br>A+# of patients–Ambulatory (sitting) | From evaluation of patient(s) | Medic or senior person present | Required so that the appropriate number of evacuation vehicles may be dispatched to the pickup site. They should be configured to carry the patients requiring evacuation. |
| 6 | Security of pickup site (wartime) | N—No enemy troops in area<br>P—Possibly enemy troops in area (approach with caution).<br>E—Enemy troops in area (approach with caution).<br>X—Enemy troops in area (armed escort required). | From evaluation of situation | Unit leader | Required to assist the evacuation crew in assessing the situation and determining if assistance is required. More definitive guidance can be furnished to the evacuation vehicle while it is enroute (specific location of enemy to assist an aircraft in planning its approach). |
| 6 | Number and type of wound, injury or illness (peacetime) | Specific information regarding patient wounds by type (gunshot or shrapnel). Report serious bleeding, along with patient's blood type if known. | From evaluation of patient(s) | Medic or senior person present | Required to assist evacuation personnel in determining treatment and special equipment needed. |
| 7 | Method of marking pickup site | Encrypt the brevity codes.<br>A—Panels<br>B—Pyrotechnic signal<br>C—Smoke signal<br>D—None<br>E—Other | Based on situation and availability of materials | Medic or senior person present | Required to assist the evacuation crew in identifying the specific location of the pick up. Note that the color of the panel or smoke should not be transmitted until the evacuation vehicle contacts the unit (just prior to its arrival). For security, the crew should identify the color and the unit verifies it. |
| 8 | Patient | The number of patients in each category need not | From evacuation | Medic or | Required to assist in |

| | | | | | |
|---|---|---|---|---|---|
| | nationality and status | be transmitted. Encrypt only the applicable brevity codes.<br>A—US military<br>B—US citizen<br>C—Non-US military<br>D—Non-US citizen<br>E—enemy prisoner of war (EPW) | platform | senior person present | planning for destination facilities and need for guards. Unit requesting support should ensure that there is an English speaking representative at the pickup site. |
| **8** | CBRN contamination (wartime) | Include this line only when applicable. Encrypt the applicable brevity codes.<br>C—Chemical<br>B—Biological<br>R—Radiological<br>N—Nuclear | From situation | Medic or senior person present | Required to assist in planning for the mission. (determine which evacuation vehicle will accomplish the mission and when it will be accomplished) |
| **9** | Terrain description (peacetime) | Includes details of terrain features in and around proposed landing site. If possible, describe relationship of site to prominent terrain feature (lake, mountain, tower). | From area survey | Personnel present | Required to allow evacuation personnel to assess route/avenue of approach into area. Of particular importance if hoist operation is required. |

## Chapter 5

# Medical Evacuation in Specific Environments

This chapter addresses medical evacuation in specific environments or under special circumstances. The medical evacuation effort must be well-planned and its execution synchronized to be effective. Further, medical evacuation personnel must be flexible and ready to improvise, if needed, to meet the demands of unique situations.

## SECTION I — MOUNTAIN OPERATIONS

5-1. In the past, armies have experienced great difficulty in evacuating patients from mountainous areas because they are extremely diverse in nature. Some mountains are dry and barren with temperatures ranging from extreme heat in the summer to extreme cold in the winter. In tropical regions, mountains are frequently covered by lush jungles and heavy seasonal rains occur. Many areas display high rocky crags with glaciated peaks and year-round snow cover. Elevations can also vary from as little as 1,000 feet to over 16,000 feet with drastic and rapidly occurring weather changes.

5-2. Operations in mountainous terrain require some procedure modifications. This is due to the environmental impact on personnel and equipment. Important physical characteristics and considerations that influence medical evacuation are—

- Rugged peaks, steep ridges, and deep valleys.
- Limited number of trafficable roads.
- Reduced communications ranges.
- Unpredictability of and severe changes in weather.
- Decreased partial pressure of oxygen.
- Limited availability of LZs.

5-3. In order to effectively support the tactical plan, the HSS/FHP plan must provide maximum flexibility. The HSS planner should consider using all means of evacuation. Due to the length of evacuation times and the limited means of ground evacuation, it is important to triage and prioritize patients prior to movement.

- The availability of improved, hard-surfaced roads is extremely limited, if they exist at all. Usually, improved roads are only found in valley corridors. Such roads are often dependent upon a system of narrow bridges spanning mountain streams and ravines. They may also twist along ridgelines and cling to steep shoulders.
- Secondary roads and trails may be primitive and scarce. However, they may provide the only routes capable of vehicular traffic. Cross-compartment travel between adjacent valleys may be impossible by ground vehicle. Off-road travel requires detailed planning, even for short distances.
- Because of rough terrain, the Role 2 MTF may not be able to reach the BAS by ground vehicle. An ambulance shuttle system is established with an AXP for air and ground evacuation vehicles to meet litter bearers. Litter bearers and beasts of burden may be the only means of evacuation available. Any available personnel may be used as litter bearers (nonmedical personnel from supported units may be required to augment the litter bearer teams). Close coordination between Role 2 MTFs and BASs in establishing CCPs or AXPs is necessary to—
  - Reduce distance traveled by litter bearers.
  - Reduce evacuation time.

  - Conserve personnel.
  - Locate the best potential LZs for AAs.
- In mountainous areas, evacuation of patients by air is the preferred means. Air ambulances permit the rapid movement of patients over rugged terrain. For example, to travel a distance of only 6 kilometers on foot could take up to 2 hours, while flying time could be less than 2 minutes.
- Frequency-modulated radios are the principal means of communication in this environment. The ability to transmit is hampered by the limitations of line of sight transmissions.
- The briefing of ambulance drivers needs to be extensive, including detailed strip maps and overlays. Further, specific instructions on what to do in various situations should be covered (such as if the vehicle breaks down or the unit moves).

5-4. The mountain environment, with its severe and rapidly changing weather, impacts on aircraft performance capabilities; accelerates crew fatigue; and requires special flying techniques. Having to rely on continuous aviation support for a successful mountain operation is risky.

- Flying in mountainous areas requires special training. Both the terrain and the weather influence basic flying techniques and operational planning. Rugged, mountainous terrain complicates flight route selection. Direct routes can seldom be flown without exposing the aircraft to detection and destruction by the enemy.
- Important considerations for AA operations in mountainous areas are—
  - Density altitude which is the most important factor affecting aircraft performance. Density altitude combines temperature, humidity, and pressure altitude, and provides the basis for lift capability. Density altitude can vary significantly between the pickup point and the LZ because of the time of day and changes in elevation. Frequent performance planning updates are essential.
  - Unpredictable winds which can produce significant turbulence, wind sheers, updrafts, and downdrafts. This further increases the risk of a catastrophe in a seemingly routine mission. Adverse winds along with high density altitude demand current and accurate performance planning. Pilots must plan for greater margins of safety.
  - Ice can clog intake ports, thus starving the engine of air, or it can collect on rotor blades resulting in a significant loss of lift. Asymmetrical shedding can cause severe out-of-balance rotor conditions.
  - Visibility can be reduced by low clouds or fog greatly decreasing the ability to navigate or to avoid obstacles.
  - Lack of LZs is critical because the characteristics of mountain terrain do not usually afford adequate LZs. The terrain may only allow the aircraft to hover while loading patients on board.
  - Hoist operations can be expected in mountainous terrain requiring the use of the internal or external rescue hoist. Mounting the rescue hoist on the aircraft as standard equipment in mountain operations may be required. When possible, orientation and training sessions with supported troops should be conducted to help minimize the difficulty of such missions. Depending on the terrain, the forest penetrator may also be needed to accomplish the mission.
  - Enemy air defenses must be considered because when enemy air defense capabilities preclude using AAs in forward areas, they should be used to evacuate patients from AXPs or from Role 2 MTFs.
  - Ambulatory patients may be reported as litter patients in mountainous terrain because these patients may be unable to move unassisted over the rugged terrain. Once placed on the AA, their status may be upgraded.
  - Additional crew training for ground and air evacuation crews should include training and orientation in mountaineering skills, handling patients, and survival skills. Such training might include cold weather survival training (cold injury prevention, mountain [rock] climbing, and the use of ropes and vertical rescue techniques). Training may also include individual and unit

movement at high altitudes and techniques of patient evacuation by litter, emphasizing the use of pack animals (if available from the host country), and the improvised travois litter.

- Patient loading can be hazardous and care must be taken when loading patients where there is a great deal of slope to the LZ. Emphasis on approaching and loading the aircraft from the down-slope side of the aircraft must be reinforced.

**DANGER**

**Approaching the aircraft from the up-slope side is hazardous.**

5-5. Troops operating in mountainous areas are exposed to other injuries and illnesses that frequently occur in this environment. These conditions include—

- An increased rate of fracture, sprain, and dislocation injuries.
- Incidents of acute mountain sickness, high-altitude pulmonary edema, and cerebral edema caused by rapid ascent to heights over 7,500 feet.
- Cold weather injuries.
- Dehydration and heat exhaustion.
- Sunburns and snow blindness.
- Aggravated sickle cell anemia. Although this condition is not considered a mountain illness, personnel with the sickle cell trait can be seriously affected by the decrease in barometric pressure and lower oxygen levels found at higher altitudes.

5-6. The proportion of litter cases to ambulatory cases is increased in mountainous terrain, for even the slightly wounded may be unable to move unassisted over rough terrain. Litter relay stations may be required along the evacuation route to conserve the energy of litter bearers and to speed evacuation. It is important to be able to predict the number of patients that can be evacuated with available personnel. When the average terrain grade exceeds 20 degrees, the four-man litter team is no longer efficient and should be replaced by a six-man team. The average mountain litter team should be capable of climbing 120 to 150 vertical meters of average mountain terrain and return with a patient in approximately 1 hour.

5-7. Mountain operations may require medical personnel to carry additional equipment. Items such as ropes, pulleys, pitons, piton hammers, and snap links are all necessary for evacuating patients and establishing BASs. All unnecessary items of equipment including those for which substitutes or improvisations can be made should be left behind. Heavy tentage, bulky chests, extra splint sets, excess litters, and nice-to-have medical supplies should be stored. Such medical supplies, if stored, should be readily available for airdrop or other means of transport. Medical items that are subject to freezing should be safeguarded; they should not be exposed to the low temperatures experienced in mountainous areas.

5-8. For further information on mountain operations, refer to FM 3-97.6.

## SECTION II — JUNGLE OPERATIONS

5-9. Health service support elements in a jungle environment retain the same basic capabilities as in other environments. Jungle operations, however, subject personnel and equipment to effects not found in other environments. The jungle environment degrades the ability to maneuver. Security problems are also increased and affect medical evacuation operations as much as they do the combat forces.

5-10. In jungle operations, combinations of air and ground evacuation units are used to maximize the patient evacuation potential. Using this dual system of evacuation ensures that the inherent limitations of one system can be compensated for by the other. Jungle variations affect the organizing, positioning, and securing of HSS/FHP. Due to the terrain, aerial resupply is usually a common practice. The responsiveness provided by aerial resupply requires fewer supplies to be stockpiled in the combat trains.

5-11. Jungle combat operations are characterized by ambushes and other unconventional warfare operations. The security threat caused by infiltrators requires that LOC be patrolled often and that convoys be escorted. It is, therefore, essential that HSS be performed as far forward as the tactical situation permits. Positioning assets forward—

- Improves response time.
- Reduces road movement.
- Allows the HSS elements to take advantage of the security offered by combat units.

5-12. The thick foliage often makes evacuation by ground more difficult than in other types of terrain. Factors such as the threat, limited road network, and reliance on nonmedical personnel for convoy security make air evacuation the preferred means. By using the ambulance shuttle system, patients can be transferred from forward operating ground ambulances to either ground or AAs operating further to the rear. In situations where evacuation assets are delayed by various factors (weather or terrain), patients are held for longer periods of time at forward locations. This will dictate the need for additional medical supplies. Health service support planners must try to anticipate these delays whenever possible. The increased disease and infection incidences associated with the jungle environment may worsen the patient's condition; therefore, timely evacuation is essential.

5-13. In some remote and densely foliaged jungles, the only means of evacuation may be by litter. Ambulances may not be practical on trails, unimproved muddy roads, or in swamps. As in mountain operations, there is a higher proportion of litter cases than usual. In the jungle even a slightly wounded Soldier may find it impossible to walk through dense undergrowth. At best, litter teams can carry patients only a few hundred meters over rough jungle terrain before needing rest or relief. Litter carries should be kept as short as possible and medical elements prepositioned and retained forward.

5-14. Other special planning considerations in jungle operations include—

- Water is vital in the jungle and is plentiful. Water from natural sources, however, should be considered contaminated. Water purification procedures must be taught to all Soldiers. (Refer to FMs 21-10 and 4-25.12 for additional information.)
- Due to the tropical climate, units should pack hot weather clothing when deploying to jungle areas. Insect (mosquito) nets, insect repellent, permethrin treated uniforms and sunscreen should be issued to all Soldiers operating in this environment.
- The jungle environment is ideal for the transmission of a large number of diseases. The rate of DNBI casualties is potentially the highest in this climate. The heat, humidity, and terrain places the troops at high risk for dehydration, heat injury, skin diseases, endemic diseases, and immersion foot. Small wounds can rapidly become infected and lead to loss of effectiveness and possibly require evacuation. High standards of personal hygiene must be taught, encouraged, and maintained by the command. Mosquitoes and other arthropods that carry disease flourish under jungle conditions. Use of all possible personal protective measures must be ensured. Food- and waterborne diseases leading to diarrhea or other symptoms will abound. Food service sanitation measures must be strictly followed. The potential for contamination of food and water increases with each time they are handled, stored, or transported. Soldiers must be encouraged to consume adequate amounts of water that has been purified and to eat only approved foods. In the jungle it is necessary for the commander to pay meticulous attention to the details of PVNTMED measures to maintain an effective fighting force. For additional information on PVNTMED measures, refer to FM 21-10 and FM 4-25.12.
- Health service support personnel should be trained in survival and support techniques in jungle environments. For example, training should be conducted in—
  - Hot weather acclimatization and survival.
  - Prevention, early detection, and treatment of arthropod-, food-, and waterborne diseases.
  - Land navigation in a jungle environment.
  - Field sanitation and other PVNTMED measures.
  - Care and maintenance of equipment and supplies.

- Due to the increased heat and humidity, vehicles and equipment require additional maintenance. Equipment tends to rust quickly and must be cleaned and oiled more frequently. Canvas items rot and rubber deteriorates much faster than in more temperate climates.
- The range of FM communications in the jungle is significantly reduced due to the dense undergrowth, heavy rains, and hilly terrain. The range of a radio set operated in the jungle may be reduced by 10 to 25 percent. The heavy rain and high humidity of the tropics also reduce the range (about 20 percent) and reliability of wire communications. The transmission range can be extended by using additional radio relays and field expedient antennas.
- Utility helicopters are not able to lift the same size loads that can be lifted in more temperate areas. This results in a reduced patient load in some evacuation aircraft. Again, frequent and accurate performance planning is essential for mission accomplishment.
- There may be few suitable LZs. Many LZs will only be large enough to support one or two helicopters at a time.
- Hoist operations may be required more frequently in the thick jungle vegetation where LZs are not available. The forest penetrator should be carried on all operations.
- For aviation-specific information, refer to training circular (TC) 1-400.

## SECTION III — DESERT OPERATIONS

5-15. Deserts are arid, barren regions of the earth incapable of supporting normal life due to a lack of fresh water. Although deserts are often thought of as hot climates, it is important to note that temperatures range from over 136 degrees (°) Fahrenheit (F) in some deserts, to bitter cold in others. Day to night fluctuations in temperature can exceed 70° F. Desert terrain can have mountains, rocky plateaus, or sandy dunes; some desert areas may contain all of these characteristics. Rain, when it falls, often causes flooding in low-lying areas. Winds can have a devastating effect upon AHS operations by destroying equipment and supplies and causing dust storms. Dust storms make navigation and patient treatment difficult. Since deserts vary considerably in the type of terrain and temperature, and in their cultural makeup, current medical intelligence should be obtained prior to deployment on operations conducted on desert terrain.

5-16. People have lived and fought in desert areas for thousands of years. However, the environmental effects on personnel can be extreme, especially for Soldiers not prepared for these operations.

- To be effective, Soldiers must be properly acclimatized to the desert. Two weeks are usually required to satisfactorily acclimatize troops to hot environments, using progressive degrees of heat exposure and physical exertion. Other potential acclimatization problems that may be encountered are the effects of dry air and altitude on the respiratory system. Since many desert areas are located in mountainous terrain, Soldiers must be acclimatized for both the altitude and the temperature. In some areas of the world, such as the Gobi Desert in East Asia, people must be acclimatized to the cold in addition to the dryness. (For additional information, refer to FM 21-10, FM 8-250, and TB MED 507.)
- Units deployed in desert areas typically have long LOC and are widely dispersed. As unit elements become more dispersed, commanders must rely more heavily on junior officers and NCOs to ensure that C2 is maintained and that discipline and PVNTMED programs are enforced. For a unit to be effective, a high level of discipline must exist at all levels of the organization.
- Water is the most basic need in a desert. Without it, Soldiers cannot function effectively for more than a few hours. Thirst is not an adequate indicator of the need for water. It is necessary for each commander to establish and enforce a supervised drinking program. Experience has shown many times that Soldiers do not drink enough fluids unless forced to do so. It is important to cool the water, if at all possible, to make it more appealing. Water supplies should be carefully guarded against accidental loss, sabotage, or contamination. Extra water must be carried by AHS vehicles for patients to drink and to cool heat casualties.
- Soldiers deployed in the desert are susceptible to endemic diseases and environmental injuries. Proper water discipline, vaccines, prophylactic measures, field sanitation measures, personal

hygiene, and other PVNTMED measures can reduce these risks. Cold weather injuries, heat injuries, and respiratory disease can also be prevalent. Proper clothing, equipment, and a water discipline program must have command emphasis in desert operations.

- Winds may very easily damage materiel such as aircraft, antennas, and tents. Equipment is protected by using covers, tie-downs, and shelters. Terrain helps shield equipment from the wind if site selection is done carefully. In some cases, special tools, such as extra long metal tent stakes, are necessary.
- The effects of wind and sand are interrelated. Desert sand starts to become airborne when the wind reaches about 20 knots. Sandstorms—
    - Restrict visibility.
    - Pose a hazard to eyes (especially for Soldiers wearing contact lenses).
    - Can contaminate water supplies (if they are not protected).
    - Make navigation difficult.
- The sun may cause sunburn of the skin and eyes (if protection is not used).

5-17. Seven characteristics of the desert environment that may adversely effect equipment are—

- Terrain. Trafficability varies with the type of terrain covered. Open, flat, and rocky terrain affords higher trafficability than do mountainous areas, lava beds, or salt marshes. Drivers must be well-trained in judging the terrain over which they are driving to select the best alternative routes.
    - Tracked vehicles are best suited for desert operations. However, they can throw tracks when traversing a rocky area. Their use is also limited in rough terrain with steep slopes.
    - Wheeled vehicles may be used in desert operations; however, they normally have a lower average speed than tracked vehicles and a higher incidence of damage and malfunction. Wheeled vehicles often bog down in sandy areas and cannot traverse many of the rougher areas.
    - In planning for desert operations, vehicles should carry extra repair parts (fan belts, tires, and other items apt to malfunction).
- Excessive heat causes vehicles to overheat, leading to greater than normal wear. The frequency of leaks on vehicles and aircraft is greater than in some other environments. Engine and transmission seals tend to dry out and crack; fuel lines wear out quickly; and water requirements for cooling vehicle engines are greater. Loss of water, through evaporation, must be included in logistical planning. Using vehicle and aircraft covers reduce the effects of heat while vehicles and aircraft are not in use. Refer to FM 1-202.
    - Batteries do not hold their charge efficiently in intense heat. Dry battery supplies should be increased to compensate for a higher usage rate.
    - Communications equipment must be protected from the heat in the desert. Dust covers are used on this type of equipment. If the equipment has ventilating ports, these should be cleaned regularly to avoid clogging.
    - Medical supplies must be protected from the heat to prevent deterioration. The shelf life of some medical supplies decreases when stored in hot climates.
    - Medical equipment must be protected from the effects of heat. It may be protected using the same techniques as those used to protect communications equipment.
- The sun burns unprotected skin and it may damage unprotected eyes. Soldiers should dress in loosely fitting clothing, use sunburn cream or oils to protect exposed skin, and wear sunglasses or goggles to protect their eyes. Soldiers should remain fully clothed. Removing clothing increases direct exposure of the skin to the sun and eliminates the beneficial cooling effects of the moisture trapped in clothing. Radiant light or its heat effects may be detrimental to plastics, lubricants, pressurized gasses, rubber, and some fluids. All vehicles and aircraft should be kept well-ventilated, and windshields should be covered to reduce heat buildup inside. Supplies of all types should be stored in a well-ventilated, shady area. Placing supplies in covered holes in the ground may reduce adverse heat effects.

- Dust and sand present one of the greatest dangers to the proper functioning of equipment. Sand mixed with lubricants forms an abrasive paste. Lubrication fittings, bearings, and filters should be inspected frequently and changed when required.
  - Aircraft should not be exposed to dust and sand any more than is absolutely necessary. Ground handling instead of hovering reduces sand ingestion. Dust and sand increase failure of microphone switches, signal distribution panels, and circuit breakers, and cause electrical motors and generators to burn out. Wheel and flight control bearings require more frequent cleaning; engines should be flushed frequently.
  - Medical and communications equipment may be adversely affected by dust and sand. Over a period of time, electrical insulation is damaged by windblown sand. When combined with the effects of lubricants on the insulation, dust and sand can become a major communications problem. Special care should be taken to brush dust off radio equipment and to keep ventilating ports and channels clear.
  - Sand can accumulate in airframes, on the bottom of armored vehicles, and in bearings on all types of equipment. This accumulation, combined with oil and condensation, adds extra weight to aircraft and vehicles as well as jamming their control linkages. Sand and grease buildups must be removed from bearings to ensure safe operation and control of aircraft and vehicles.
  - Dust trails created by hovering aircraft or ground vehicles can be seen in excess of 10 miles on a relatively flat desert. This exposes these assets to direct and indirect enemy fires. Ground vehicles should reduce their speed to the point that they do not create a dust signature.
- Humidity is a factor in some desert areas of the world, especially in the Middle East. Humidity can become a problem for short periods of time in other desert areas. Light coats of lubrication can help prevent rust; however, these benefits should be weighed against the dust-gathering qualities of oil. Demisting equipment is used on optics and night vision equipment to combat the effects of humidity.
- Temperature variation can cause condensation in humid desert areas affecting optics, fuel lines, air tanks, and weapons. Expansion and contraction of air and fluids cause tires to over inflate during the day and under inflate at night. Fuel tanks may overflow during the day causing a fire hazard. Oil fluid levels become overfull and cause leaks during the day, or insufficient lubrication occurs when the oil cools. Vehicle operators and crew chiefs must ensure that the effects of temperature variations do not become a significant problem.
- Static electricity is prevalent in the desert. This is important to remember during refueling operations and when oxygen is being used onboard vehicles or aircraft. Proper refueling procedures must be followed. Static electricity also causes severe shock to ground personnel in sling load and hoist operations.

**CAUTION**

During hoist and sling load operations, the load must touch the ground before the ground crew can handle it.

5-18. To ensure success in desert operations, detailed planning is required. Factors to consider include—

- Additional quantities of water are required for HSS/FHP operations for the survival of both medical personnel and their patients. Load plans for all vehicles and aircraft must include water. Water is as mission essential as any piece of unit equipment. It should be a priority item when load plans are developed.
- Prescribed load lists are expanded to carry sufficient quantities of repair parts easily degraded by the environmental factors. For example, rubber and plastic fittings and tubes, or spare parts for communications equipment.

- Plan for the effects of wind, sand, and sun on all equipment. All plastic and glass surfaces on vehicles, aircraft, and other equipment should be covered when not in use. Covers should be ordered or made prior to deployment.
- Fuel planning is critical due to power limitations, extended range requirements, and increased vulnerability of refueling sites in the relatively open desert terrain. Careful planning of FARPs is essential for mission accomplishment.
- Units should plan to pack both hot and cold weather clothing when deploying due to varying temperatures between day and night.
- Petroleum, oils, and lubricants (POL) should be of the proper viscosity for desert operations. Maintenance services are also performed more frequently on ground vehicles and aircraft, thus requiring a larger amount of POL than normal.
- Extra filters of all types are planned for due to a higher consumption rate.

5-19. Training for desert operations is not significantly different than training for operations in other areas except for the following:

- Because many desert areas are in mountainous terrain and because high temperatures increase density altitude, aeromedical evacuation units should conduct mountain training to prepare for contingencies in desert areas. Further, procedures and techniques for evacuation in mountainous terrain must be practiced by all AHS personnel. Special equipment requirements (paragraph 5-7) must also be considered.
- Navigation in desert terrain varies from simple to extremely difficult. Factors affecting navigation are the type of desert and the scale and quality of the available navigational charts. At times, aircraft may have to use dead-reckoning navigational techniques (time, distance, and heading). Ground vehicles must have compasses available, as they have to rely on compass headings and odometer readings to navigate. Ground and AA crews should be able to interpret navigational charts and maps of all types and scales. Use of convoys is a viable technique to improve security and to ensure that ground vehicles do not get lost. Aircraft may be used to assist in navigation by convoys in those areas in which there are poor road networks and the terrain offers no distinctive features by which to navigate.

5-20. In principle, medical evacuation operations in the desert do not differ greatly from these operations in other environments.

5-21. Helicopter landing sites should be chosen with care. Common mistakes made by many units when establishing the LZ are—

- Locating the pad relative to the patient and tents, vehicles, and other obstacles. A common tendency is to locate the helipad downwind of MTFs so that approaches may be made into the wind towards the facility. In mountainous deserts, winds normally channel down the valleys and are more predictable along valley floors. A better site selection for an LZ is with the MTF, alongside the approach and takeoff zone. Thus, the landing direction is up or down the valley, depending on the airflow, and the MTF is not over flown.
- Situating landing sites in washes, small confined areas between large rocks, or close to moving tracked vehicles. Map coordinates are rarely accurate unless the site is beside a major terrain feature. Therefore, LZs should be located next to major terrain features or on higher ground where they can be seen from the air at a distance of 2 to 3 kilometers, if possible. Lack of distinctive features in the open desert and on large-scale maps makes pinpoint navigation difficult, especially at night.
- Marking of helicopter LZs is done so that the pad can be seen from the air, but the markings should not be a hazard in themselves. If engineer tape is used, it should be firmly secured to prevent it from blowing loose. Panel markers are not a good tool to use as they are difficult to see. If panel markers are used, they need to be secured. If used, flares or marker smoke should not be deployed on or directly upwind from the pad. Smoke grenades or flares should not be thrown under the aircraft as it lands. Avoid using white smoke to mark the LZ. Colored smoke is probably the best daylight marking method. It is difficult to detect a smoke grenade more

than 2 to 3 kilometers away, but an aircraft in the general vicinity can normally see it. Radios are used to guide aircraft to the LZs, but this creates an electronic signature. Units requesting medical evacuation must be prepared to signal the evacuation aircraft upon its arrival. Normally, map coordinates will guide the aircraft to within 2 to 3 kilometers of the LZ. Even from nap-of-the-earth (NOE) altitudes, the aircrew may be able to see several units in the area. The requesting unit must signal the aircraft to ensure the designated LZ is used.

5-22. When planning for night flights, considerations include the following—

- Moonlight aids the medical evacuation pilot by providing him with the light to see with either unaided vision or night vision goggles (NVGs). When adequate ambient light exists, medical evacuation crews function almost as effectively at night as they do during daylight. The small arms threat is somewhat reduced at night, although it still exists from radar-guided weapons, infrared-sited weapons, and passive night vision device-equipped weapons systems. Flying into a bright moon with NVGs on can be compared to flying into the sun during the day. The goggles darken and visibility becomes extremely poor. Flight routes should not be planned to fly directly into a bright moon if NVGs are to be used.
- The lack of visual cues over sand is similar to that over water. It is very easy for pilots to become disoriented and fly into the ground. Reliance on radar altimeters is a must over flat sandy areas of the desert.
- Frequently, desert areas do not have sufficient ambient light to allow adequate night vision, even with the aid of NVGs. A pilot wearing NVGs is often unable to see the ground at an altitude of 100 feet using a landing light equipped with a pink light filter. Under these conditions, dead reckoning is the only effective navigation method unless global positioning satellites, Doppler equipment, or NAVAIDS are available. Unfiltered light can be used with or without NVGs; however, this increases the risk of exposing the aircraft's position to the enemy.
- Desert warfare is usually characterized by extended battle zones which increase evacuation distance and time. Combat health support units are located further to the sustainment area in the desert. Establishing an ambulance shuttle system or CCPs is useful. Combat health support units require a greater number of vehicles for operating in deserts than in other environments. Air evacuation by fixed and rotary-wing aircraft is the preferred method due to their speed and range. Further, using aircraft reduces the load on ground vehicles. Augmentation from higher echelon HSS may also be required to meet the extended evacuation needs.
- Smoke is used extensively on the modern battlefield by both sides. It can be effectively used to mask friendly actions to include medical evacuation. (Refer to Appendix E for further information.)
  - Smoke can be a major hazard, especially to medical evacuation helicopters. Smoke reduces visibility and forces an aircraft higher where it can be acquired by threat weapons systems. The phenomenon of inversion occurs often in the desert. When this happens, medical evacuation vehicles and aircraft may be able to work underneath the smoke using the smoke layer for overhead concealment.
  - Medical units must coordinate closely with supported organizations on smoke operations. Smoke can either help or hinder the evacuation mission, depending upon how it is used.
- Communications in the desert are affected by a number of factors. Atmospheric interference and the skip of signals occur frequently. Mineral deposits in the desert may unexpectedly disrupt communications. Many of these problems can be overcome by using additional radio relays, pre-established control measures, and visual signals.
- Artificial lights may be used at times in the desert. They are very easily detected. Even with blackout lights, vehicles using lights can be detected for miles with NVG. Serious consideration should be given to driving without using lights when the tactical situation dictates. Ground guides are used to help vehicles navigate through areas that are not clearly marked or through area where troops are present.
- Wind is one of the most significant environmental factors affecting medical evacuation in the desert. Wind can be destructive to both structures and equipment; tents, antennas, and aircraft

can be easily damaged. Wind direction and speed vary greatly within the space of a few miles. Velocity is substantially increased when wind channels between hills and direction changes due to interference of terrain features. The wind frequently makes aeromedical evacuation impossible by exceeding the operating limitations of the aircraft. At other times, it may limit the use of some potential LZs. Blowing sand and dust can slow down the evacuation system by making navigation by either ground or AA difficult, if not impossible. High winds are predictable to a certain extent. For example, at certain times of the year in the Mojave Desert high winds occur every day at dusk and last for 3 to 4 hours. At other times, high winds, based on frontal weather patterns, can remain for several days at a time. These factors should be considered by HSS planners, and medical evacuation assets should be massed or relocated accordingly.

- The desert provides little or no protection from enemy air defenses except in mountainous terrain. Aircraft may have to be flown above NOE altitudes to prevent a dust signature. These factors cause increased exposure and vulnerability of AAs to enemy air defenses and may limit their employment.

5-23. Further Information. Refer to FM 90-3 for additional information on desert operations.

## SECTION IV — EXTREME COLD WEATHER OPERATIONS

5-24. Operations in the extreme cold have many of the limiting factors found in desert operations. The tundra and glacial areas are harsh, arid, and barren. Temperatures may reach lows of -80° F to -100° F which, combined with gale force winds, make exposure unsurvivable.

- The greatest environmental detriment to operations is blowing snow. This results in a loss of depth perception from total white conditions. Blowing snow is caused by the wind or by the rotor wash of helicopters; its effect reduces visibility to zero.
- Other environmental considerations are as extreme but easier to circumvent. Solid footing is suspect in both the dead of winter and in the summer. Snow and ice cover crevasses, holes, and otherwise unstable ground. In traversing suspect ground situations, consider linking Soldiers by rope. During the summer, ground transportation is more restricted than in any other environment due to the marsh and muskeg composition of the arctic tundra. Patients must be sustained for a longer duration due to terrain delays and the lack of direct lines of evacuation.
- Greater responsibility has to be placed on each Soldier, especially for maintenance of nutrition and water consumption. It is imperative to stress that leadership and training are important in the prevention of cold weather injury. Strict adherence to the guidelines found in FMs 21-10 and 31-70 assures an effective fighting force. Water conservation is essential; however, adequate consumption by the individual should be enforced.
- Factors to consider for conducting evacuation in arctic operations include the following:
  - Arctic warfare is usually characterized by extended battle zones that increase evacuation distance and time. Establishing an ambulance shuttle system or CCPs is useful. Augmentation from higher echelon AHS may also be required to meet the extended evacuation needs.
  - Additional supplies of water should be carried by ambulances and maintained at CCPs, if possible.
  - Due to the decreased temperature and frozen environment, ambulance maintenance requirements are increased. Lubricants must be of the correct viscosity for the temperature. In extreme cold, batteries perform less efficiently. Consult the specific aircraft maintenance manual for the recommended battery and procedures for extreme cold weather operations.

---

***Note.*** All ambulances are considered deadlined without a functional heater for the patient compartment.

---

  - The proper storage of medical supplies is essential to prevent loss from freezing.

- There are few terrain features or road networks; therefore, evacuation routes must be surveyed and marked over open terrain. At extreme latitudes, operations during the winter months are conducted in extended hours of darkness. The use of NVG may be required. Compass accuracy is inconsistent due to a geomagnetic phenomenon. Beacons and homing devices are essential for air navigation.
- Weather is extremely unpredictable. There are too few observers to allow for accurate assessment of weather patterns. Unfavorable weather conditions cause unexpected delays; therefore, medical personnel must be prepared to provide survival measures for their patients and themselves.
- Landing zones must be chosen with extreme care in both winter and summer. Blowing snow mandates instrument-assisted takeoffs and running landings. Landing areas must be correspondingly larger. The full weight of the aircraft cannot be allowed to settle on the skis until after firm ground conditions are established. Movement of patients to and from the aircraft is difficult. Where an aircraft lands is where it stays. A rocking motion, to free the skis prior to lift off, is performed using the cyclic and antitorque controls.

- Thorough planning and strict preparation are the keys to survival. Factors to consider include the following:
  - Mud obstacles at noon may become an avenue of approach at midnight.
  - Snow complicates all work. Snow-covered terrain hampers reinforcements, muffles noise, makes cross-country driving hazardous, and creates different camouflage requirements.
  - Because of thermal sights, a complete reappraisal of concealment is required.
  - Tracks in the snow destroy concealment.
  - No Soldier is assigned to any job alone. The buddy system is used at all times.
  - Anticipate that all maintenance tasks will take twice as long.
  - Bare metal can stick to skin or wet garments in subfreezing temperatures.
  - Fuel spilled on skin or garments increases the freezing factor; it is one of the greatest causes of injury in winter operations.
  - When operating in the cold, anticipate increased POL needs. Fuel consumption can rise as much as 25 percent for vehicles operating in deep snow, slush, or mud.
  - The recommended fuel for Yukon stoves is diesel.
  - Make every effort to warm gearboxes and engines before starting.
  - The first consideration in the AO is heat; followed by shelter for sustained work.
  - Soldiers need to stand clear of taut cables; steel tends to be brittle and breaks in extremely cold temperature.
  - Fire extinguishers are winterized by adding 15 percent nitrogen to the carbon dioxide.
  - Degradation of battery life requires changes as much as six times more frequently than in a more temperate environment.
  - Radio sets are warmed up prior to transmission. The sets may be turned on but should not transmit for at least one-half hour.

---

***Note.*** Single-channel ground and airborne radio system (SINCGARS) radios do not require a warm up period.

---

  - Frost shields (such as using the plastic bag in which the batteries are packed) should be placed over microphones.
  - Grounding rods have to be buried horizontally instead of pounded in vertically. Recovery of stakes and rods placed in the ground is significantly more difficult.
  - Flooring is needed in heated areas because of the thawing of the tundra.
  - Soldiers must take breaks for water and warmth.
  - Static electricity presents a serious safety hazard especially around flammable materials.

5-25. For additional information, refer to FM 31-70.

## SECTION V — CHEMICAL, BIOLOGICAL, RADIOLOGICAL, AND NUCLEAR ENVIRONMENTS

5-26. Medical evacuation and treatment are conducted continuously throughout operations conducted in a CBRN contaminated environment. The AHS commander must have a comprehensive plan which is rehearsed on a periodic basis to ensure the timely evacuation and treatment of casualties in a CBRN environment. Techniques and procedures which are essential for operating in a contaminated environment should be contained in the unit TSOP. The number of casualties and their medical condition, type of contaminant, the size of the land area contaminated, the expected duration of operation, risk assessment and acceptable level of risk, and the number of AHS assets (medical personnel, medical units, and evacuation vehicles and aircraft) initially contaminated will determine the quantity and type of uncontaminated AHS resources, if any, which will be introduced into the contaminated environment to ensure timely medical treatment and evacuation occur.

5-27. The Commander must take into consideration the number of assets he is willing to commit during evacuation operations in a CBRN contaminated environment. Since the combinations of evacuation methods are nearly endless, the commander has greater flexibility in tailoring an evacuation plan to meet his particular tactical situation in a CBRN contaminated environment.

5-28. On the modern battlefield there are three basic modes of evacuating patients (personnel, ground vehicles, and aircraft).

- In using personnel to physically carry the casualties, the commander must realize the inherent stress involved. Cumbersome mission-oriented protective posture (MOPP) gear needed in a contaminated environment (added to climate, increased workloads, and the fatigue of battle) greatly reduces the effectiveness of unit personnel.
- If the commander must send evacuation personnel into a radiological contaminated area, he must establish operational exposure guidance (OEG) for the medical evacuation operation. Radiation exposure records are maintained by the unit CBRN NCO and are made available to the commander, staff, and surgeon. Based on OEG, the commander decides which medical evacuation assets to send into the contaminated environment.

5-29. Commanders should make every effort to limit the number of evacuation assets which are contaminated while still maintaining a timely and effective medical treatment and evacuation operation.

- It is expected that a certain number of both ground and AAs will become contaminated in the course of battle. The commander can, therefore, segregate the contaminated ones. This results in the smallest impact on his available assets and the greatest possibility for continuing the patient evacuation mission. Optimize the use of resources, medical or nonmedical, which are already contaminated before employing uncontaminated resources.
- Once a vehicle or aircraft has entered a contaminated area, it is highly unlikely that it will be able to be spared long enough to undergo a complete decontamination. This depends upon the contaminant, the tempo of the battle, and the resources available. Normally, contaminated vehicles (air and ground) have restricted use and are confined to dirty environments.
- Introducing uncontaminated aircraft into a contaminated area should be avoided, whenever possible. Ground ambulances should be used instead of AAs as long as their use does not adversely affect the patient's medical condition. Ground ambulances are more plentiful and are easier to decontaminate. This does not, however, preclude using aircraft in a contaminated environment or in the evacuation of contaminated patients.
- The relative positions of the contaminated area, FLOT, and threat air defense systems determine if and where helicopters are to be used. The commander may choose to restrict one or more helicopters to the contaminated areas and use ground vehicles to cross the line separating contaminated and clean areas. The ground ambulance can proceed to the receiving MTF with a

patient decontamination station. The patient can then be transferred to a clean ground or AA if further evacuation is required. The routes used by ground vehicles to cross between contaminated and clean areas are considered dirty routes and should not be crossed by clean vehicles. The effects of wind and time upon the contaminants must also be considered.

- The rotor wash of the helicopters must always be kept in mind when evacuating contaminated casualties. The intense winds disturb the contaminants in the area and further aggravate the condition by additionally spreading the contaminants. Ideally, the aircraft must be allowed to land and reduce to a flat pitch prior to bringing any patients near. This will be dictated by the tactical situation, but allows some reduction in the effects of the downwash. A helicopter must not land too close to a decontamination station (especially upwind) because any trace of contaminants in the rotor wash will compromise the decontamination procedure.
- Immediate decontamination of aircraft and ground vehicles should be accomplished to minimize crew exposure. Units should develop their own procedures for thorough decontamination and document them in their TSOPs. A sample aircraft decontamination station that may be tailored to a particular unit's needs is provided in FM 3-11.5.
- Evacuation of patients must continue even in a contaminated environment. The commander must recognize the constraints placed upon him by resources and plan and train to overcome deficiencies.
- Refer to FM 4-02.7 for additional information on AHS operations in a CBRN environment.

## SECTION VI — SHORE-TO-SHIP EVACUATION OPERATIONS

5-30. It is imperative that Army aeromedical evacuation units be able to interface with USN air-capable ships. Lessons learned from past operations, such as Vietnam and Grenada, have shown that US Army helicopters should be able to operate to and from USN air-capable ships. An interservice agreement between the Army and the Navy allows for deck-landing qualification of Army pilots. (Refer to FM 1-564 for additional information.)

- It is important that units having contingency missions requiring Navy support establish training requirements to obtain naval-operations orientation, water egress training, water survival, and deck-landing qualification. This enhances the successful accomplishment of the aeromedical evacuation mission to naval vessels.
- In past joint operations, communications have been burdensome for both Army and Navy elements. Commonality of communication requirements should be established during training exercises. Communication equipment and frequencies for medical evacuation to Navy vessels must be established. This will provide smooth integration of Army helicopters into the Navy airspace management system during actual operations.
- As the Navy vessels may operate relatively long distances from the ground combat operations, Army aeromedical evacuation units need to be proficient in over water navigation. The use of NAVAIDS from the Navy element in support of the operation is the first priority for over water navigation. Basic dead-reckoning remains a secondary measure.

5-31. An important aspect of joint operations is the medical capabilities of Navy vessels servicing the CZ. Knowledge of ship's medical capabilities assists the MRO to direct patients to proper treatment sites. There are many classes of ships which can meet the medical needs of ground forces. Destroyer tenders and aircraft carriers have helicopter landing areas, one operating room (OR) and, at a minimum, one medical officer. Amphibious ships have the most extensive medical facilities of any Navy combat ship. The Navy has 59 amphibious ships in active commission plus two tank landing ships, which are operated by the Navy Reserve Force. The primary mission of the amphibious ships is to transport and support the Fleet Marine Force. The ships have the additional duty of casualty receiving and treatment ships (CRTS). During normal operations, the medical staff is kept to a minimum. The medical staff is augmented when expanded capabilities are needed. Current information regarding landing requirements and medical capabilities should be obtained during training periods with the Navy. Casualty care is secondary to the combat mission of all US combat ships.

- Amphibious assault ships "WASP" class are designated by the Navy as amphibious helicopter assault carrier dock (LHD) (followed by a number) and have the largest patient care facilities on any US combat ship. The WASP class ships have six main ORs, four dental ORs, bed capacity that can be expanded to 600, and it carries 1,500 pints of frozen blood. This ship can receive casualties from helicopters or landing craft.
- Amphibious assault ships "TARAWA" class are designated by the Navy as amphibious helicopter assault carrier (LHA) (followed by a number). The TARAWA class ships have three main ORs, two dental ORs, an overflow bed capacity of 300, and carries 1,500 pints of frozen blood. The ship can receive casualties from helicopters or landing craft.
- Amphibious assault ships "IWO JIMA" class are designated by the Navy as amphibious assault helicopter carrier—aviation (LPH) (followed by a number). These ships were specifically designed to operate helicopters. The IWO JIMA class ships have two ORs and an overflow bed capacity of 200.
- The amphibious transport dock (LPD) (followed by a number) has less medical capabilities than the LHD, LHA, or LPH ships. It can be designated as secondary casualty receiving ship.
- The older dock landing ship (LSD) (followed by a number) can be used as a secondary CRTS when augmented. The newer class of LSD currently under construction can be used as a casualty receiving ship with a capacity for 50 wounded.
- The tank landing ship (LST) (followed by a number) is another type of ship used in amphibious operations. It is designed with a helicopter platform and a stern ramp. Patients can be delivered by air or boat when required by tactical or mass casualty situations. When the LST is augmented with medical personnel and materiel, it can be used for the emergency treatment and evacuation of patients.
- The troop transport (AP) (followed by a number) is not in active service. When available, the troop transport can be outfitted with special medical facilities and carry sick, injured, and wounded personnel.

5-32. The Military Sealift Command operates two hospital ships. The United States Navy Ship (USNS) MERCY T-AH 19 and the USNS COMFORT T-AH 20. One ship is based on each coast and, when needed, will be assigned medical staffs from military hospitals, getting underway within 5 days. The hospital ships MTFs were designed for a total capacity of 1,000 casualties, including 500 acute care beds and 500 recuperation beds. The hospital ships have 50 trauma stations in the casualty receiving area; 12 operating rooms; a 20-bed recovery room; 80 intensive care beds; and 16 intermediate, light, limited care wards. The maximum patient flow rate, for which the helicopter facility and the casualty reception area were designed, is 300 patients per 24 hours. There is a limited capability to receive casualties from boats.

5-33. The US Army has the shore-to-ship medical evacuation mission on an area support basis for Marine forces deployed on land.

## SECTION VII — AIRBORNE AND AIR ASSAULT OPERATIONS

5-34. Airborne and air assault operations are some of the most complicated military operations to undertake, not only from an operational perspective, but also from a sustainment and conservation of the fighting strength perspective. A number of personnel within the medical chain that need to coordinate, not only amongst the AMEDD, but also include the aviation community as what proponent own the aircraft (for example, BAE, BSS, medical platoon, and such). Manual evacuation techniques may have to be used and planners should plan for redundant communications, only ground medical evacuation may or may not be available.

5-35. The airborne and air assault operational forces are specialized forces employed to maximize their design characteristics. Airborne units are a flexible force that can be strategically or tactically deployed. They can be inserted rapidly anywhere in the world as either a deterrent or strike force. Air assault units are flexible and lethal fighting organizations. They are ideally suited for rapid employment to critical areas beyond the reach of ground forces.

5-36. After airborne forces have landed in the objective area, they reorganize and maneuver to seize objectives. When it is necessary for assault aircraft to land in the drop zone, they are parked and unloaded rapidly. Then, they may be used to transport Soldiers injured during the parachute assault. It must be understood that organic medical units may experience an overload of patients during the early phases of an airborne assault. These units have to hold the patients until either ground link-up is made or evacuation can be established at airheads. Aeromedical evacuation from the airhead is accomplished using tactical and strategic USAF aircraft.

5-37. The air assault division's organic aircraft have the ability to attack from any direction, over fly obstacles, and bypass enemy positions. Evacuation of patients in the assault phase is accomplished by AAs. Air ambulances may accompany the air assault task force (AATF) or respond from laager sites once the initial assault has taken place. If AAs are providing on-call support, it will be necessary to fly secure air avenues of approach.

5-38. When both airborne and air assault divisions have been employed and become a part of other conventional forces, their operations are similar to that of light infantry forces. During initial deployment, division medical evacuation assets may be used to evacuate patients to the airhead for air evacuation directly to division/corps hospitals.

## Chapter 6
# Medical Regulating

## GENERAL

6-1. Medical regulating is the coordination and control of moving patients to MTFs which are best able to provide the required specialty care. This system is designed to ensure the efficient and safe movement of patients.

## PURPOSE OF MEDICAL REGULATING

6-2. Medical regulating entails identifying the patients awaiting evacuation, locating the available beds, and coordinating the transportation means for movement. Careful control of patient evacuation to appropriate hospitals is necessary to—

- Effect an even distribution of cases.
- Ensure adequate beds are available for current and anticipated needs.
- Route patients requiring specialized treatment to the appropriate MTF.

6-3. The factors that influence the scheduling of patient movement include the following:

- Patient's medical condition (stabilized to withstand evacuation).
- Tactical situation.
- Availability of evacuation means.
- Locations of MTFs with special capabilities or resources.
- Current bed status of MTFs.
- Surgical backlogs.
- Number and location of patients by diagnostic category.
- Location of airfields, seaports, and other transportation hubs.
- Communications capabilities (to include radio silence procedures).

## MEDICAL REGULATING TERMINOLOGY

6-4. As medical regulating may include coordination with other services, it is necessary to use the correct terminology. These terms include—

- Intracorps Medical Regulating—This is the system by which patients are transferred or evacuated from an FSB or main support battalion (MSB) to a combat support hospital (CSH).
- Intratheater Medical Regulating—This is the system by which patients are transferred or evacuated from one hospital to another within the AO. This includes evacuations between CZ hospitals, between EAC hospitals, or from CZ hospitals to EAC hospitals.
- Intertheater Medical Regulating—Moving patients between, into, and out of the different theaters of the geographic combatant commands and into the CONUS or another supporting theater. (JP 4-02.)
- Patient Administrator—The patient administrator (PAD) accomplishes the medical regulating function at the hospital level in addition to his normal duties. His medical regulating functions include consolidating all evacuation requests within the hospital and forwarding an evacuation request to his next higher headquarters for action. The PAD is also responsible for keeping his next higher MRO apprised of the current beds available and the OR status.

- Medical Regulating Officer—The MRO functions as the responsible individual at C2 headquarters for receiving and consolidating evacuation requests. These requests are initiated by the DSS or subordinate hospitals. The MRO also maintains the current patient status, bed status, and the surgical backlog at subordinate hospitals. His duties include—
  - Managing what patient classes are regulated into his facility.
  - Determining what resources are available to move the patients and coordinating for the use of these assets.
  - Maintaining accountability of patients within the MTFs.
  - Preparing reports as required.
- Theater Patient Movement Requirements Center—The TPMRC is a joint agency normally located at or near the unified theater headquarters. The theater surgeon supervises the functions of this office. These functions include—
  - Maintaining direct liaison with the Global Patient Movement Requirements Center (GPMRC), the MROs of component services, and the transportation agencies which furnish the means for evacuation.
  - Obtaining periodic reports of available beds from the services MROs providing hospitalization.
  - Selecting hospitals based on the reported bed availability to receive patients within EAC.
- Global Patient Movement Requirements Center—The GPMRC is a joint agency located in CONUS and established by the USTRANSCOM. The GPMRC receives requests from the TPMRCs. The primary role of the GPMRC is to apportion intertheater assets to the TPMRCs, collaborate and integrate proposed TPMRC intertheater plans and schedules, and communicate lift and bed requirements. The destination hospital is determined based on the patient's medical needs, the available transportation resources, and MTF capabilities.
- Joint Patient Movement Requirements Center (JPMRC)—Joint patient movement requirements center provides theater patient movement requirements center-type domain, automatic information system support and operations for a joint task force operating within a unified command area of responsibility. The theater patient movement requirements center maintains overall responsibility for theater patient movement operations, but the joint patient movement requirements center is responsible for patient movement operations within its area scope of responsibility and coordinates with the theater patient movement requirements center for intra-theater patient movement and the GPMRC for inter-theater patient movement. Also called JPMRC.
- Theater Aeromedical Evacuation System (TAES)—The TAES is a functional organization which is provided by the USAF and performs the mission of theater AE. Significant components of the TAES include the following:
  - Aeromedical evacuation command squadron.
  - Aeromedical evacuation control team (AECT).
  - Mobile aeromedical staging facility.
  - Contingency aeromedical staging facility (CASF) (under the C2 of the medical group commander).
  - Aeromedical evacuation crews.
  - Critical care air transport teams (CCATTs) (not necessarily Air Force).
- Aeromedical Evacuation Command Squadron—The AE command squadron provides C2 of all assigned TAES forces and can deploy in advance of other AE components to arrange support requirements for AE forces. The command squadron advises commanders and other personnel/agencies on AE operations, capabilities, and requirements and provides procedural and technical guidance for attached and transiting AE elements.
- Aeromedical Evacuation Control Team—The AECT is a CASF (under the C2 of the medical group commander) and is an USAF element responsible for linking validated patient movement requests with the appropriate airlift platform. The AECT is normally located within the Air

Mobility Division of the theater air operations center (AOC) for intratheater and the tanker airlift control center (TACC) for intertheater coordination of AE operations using USAF aircraft. The CONUS and strategic AOC for the USAF is the TACC located at Scott Air Force Base, Illinois. The AECT coordinates with and receives requirements from the PMRC for execution.

- Aeromedical Evacuation Liaison Team (AELT)–The AELT provides a direct communications link and immediate coordination between the user service and the TAES. The AELT is located at any level where Air Force fixed-wing requests are initiated, verifies and coordinates with the AECT/patient movement requirements center (PMRC) the physiology of flight issues and patient flight/movement requirements, assists with patient preparation for flight, and directs patient on-load activities.
- Mobile Aeromedical Staging Facility—The MASF is a USAF staging facility employed at forward airfields in the CZ to provide a temporary staging capability for preparation of patients being evacuated from corps to EAC hospitals. The MASF is employed to ensure patients are prepared for aircraft loading with the main focus of reducing AE aircraft ground time.
- Contingency Aeromedical Staging Facility—The CASF, similar to the MASF, is a USAF staging facility. The CASF, which provides greater capability and longer holding periods than the MASF, is a fixed facility located at theater AE strategic hubs and interfaces with the TAES. As opposed to the TAES elements, which are under the C2 of the AE command squadron, the CASF reports to the medical group commander.
- Military Sealift Command (MSC)—The MSC is the USN element responsible for coordinating movement of supplies, equipment, and personnel into the JOA by Navy ships. Further, it coordinates, through the TPMRC, the medical evacuation of patients by ship from the JOA to the support base or hospital ship, as required.
- Corps Movement Control Center (CMCC)—The CMCC is the corps movement control organization. It provides centralized movement control and highway regulation for movement of personnel and material into, within, and out of the corps area. When USAF capabilities are exceeded, the CMCC coordinates requests for additional air and ground resources. It also obtains the necessary clearances to support the medical evacuation mission from the CZ.
- Movement Control Battalion (MCB)—The MCB mission is to provide movement management services and highway traffic regulations and to coordinate for personnel and material movements into, within, and out of the theater. The MCB coordinates with allied and HN movement control agencies. It also coordinates with the USTRANSCOM and its subordinate units (such as the Air Mobility Command and MSC) and prepares movement and port clearance plans and programs.
- Joint Military Transportation Board (JMTB)—The JMTB is a joint staff composed of members of the Army, USAF, and USN that coordinates transportation requirements for patients requiring intertheater evacuation.
- Transportation Command Regulating and Command and Control Evacuation System—The Transportation Command Regulating and Command and Control Evacuation System is the single DOD automated information system that supports medical regulating, patient movement, and patient in-transit visibility requirements.
- Automated Patient Evacuation System—The Automated Patient Evacuation System is the system that automates the patient movement portion of medical evacuation.

# MEDICAL REGULATING FOR THE DIVISION

6-5. Medical regulating in and from the division is the responsibility of the DSS. Medical regulating in the division is not as formalized as the rest of the medical regulating system. It is usually operated procedurally so as not to depend solely on communications to effect rapid evacuation. The medical regulating function in the DSS is concerned primarily with.

- Tracking the movement of patients throughout the division MTFs and into the corps facilities.
- Monitoring the use of ambulance assets.

6-6. Air and ground ambulances placed in GS of the BCTs can be field sited in the sustainment area when the mission directs. When these assets go forward to the BSMC or the ASMC to evacuate patients to corps MTFs, they have corps MTF destinations predetermined (blocks of beds). The DSS, in coordination with the medical group/brigade MRO, establishes the number of patients a supporting corps hospital can accept during a particular period of time. These blocks of available beds are then provided to the GS ambulances prior to the call for missions.

***Note.*** Under the MRI force structure, the medical group functions were absorbed by the medical brigade.

- Once an evacuation mission is completed, the originating division MTF contacts the patient disposition section of the BSS and provides—
  - Patient numbers by category and precedence.
  - Departure times.
  - Modes of transportation.
  - Destination MTFs.
  - Any other information required by TSOP.
- The DSS, in turn, notifies the medical group/brigade MRO via the patient administration net, which is monitored by the corps MTFs. Since corps ground ambulances have no onboard communications ability and AAs have no amplitude-modulated high-frequency capability at present, all patient information is passed to the gaining MTFs via the patient administration nets. To reduce the turnaround time for ground ambulances and to move more serious patients to the CSHs in the sustainment area—
  - Air ambulances are given blocks of beds in the corps hospitals farther to the rear.
  - Ground ambulances are normally given blocks of beds in the more forward deployed CSHs.

6-7. Medical evacuation can be effected immediately, procedurally, and under conditions of communications silence without interrupting the continuum of care by—

- Preparing patient estimates.
- Prioritizing and task-organizing ambulance support.
- Assigning blocks of hospital bed designations prior to the start of the mission.
- The assistant chief of staff (personnel) (G1) is also involved in medical regulating as the G1 is responsible for personnel accountability and casualty reporting. Casualty liaison teams (CLTs) will normally be colocated at CSH to assist and coordinate patient tracking and casualty reporting.

# MEDICAL REGULATING WITHIN THE COMBAT ZONE

6-8. The requirement to transfer patients from one hospital to another within the CZ occurs. This results from—

- Surgical backlogs.
- Mass casualty situations.
- Specialty care requirements.
- Planned movement of an MTF.

6-9. When it is necessary to transfer a patient, the attending physician notifies the hospital PAD. The PAD consolidates all such requests from the hospital and requests movement authority from the IPMC.

6-10. If the medical IPMC can transfer the patient or patients to its subordinate hospitals, he designates the hospitals to receive the patients and notifies both the requesting and receiving hospitals of the transfer. The IPMC also tasks subordinate medical evacuation units for the assets to transfer the patients.

6-11. If the MEDBDE cannot provide the needed hospitalization within its own resources, the MRO forwards the request to the medical TPMC for action. The TPMC then designates the receiving hospitals and notifies the subordinate IPMCs. The medical IPMCs disseminate the information to the hospital PAD and coordinate the evacuation resources for the transfer. The MRO also coordinates the regulation of patients to—

- Other US military service hospitals and naval hospital ships.
- Allied nations' military hospitals.
- Other authorized supporting facilities.

# MEDICAL REGULATING FROM THE COMBAT ZONE TO ECHELONS ABOVE CORPS

6-12. Hospital attending physicians and oral and maxillofacial surgeons submit daily reports to the hospital PAD listing the patients requiring evacuation. The PAD assembles this information and transmits the report to the medical brigade headquarters. This report is a request for transportation, as well as a notification of the number of patients requiring evacuation. The report classifies the patients according to—

- Diagnostic category.
- Desired on-load points.
- When the patients will be available for evacuation.

6-13. The medical brigade TPMC consolidates these reports from each hospital attached to the medical brigade and forwards his report to the corps TPMC. The medical brigade MRO consolidates the reports and transmits the data to the theater TPMC.

6-14. If a JPMRC has been activated within the theater, the MEDCOM MRO consolidates all reports from the CZ medical brigades and forwards them to the JPMRC. The JPMRC designates hospitals in the EAC to receive the patients. The designation is based on the previously received bed status reports from all service components and available means of evacuation. The JPMRC then notifies the MEDCOM MRO of designated hospitals. The TPMC accomplishes this task if the JPMRC is not activated.

6-15. The primary means of moving patients from the CZ out of theater is USAF aircraft. With the elements of the TAES deployed, it is possible to find AELTs at each echelon and as far forward as Air Force fixed-wing requests are initiated. The PMRC monitors the MRO patient evacuation requests and passes requirements to the AELT. At the same time they pass airlift requirements to the AECT, seeking an aircraft to perform the evacuation mission. The AELT, through the MRO patient movement request, requests the PMRC/AECT to move patients. Included in the request are the originating medical facility (OMF) and the destination airfields. The airfields selected are those serving the hospitals designated to receive patients.

6-16. The AECT is a component of the TAES and performs the mission of coordinating the movement of and providing in-flight medical care to patients while under the USAF control. The AECT receives patient movement requirements from the TPMRC, then works with the airlift control team (ALCT) in the AOC to meet the evacuation requirements.

6-17. The AOC coordinates the forward movement of cargo and personnel aboard USAF aircraft with other USAF units, Army transportation representatives, and USN agencies. Certain of these aircraft are scheduled to evacuate patients on their return trips. These aircraft seldom go forward solely to evacuate patients; however, these missions may be used for retrograde airlift and mixed cargo/AE missions is the primary means of AE due to limited aircraft and the need to maximize the use of this asset.

6-18. After the schedules have been arranged, the AECT returns the detailed flight schedule to the JPMRC who passes the information to the TPMC/AELT and the parent AE element.

6-19. The MEDCOM MRO, in coordination with the AELT, issues these instructions to the IPMCs (with the authority to move patients in Army CZ facilities) and the receiving hospitals. The hospitals must prepare to receive the patients at the destination airfields. (This may be accomplished by collocating an ASMC at the airfield to receive the incoming patients.) The patients are sorted by destination hospital and moved by Army medical evacuation means. The instructions mentioned above include, as a minimum, the—

- Number of patients to be moved.
- On-load airfield.
- Destination airfield.
- United States Air Force aircraft mission number.
- Estimated time of arrival at the destination airfield.

6-20. The TPMC issues the flight and movement instructions to its subordinate medical IPMCs. The IPMCs then direct the evacuation units and hospitals within their AOs to move the patients to the on-load airfield according to the arrival time of the aircraft. This movement must be closely controlled, as a MASF can accommodate up to 50 patients. The patients cannot be delivered to the MASF earlier than 6 hours prior to arrival of the aircraft and no later than 1 hour prior to arrival.

## MEDICAL REGULATING WITHIN ECHELONS ABOVE CORPS

6-21. Medical regulating within the EAC is similar to the system used within the CZ. Attending physicians within the Role 3 hospitals notify the hospital PAD of patients requiring evacuation to CSHs. The PAD then consolidates the requests from the hospital and forwards the consolidated requests to the IPMC. The IPMC, in turn, consolidates the requests and forwards them to the TPMC.

6-22. The TPMC, based on periodic bed status and availability reports from subordinate hospitals, designates specific hospitals to receive the patients. The hospitals are designated based on bed availability, to include specialty beds, to support the specific patient. The TPMC then notifies the requesting medical IPMC of the designated hospitals and, in turn, notifies the designated hospitals.

## INTERTHEATER MEDICAL REGULATING

6-23. The patients who are evacuated to EAC are treated there and then further evacuated to CONUS. The attending physicians at the hospital notify the PAD. The PAD then consolidates these requests and forwards them to the IPMC. This IPMC forwards the consolidated request to the TPMC who, in turn, consolidates and forwards a request to the JPMRC (if established) or the TPMRC.

6-24. Upon request of the J/TPMRC for authority to evacuate patients to CONUS, the GPMRC directs the distribution of these patients into hospitals throughout the CONUS; advises the J/TPMRC of the destination hospital; and provides the authority for such movement. As a rule, the destination hospitals are military facilities. Civilian national disaster medical system member hospitals and other federal hospitals may also receive patients. The Department of Veterans Affairs hospitals, for example, may receive patients who are expected to be discharged from service. The GPMRC validates the TPMRC patient movement requirements and, if moving by air, tasks the TACC to plan, schedule, and execute the intertheater evacuation.

6-25. When the J/TPMRC receives the authorization to move patients, it notifies the TPMC of destination hospitals in CONUS. The TPMC coordinates with the JMTB to arrange movement of CONUS-bound patients. The TPMC then authorizes the movement to ASFs/aeromedical staging squadrons (ASTSs) that are located on or near air bases or airstrips capable of handling long-range aircraft. Transportation is arranged, within Army channels, to move patients from the hospitals to the staging facilities. The medical brigade, in coordination with the AELT, then notifies the subordinate CSHs of the flight schedule and the evacuation arrangements for movement to strategic airheads. At strategic airheads, there is an established ASF/ASTS. When the patients are delivered to the USAF, the responsibility for those patients is transferred from the Army hospital to the TAES. Upon arrival in CONUS further movement is the responsibility of the GPMRC.

6-26. All patients may not be able to be moved by air from the theater to CONUS. In that event, the MSC is used to move them by surface means. The movement authority also comes from the G/T/JPMRC or TPMC which has arranged with the Navy Service Component Command for the movement of patients by hospital ships. When the patients are moved by ships, the MEDCOM has to provide holding facilities at the port (collocating an ASMC can provide this support). Patients are delivered to these holding facilities and held there until loaded aboard the ships.

# MOBILE AEROMEDICAL STAGING FACILITY

6-27. The MASF is a 13-person, mobile, tented, temporary staging facility deployed to provide supportive patient care and administration. Each MASF is capable of a throughput of 40 patients per 24 hours and should be able to sustain this tempo for 72 hours before augmentation is required and is not intended to hold patients overnight or for an extended period.

6-28. This theater system is used to evacuate patients from—

- United States Air Force operational locations within the CZ to hospital facilities outside the CZ.
- Airhead or airborne objective areas where airborne operations include USAF forward logistics support.

6-29. Mobile aeromedical staging facilities—

- Provide supportive medical care when not augmented by the CCATTs. (When augmented by CCATTs, continued stabilization of patients can be accomplished. Aeromedical evacuation crews and CCATTs fly airlift missions to provide in-flight patient care.)
- Confirm sending facility has prepared patients for evacuation.
- In coordination with the OMF, ensure patient evacuation manifests are completed. (If the OMF is an Role 2 facility [such as an FSMC augmented with an FST], the forms required to complete the manifest may not be available and the MASF will be required to complete the appropriate forms.)
- Identify patient baggage tags.

6-30. Upon deployment, the OMF provides an adequate quantity of medications for patients' transit time to the regulated destinations.

6-31. The MASF staff also establishes liaison with the OMF. The AELT is composed of one medical service corps officer for administrative assistance and a flight nurse. The team provides the initial interface between the user service and the TAES. The AELT is located at any level of the combat forces where Air Force fixed-wing requests are initiated.

# LIMITATIONS OF THE UNITED STATES AIR FORCE THEATER AEROMEDICAL EVACUATION SYSTEM

6-32. There are a number of limitations that are inherent in the current system. These include the following:

- Absence of biological weapon and chemical weapon agent decontamination ability.
- The MASF does not have the capability to provide patient meals.
- The AECT ensures the initial 30-days medical resupply package arrives at the MASF. The MASF/AELT/aeromedical evacuation operations team rely on the user service for all other logistical support.
- It is the Army's responsibility to provide food and other logistical support required including moving patients back to Army facilities should USAF AE support be delayed.

## ORIGINATING MEDICAL FACILITY'S RESPONSIBILITIES

6-33. Once the authorization to move the patient has been given, the OMF must complete the following administrative procedures prior to entering the patient into the TAES:

- The patient's baggage tag, patient evacuation manifest, and patient evacuation tag are the specified evacuation forms for all services and are completed as required by triservice regulation.

*Note.* If the OMF is a Role 2 facility (such as the BSMC), the forms may not be available; in which case, the forms will need to be completed by the MASF.

- All of the patient's medical records must be collected together and packaged. The dental records are forwarded separately in the event they are needed for identification.
- At the appropriate time, the OMF provides transportation to the MASF and assists in the offload.
- The OMF must provide the necessary medications, medical supplies, and equipment to support the patients' travel time to the regulated destination.
- Any requirements for armed guards must be met by the echelon commander.

*Note.* Medical units do not provide guards for prisoners or EPW in their care. When guards are required, they are provided by the echelon commander. The OMF will coordinate for this support when needed.

- A limited amount of personal baggage is authorized if each piece is properly tagged and delivered to the MASF with the patient. Patients will always be evacuated with CBRN-protective equipment, less the protective overgarment.
- Each patient must be clearly identified with a wristband or equivalent identification and properly classified as to his medical condition.
- The OMF must ensure that each patient is properly briefed and prepared for his evacuation prior to his arrival at the MASF.

## MEDICAL REGULATING FOR ARMY SPECIAL OPERATIONS FORCES

6-34. As in medical evacuation, the medical regulating plan must be integrated with the ARSOF operational and logistic plan. Maximum use of opportune (operational and logistics) aircraft and command and logistics communications nets must be coordinated to expedite mission requests and ensure success.

6-35. The ARSOF medical planner must constantly coordinate with the battalion or group operations and logistics sections to obtain up-to-date information of opportune transportation assets to be used for evacuation. In an operation, or when the theater is not sufficiently developed to allow the TAES to be used effectively, the primary means of air evacuation will be those SOA or USAF SOF airframes conducting the clandestine mission. It is essential that coordination is made through the Theater Special Operations Command (TSOC) or the highest C2 element for flight medics or para rescue men to accompany the flight when backhauling the casualties. Otherwise, a medic from the SOF unit being supported may have to accompany the patient, leaving the mission without proper medical support, or the casualty may have to be transported without en route care.

6-36. For all other special operations, the supporting medical evacuation unit provides air and ground ambulances according to standard doctrinal procedures. United States Air Force MASFs or AELTs may be collocated at SOF support bases, or C2 bases, particularly during contingency operations where the build-up phase allows for prepositioning of assets.

6-37. During sustained special operations missions, the TSOC cannot afford to lose the services of low-density ARSOF skilled Soldiers who become casualties. Every effort must be made to preserve ARSOF capabilities and combat power in theater. A determination must be made as to who can be treated and returned to duty at hospitals within the EAC. As an exception to the theater evacuation policy, the combatant command may retain injured or wounded ARSOF in theater where they can be returned to limited duty without jeopardizing their recovery and health. There they can assume the support duties performed by other ARSOF Soldiers, freeing the latter for operational duties.

# Appendix A
# Geneva Conventions and Medical Evacuations

The conduct of armed hostilities on land is regulated by customary international law and lawmaking treaties such as The Hague and Geneva Conventions. The rights and duties set forth in the Conventions are part of the supreme Law of the Land. The US is obligated to adhere to these obligations even when an opponent does not. It is a DOD and Army policy to conduct operations in a manner consistent with these obligations. An in-depth discussion of the provisions applicable to medical units and personnel is provided in FM 4-02. This appendix discusses only those articles or actions which affect medical evacuation operations.

## DISTINCTIVE MARKINGS AND CAMOUFLAGE OF MEDICAL FACILITIES AND EVACUATION PLATFORMS

**This paragraph implements STANAG 2454.**

A-1. All US medical facilities and units, except veterinary, display the distinctive flag of the Geneva Conventions. This flag consists of a Red Cross on a white background. It is displayed over the unit or facility and in other places as necessary to adequately identify the unit or facility as a medical facility.

### AUTHORIZED EMBLEM

A-2. The Geneva Conventions authorizes the use of the following distinctive emblems on a white background: Red Cross; Red Crescent; and Red Lion and Sun. In operations conducted in countries using an emblem other than the Red Cross on a white background, US Soldiers must be made aware of the different official emblems. Although not specifically authorized as a symbol in lieu of the Red Cross, enemies of Israel in past wars have recognized the Red Star of David and have afforded it the same respect as the Red Cross. This showed compliance with the general rule that the wounded and sick must be respected and protected when they are recognized as such, even when not properly marked.

A-3. United States forces are legally entitled to only display the Red Cross. However, commanders have authorized the display of both the Red Cross and the Red Crescent to accommodate HN concerns and to ensure that confusion of emblems would not occur. Such use of the Red Crescent must be in a smaller size than the Red Cross.

**This paragraph implements STANAG 2931.**

#### Camouflage of the Distinctive Emblem

A-4. Camouflage of medical facilities (medical units, medical vehicles, and medical aircraft on the ground) is authorized when the lack of camouflage might compromise tactical operations. The marking of facilities and the use of camouflage are incompatible and should not be undertaken concurrently.

A-5. If failure to camouflage endangers or compromises tactical operations, the camouflage of medical facilities may be ordered by a NATO commander of at least brigade level or equivalent. Such an order is temporary and local in nature and is rescinded as soon as circumstances permit.

***Note.*** There is no such thing as a *camouflaged* Red Cross. When camouflaging a medical unit or ambulance, either cover up the Red Cross or take it down. A *black cross* on an olive drab or any other background is not a symbol recognized under the Geneva Conventions.

### MEDICAL AIRCRAFT

A-6. Medical aircraft exclusively employed for the removal of wounded and sick and for the transport of medical personnel and equipment shall not be attacked, but shall be respected by the belligerents, while flying at heights, times, and on routes specifically agreed upon between the belligerents concerned.

A-7. The medical aircraft shall bear, clearly marked, the distinctive emblem together with their national colors on their lower, upper, and lateral surfaces.

A-8. Unless agreed otherwise, flights over enemy or enemy-occupied territory are prohibited.

A-9. Medical aircraft shall obey every summons to land. In the event that a landing is thus imposed, the aircraft with its occupants may continue its flight after examination, if any.

A-10. In the event of involuntary landing in enemy or enemy-occupied territory, the wounded and sick, as well as the crew of the aircraft, shall be prisoners of war; medical personnel will be treated as prescribed in these Conventions.

## SELF-DEFENSE AND DEFENSE OF PATIENTS

A-11. When engaging in medical evacuation operations, medical personnel are entitled to defend themselves and their patients. They are only permitted to use individual small arms.

A-12. The mounting or use of offensive weapons on dedicated medical evacuation vehicles and aircraft jeopardizes the protections afforded by the Geneva Conventions. These offensive weapons may include, but are not limited to machine guns, grenade launchers, hand grenades, and light antitank weapons.

A-13. Medical personnel are only permitted to fire in their personal defense and for the protection of the wounded and sick in their charge against marauders and other persons violating the Law of War.

## ENEMY PRISONERS OF WAR/DETAINEES

A-14. Sick, injured, wounded enemy EPW, and detainees are treated and evacuated through normal medical channels but are physically segregated from US, allied, or coalition patients.

A-15. Personnel resources to guard EPW or detainee patients are provided by the echelon commander. Medical personnel do not guard EPW or detainee patients.

## COMPLIANCE WITH THE GENEVA CONVENTIONS

A-16. The US is a party to the 1949 Geneva Conventions. Two of these Conventions afford protection for medical personnel, facilities, and evacuation platforms (to include aircraft on the ground). All HSS personnel should thoroughly understand the provisions of the Geneva Conventions that apply to medical activities.

### CONSEQUENCES OF VIOLATIONS

A-17. Violation of these Conventions can result in the loss of protection afforded by them. Medical personnel should inform the tactical commander of the consequences of violating the provisions of these Conventions. The consequences may include—

- Medical evacuation assets subjected to attack and destruction by the enemy.
- Health service support capability degraded.
- Captured medical personnel becoming prisoners of war rather than retained persons. They may not be permitted to treat their fellow prisoners.
- Loss of protected status for medical unit, personnel, or evacuation platforms (to include aircraft on the ground).

## PERCEPTION OF IMPROPRIETY

A-18. Because even the perception of impropriety can be detrimental to the mission and US interests, HSS commanders must ensure they do not give the impression of impropriety in the conduct of medical evacuation operations. For example, if a medical commander included in the unit's TSOP rules governing the use of automatic or crew-served weapons, it would give the impression that the unit possessed and intended to use these types of weapons. Under the provisions of the Geneva Conventions, medical units are only authorized individual small arms for use in the defense of the patients under their care and for themselves. Even though the unit does not possess these types of weapons, the entry in the TSOP could be misinterpreted and a case made that the commander intended to use these weapons in violation of the Geneva Conventions.

# Appendix B
# Legacy Units

Currently there are two units that are still active but will be transitioning out of the Army's inventory. They are the medical evacuation battalion and the medical company, AA (UH-1V).

## SECTION I — MEDICAL EVACUATION BATTALION

B-1. The HHD, medical evacuation battalion, serves as the central manager of ground and air evacuation assets within the corps and EAC.

B-2. Assignment. The medical evacuation battalion is assigned to the MEDCOM in the EAC or to the medical brigade in the corps. It is normally further assigned to a medical group for C2.

> ***Note.*** Under the MRI design, the medical group was replaced by the MEDBDE.

B-3. Air and ground ambulance companies assigned to the MEDCOM or MEDBDE are attached to the medical evacuation battalion for C2.

B-4. The basis of allocation is one medical evacuation battalion per a combination of three to seven of the following units:

- Medical companies, AA.
- Medical companies, ground ambulance.

B-5. Mission and capabilities.

- The mission of the medical evacuation battalion is to provide C2 of air and ground medical evacuation units within the AO. It tactically locates in the area where it can best control subordinate air and ground ambulance companies.
- The medical evacuation battalion is designed to focus on C2, planning, patient evacuation, subordinate unit support, and vehicle management. Specific capabilities are—
    - Command and control, planning and supervision of operations and training, and administration of a combination of air and ground ambulance companies.
    - Staff and technical supervision of aviation operations, safety, standardization, and aviation unit maintenance (AVUM)-level maintenance within the attached AA companies.
    - Coordination of medical evacuation operations and communications functions.
    - Coordination of logistics and service support to attached units.
    - Aviation medicine and unit-level AHS.
- This unit is dependent upon appropriate elements of the corps or ASCC for—
    - Personnel service support.
    - Health service support, to include hospitalization.
    - Mortuary affairs support.
    - Laundry, shower, and clothing repair.
    - Communications security equipment maintenance.
    - Military police support.

B-6. Organization and functions.

- Medical Evacuation Battalion (Figure B-1). The headquarters and headquarters detachment (HHD), medical evacuation battalion, is organized into a—
  - Battalion headquarters section.
  - Adjutant (US Army) section.
  - Intelligence Officer (US Army)/S3 section.
  - Supply Officer (US Army) section.
  - Detachment headquarters.
  - Treatment team.
- Battalion Headquarters Section. This section provides C2 of the assigned and attached air and ground ambulance companies. It also assists the commander on all military intelligence matters (to include the health threat), organization, training, operations, planning, personnel service support, and logistics support. Further, it provides information on the health of the command and aviation medicine expertise, as well as providing supervision over technical and flight aspects of administration, training, and safety within subordinate aviation units.
- Adjutant (US Army) Section. This section operates according to METT-TC and is the principle coordinating staff officer responsible for the delivery of human resources support. The S1 is responsible for the execution of all of the HR core competencies to include —PRM; PASR; PIM; R5 operations; casualty operations; EPS; and postal operations. It also includes MWR operations and HR planning and staff operations. The S1 also coordinates with elements of supporting agencies for finance, legal, religious, and administrative services.
- Intelligence Officer (US Army)/S3 Section. This section assists the S2/S3 officer in the execution of his duties and is capable of sustained 24-hour operations. This section remains abreast of the tactical situation and determines future medical evacuation requirements. It plans for ground and air evacuation operations, coordinates command post operations, and maintains the status of the AA units and plans for their employment. Further, this section maintains communications systems and nets, determines intelligence requirements, coordinates with movement control elements, and prepares orders and overlays.
- Supply Officer (US Army) Section. The S4 section assists the S4 officer in the execution of his duties. This section plans, coordinates, and supervises the requisitioning, receipt, storage, issue, and accounting for all classes of supply. Further, it monitors and keeps the commander informed on all matters pertaining to maintenance on assigned aircraft, ground vehicles, and medical equipment. This section also serves as the interface with the supporting medical battalion, logistics (forward/rear) for medical nonexpendable and durable item supply transactions.
- Detachment Headquarters Section. The detachment headquarters section provides, C2, administration, and logistics support for assigned personnel. It is also responsible for company supply and armament functions, food service operations, maintenance operations, and unit administration.
- Treatment Team. The treatment team provides unit-level AHS to assigned and attached elements collocated with the detachment headquarters and to adjacent units on an area support basis. The physician is a flight surgeon and provides staff assistance to the battalion commander on all matters pertaining to aviation medicine. The flight surgeon provides care and treatment for all assigned and attached aircrew members. This physician is dual-hatted as the battalion surgeon.

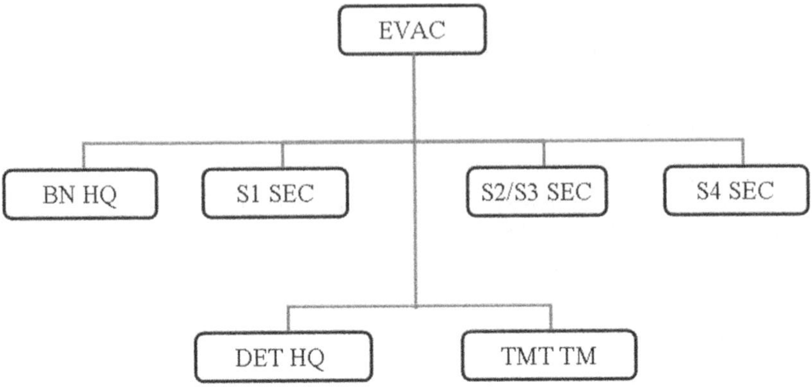


**Figure B-1. Medical evacuation battalion**

## SECTION II — MEDICAL COMPANY, AIR AMBULANCE

B-7. The medical company, AA, provides aeromedical evacuation for all categories of patients consistent with evacuation precedence and other operational considerations. Medical evacuation is effected from as far forward as possible in the tactical AO to division- and corps-level MTFs.

B-8. Assignment.

- The medical company, AA, is normally assigned to the MEDCOM or MEDBDE and attached to the medical evacuation battalion for C2.
- The basis of allocation is one unit in support of each division or equivalent force supported. Further, one unit is in GS in the corps per two division or fraction thereof; or .333 units per separate brigades or ACRs.

B-9. Mission and capabilities.

- The mission of the medical company, AA, is to provide.
  - Aeromedical evacuation support within the theater, either DS to the divisions or GS to the corps.
  - Emergency movement of medical personnel, equipment, and supplies including whole blood, blood products, and biologicals.
- Specific capabilities of this unit are to operate on a 24-hour-a-day basis and evacuate patients based on operational capability (dependent on type of aircraft).
  - Operate 15 AAs (UH-60s). These ambulances are each capable of carrying six litter patients and one ambulatory patient, or seven ambulatory patients, or some combination thereof. Single patient lift capability is 90 litter patients, or 105 ambulatory patients, or some combination thereof. In-flight medical treatment and surveillance of patients is provided by a flight medic.
  - Operate 15 AAs (UH-1H/V). These ambulances are capable of carrying six litter, or nine ambulatory patients, or some combination thereof. Single patient lift capability is 90 litter, 135 ambulatory, or some combination thereof. In-flight medical treatment and patient surveillance are provided by a flight medic.
- Provide internal/external load capability for the movement of medical personnel and equipment.

- Perform AVUM on all organic aircraft and organizational maintenance on all organic avionics equipment. It also performs unit-level maintenance on all organic equipment less medical.
- Provide air crash rescue support, less fire suppression.
- Operate as an area support MEDEVAC section and three FSMTs to provide flexibility in supporting division, brigade, or brigade TF equivalent operations.

B-10. This unit is dependent upon:

- Support elements of corps or ASCC for—
  - Financial management, legal, and religious support.
  - Human resources support.
  - Logistics.
  - Health service support, to include medical supply and equipment.
  - Food service support.
  - Communications security equipment maintenance.
  - Mortuary affairs support.
  - Military police support.
  - Laundry, shower, and clothing repair.
  - Engineer support for heliport/landing strip construction and maintenance.
- The supporting aviation intermediate maintenance (AVIM) organization for AVIM support.

B-11. Organization and functions.

- The medical company, AA (Figure B-2), is organized into a—
  - Company headquarters.
  - Flight operations platoon consisting of a platoon headquarters, a flight operations section, and an airfield service section.
  - Aircraft maintenance platoon consisting of a platoon headquarters, a component repair section, and a maintenance section.
  - Air ambulance platoon consisting of a platoon headquarters, an area support MEDEVAC section, and three FSMTs.

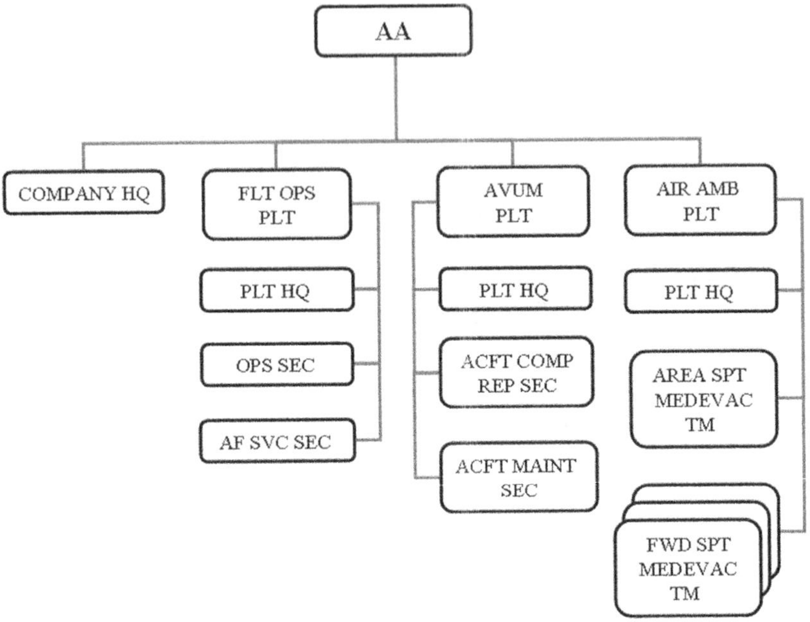


**Figure B-2. Medical company, air ambulance**

## Appendix C

# Tactical Standing Operating Procedures

This appendix provides a sample TSOP for an MMB engaged in medical evacuation missions. It should not be considered as all-inclusive. It may be supplemented with the information and procedures required for operating within a specific command or special operation.

## PURPOSE

C-1. The TSOP prescribes policy, guidance, and procedures for routine support of tactical operations of a specific unit. It should cover broad areas of unit operations, but be sufficiently detailed to provide newly assigned personnel with the guidance required for them to perform their mission. A TSOP may be modified by the TSOP and OPLANs/OPORDs of higher headquarters. It applies to a specific unit and all subordinate units assigned and attached. Should a TSOP not be in conformity with the TSOP of the higher headquarters, the higher headquarters TSOP governs. The TSOP is periodically reviewed and updated as required.

### FORMAT FOR THE TACTICAL STANDING OPERATING PROCEDURE

C-2. There is not a standard format for all TSOPs; however, it is recommended that a unit TSOP follow the format used by its higher headquarters. The TSOP can be divided into sections (specific functional areas or major operational areas) and further subdivided into annexes. An annex can be further subdivided into appendixes and then into tabs. Appendixes can be used to provide detailed information on major subdivisions of the annex, and tabs can be used to provide additional information (such as report formats or area layouts) addressed in the appendix.

C-3. Regardless of the format used, the TSOP follows a logical sequence in the presentation of material. It should discuss the chain of command, major functions and staff sections of the unit, operational requirements, required reports, necessary coordination with higher and subordinate elements for mission accomplishment, programs (such as command information, preventive medicine measures [PMM], and combat stress control) and other relevant topics.

C-4. Pagination of the TSOP can be accomplished by starting with page 1 and numbering the remaining pages sequentially. If the TSOP is subdivided into sections, annexes, appendixes, and tabs, a numbering system that clearly identifies the location of the page within the document can be used. Annexes are identified by letter and are listed alphabetically. Appendixes are identified by numbers and arranged sequentially within a specific annex. Tabs are identified by a letter and are listed alphabetically within a specific appendix. After numbering the initial sections using the standard numbering system (sequentially starting with page 1 through to the end of the sections), number the annexes and their subdivisions. They are numbered as the letter of the annex, the number of the appendix, the letter of the tab, and the page number. For example, page 4 of Annex D is written as "D-4"; page 2 of Appendix 3 to Annex D is written as "D-3-2"; page 5 of Tab A to Appendix 3 of Annex D is written as "D-3-A-5." This system of numbering makes the pages readily identifiable as to their place within the document as a whole.

C-5. In addition to using a numbering system to identify specific pages within the TSOP, descriptive headings should be used on all pages to identify the subordinate elements of the TSOP.

- The first page of the TSOP should be prepared on the unit's letterhead. The remaining pages of the major sections should include the unit identification in the upper right-hand corner of the page (for example: "Multifunctional Medical Battalion").
- A sample heading for an Annex is: "ANNEX B (Command Post) to Multifunctional Medical Battalion."

- A sample heading for an Appendix to Annex B is: "APPENDIX 3 (Command Post Security) to ANNEX B (Command Post) to Multifunctional Medical Battalion."
- A sample heading for a Tab to Appendix 3 to Annex B is: "Tab A (Tactical Operations Center [TOC] Security) to APPENDIX 3 (Command Post Security) to ANNEX B (Command Post) to Multifunctional Medical Battalion."

C-6. As the TSOP is developed there may be an overlap of material from one annex to another. This is due in part to similar functions that are common to two or more staff sections. Where overlaps occur, the material presented should not be contradictory. All discrepancies will be resolved prior to the authentication and publication of the TSOP. The TSOP is authenticated by the unit commander.

## SAMPLE TACTICAL STANDING OPERATING PROCEDURES (SECTIONS)

C-7. The information contained in this paragraph can be supplemented. It is not intended to be an all-inclusive listing. Different commands will have unique requirements that need to be included. The first section of the TSOP identifies the specific unit/headquarters that developed it.

- Scope–This document establishes and prescribes procedures to be followed by the unit identified and its assigned, attached, or OPCON units/elements.
- Purpose–This document provides policy and guidance for routine tactical operations of this headquarters and its assigned, attached, or OPCON units.
- Applicability–Except when modified by policy guidance, TSOP, or OPLANs/OPORDs of the higher headquarters, this document applies to this unit and to all units assigned, attached, or under OPCON for combat operations. These orders, however, do not replace judgment and common sense. In cases of nonconformity, the document of the higher headquarters governs. Each subordinate element will prepare a unit TSOP, conforming to the guidance herein.
- References–This paragraph can include any pertinent regulations, policy letters, higher headquarters TSOP, or any other appropriate documents.
- General information–This paragraph discusses the required state of readiness of the unit; primary, secondary, and contingency missions; procedures for operating within another command's AO; and procedures for resolution of conflicts with governing regulations, policies, and procedures.

C-8. The second section of the TSOP discusses the specific organization.

- Organization–This paragraph furnishes specific information concerning the authority for establishing the unit, such as applicable MTOE or other staffing documentation.
- Succession of command–The guidance for determining the succession of command is discussed.
- Task organization–Task organization is contingent on the mission and will be approved by the headquarters ordering deployment.
- Joint, multinational, and interagency operations–This paragraph provides guidance on any issues concerning C2 and TF organization in joint, multinational, and interagency operations. It also provides guidance on liaison officer requirements.
- Organizational charts–Contained in Annex A.

C-9. The third section of the TSOP discusses the unit functions.

- Battalion headquarters–This paragraph discusses the C2 of the assigned, attached, and OPCON units.
- Headquarters detachment–This paragraph discusses the functions of the headquarters detachment element, such as supervising movements, internal arrangements, area layout, physical security, and operation of the headquarters and staff.
- Attached, assigned, and OPCON units–This paragraph discusses the missions and functions of these units (such as providing medical evacuation of patients, emergency movement of blood and blood products, biologicals, medical logistics, and transportation of medical personnel and equipment).

- Staff responsibilities–This paragraph lists the unit's key personnel and their duties as prescribed in FM 5-0 and any command-specific duties.

C-10. The fourth section of the TSOP pertains to staff operations and is subdivided into annexes.

## SAMPLE TACTICAL STANDING OPERATING PROCEDURES (ANNEXES)

C-11. Annexes are used to provide detailed information on a particular function or area of responsibility. The commander determines the level of specificity required for the TSOP. Depending upon the complexity of the material to be presented, the annex may be further subdivided into appendixes and tabs. If the annex contains broad guidance or does not provide formats for required reports, paragraphs may be used. The annex should not require further subdivision. However, as the material presented becomes more complex, prescribes formats, or contains graphic materials, the annex may require additional subdivision. This paragraph discusses the subdivision of the annex by appendixes. It does not contain examples of subdividing the information presented in the appendixes into tabs. Applicable references such as ARs, FMs, and technical manuals should be provided in each annex. The number of annexes and their subdivisions presented below are not to be considered as an all-inclusive listing. Different commands will have unique requirements; therefore, supplementation of the information presented is permitted.

C-12. Annex A. (Organizational Charts.)

C-13. Annex B. (Command Post.)

- General. The division may operate main and/or tactical command post CPs (TAC CP) depending on the mission and tactical situation. Characteristics of the CPs include—
  - Main CP–Normally located in the sustainment area. Personnel staffing is tailored to provide planning, coordinating, and C2 of assigned/attached/OPCON units. The area location for the main CP is selected by the battalion S3; the exact site is designated by the commander in coordination with the XO and battalion S3. The XO designates work areas within the main CP. The commander, HHD, serves as the headquarters commandant. His duties include coordinating for and obtaining construction, maintenance, and logistical services and support for the CP.
  - Tactical command post CP–Normally established in the objective area or at an intermediate staging base (ISB). The ISB is the forward designated staging base at which final preparation for assault operations are controlled and is generally a safe area for support operations. The TAC CP is tailored to extend the commander's span of control and to better assess the situation on the ground and coordinate the arrival and subsequent activities of AHS organizations within the AO.
- Battalion tactical operations center.
  - Definition–The TOC is the command element of the battalion containing communications and personnel required to command, control, and coordinate HSS operations. The TOC is located within a secure, controlled area whether at a main CP or TAC CP.
  - Purpose–The purpose of the TOC is to provide the commander, in a secure environment, current evaluated information and recommendations concerning HSS/FHP operations.
  - Responsibilities–The battalion commander has overall supervision and control over the TOC. The battalion XO operates the TOC and has primary staff responsibility in the absence of the commander.
  - Operations–The TOC will operate on a 24-hour basis. It is principally staffed by each primary staff section furnishing necessary manpower as required. Secure and nonsecure telephone communications connect the TOC to other staff sections within the CP area. Access to the TOC is strictly controlled by means of an access roster, and if available, security badges. Only essential personnel and authorized visitors are allowed to enter. Each staff section will maintain an SOP on the organization and operation of its element. All elements within the TOC will, when appropriate, maintain a current situational map of their specific operations. Discussion and portrayal of tactical plans outside of the security area is prohibited.

  - Composition of the TOC–This is a listing of those personnel comprising the TOC. It normally includes the commander, XO, command sergeant major, principal staff members, and other specific staff members such as the S3 (air) or the battalion aviation maintenance officer.
  - Tactical operations center configuration–This is a schematic representation of the physical layout of the TOC.
- Responsibilities for the main CP services–The HHD commander's main mission is to support the headquarters, medical evacuation battalion. He has the personnel and resources assigned to facilitate this mission. He plans and coordinates for providing shower, laundry, transportation, maintenance, and other required services. He is also responsible for establishing and maintaining security; supplying fuel and other POL products; establishing the support area for the main CP (orderly room, supply room, motor pool, billets, and dining facilities); providing and operating generators to meet electrical requirements.
- Camouflage–This appendix discusses what camouflage procedures are required, to include type and amount of required camouflage materials (such as nets and shrubs); display of the Geneva Conventions distinctive emblem on facilities, vehicles, and aircraft on the ground (STANAG 2931); and other pertinent policies, guidance, or procedures.
- Message center–This establishes procedures for the handling of classified messages; provides delivery and service of IMMEDIATE and FLASH messages to the appropriate staff section; establishes procedures for preparing outgoing messages; and establishes a delivery service to the servicing message center for transmission of outgoing messages.

C-14. Annex C. (Administration and Human Resources.) This annex outlines procedures relating to administrative and human resources matters and associated activities.

- Personnel accountability and strength reporting.
  - Personnel daily summary (PDS)–This provides the procedures for filling out and submitting a daily personnel status report. The instructions may include requirements for encrypting the report prior to transmission, and specific guidance on time of submission, corrections, or other administrative requirements.
- Personnel readiness management.
  - Maintain the personnel readiness of unit personnel to include projected gains and losses.
- Personnel information management.
  - Maintain 100 percent accountability of the unit's military personnel files.
  - Update records directly through the officer record brief/enlisted record brief.
  - Reception, replacement, RTD, R&R, and redeployment.
  - Track the flow and duty status of the five categories of Soldiers into and out of the unit.
  - Update deployed theater accountability software database as required.
  - Coordinate the movement of the five categories of Soldiers.
  - Replacements–Individual replacements will not be readily available during the initial phases of operations. The S1 will automatically initiate replacement requests for personnel who are reported on the PDS report as WIA, missing in action (MIA), or killed in action (KIA).
- Essential personnel services.
  - Issue identification cards and tags.
  - Personnel actions–All personnel actions will be channeled through the S1. Company XOs and 1SGs will be the company points of contact. Actions will be handled expeditiously and meet suspense dates (tactical situation permitting).
  - Efficiency reports–This paragraph provides pertinent information on the completion and submission of these reports.
  - Award recommendations–This paragraph delineates the responsibilities for and guidance concerning submitting recommendations for awards and for scheduling and conducting award ceremonies.
  - Promotions–This paragraph discusses the procedures for submitting recommendations for promotion and scheduling and conducting promotion ceremonies.

- Correspondence–All correspondence addressed to higher headquarters will be submitted through the S1. Requirements for submission, preparation, and approval are also provided.

- Casualty operations.
  - Casualty Feeder Card–This report is submitted on DA Form 1156 (Casualty Feeder Card). Instructions on the completion of the form and submission requirements are included.
  - Tracking of casualties will require coordination with CLT and CSH.
- Postal operations.
  - This includes hours of operation and services available.
- Morale, welfare, and recreation operations.
  - Sporting activities and morale and welfare activities.
  - American Red Cross.
  - This includes hours of operation and services available.
  - Post Exchange services–This includes hours of operation and availability.
- Management, and HR planning and staff operations.
  - Preparation of the personnel estimate and personnel portion of plans or OPORD.
  - Coordinate external HR support with HR operations cell within the sustainment brigade.
- Other administrative support.
  - Financial management–This service includes disbursements and currency control, payday activities, currency conversion, check cashing, and the appointment of Class A agents.
  - Legal services–Information and specific guidance on administrative boards, courts-martial authority and jurisdiction, legal assistance, and general services should be provided.
  - Religious activities–Religious activities include chaplain support, services available from different faiths, schedule of services, and hospital visitations.
  - Distribution–Pick up and delivery schedules and any command-specific issues/procedures are provided.
- Graves registration (GRREG)–Commanders at all levels are responsible for the recovery, identification, and evacuation of US dead. This section discusses the responsibilities and procedures for unit-level GRREG activities for assigned and attached personnel.

---

***Note.*** This activity is for unit members only. Medical evacuation units do not accept nor transport the remains of nonunit members. Remains will not be transported in ambulances under any circumstances.

---

  - Responsibilities–This paragraph discusses both unit and battalion requirements.
  - Disposition–Specific guidance on procedures, GRREG collection points, transportation requirements, and the handling of remains are provided.
  - Hasty burials–Specific requirements for conducting hasty burials, marking, and reporting of grave sites are included.
  - Personal effects–Guidance on accounting for personal effects and requirements for burial should a hasty burial be required is contained in this paragraph.
  - Disposition of civilian and EPW remains–The local civilian government is responsible for the burial of remains of its citizens. The remains of EPW should be accomplished in separate cemeteries from US and allied personnel. If this is not possible, separate sections of the same cemetery should be used.
  - Contaminated remains–This paragraph discusses handling and disposition requirements (to include protective clothing), procedures, and marking and reporting of burial site.
- Public information–This annex contains procedures for obtaining approval on the public release of information to include the hometown news release program. In overseas locations, specific guidance on the interaction of unit members with the news media should be provided.

- Maintenance of law, order, and discipline–This appendix should provide applicable regulations, policy, and command guidance on topics such as serious incident reports, notifications, and submission formats, straggler control, confinement of military prisoners, and EPW and detainees (also discussed below).
- Enemy prisoners of war–This appendix discusses the unit responsibilities of EPW and detainees surrendered to the medical unit. These procedures do not pertain to EPW and detainee patients captured by other units. Medical personnel do not guard, search, or interrogate EPW and detainees while in the AHS; guards are provided by nonmedical personnel designated by the tactical commander for these duties. Until EPW and detainee personnel can be evacuated to an EPW collection point, medical personnel should remember and enforce the basic skills: segregate, safeguard, silence, secure, speed, and tagging. (The speed portion of evacuating EPW and detainees to designated collection points is of paramount importance to medical units.)

---

***Note.*** The treatment of EPW and detainees is governed by international and US law and the provisions of the Geneva Conventions. Personnel should be aware of these requirements and have ready access to the applicable regulations and policy guidance.

---

- Allied, coalition, and interagency personnel–This paragraph should provide guidance on the eligibility of allied, coalition, and interagency personnel for evacuation by Army ambulance and destination locations.

C-15. Annex D. (Intelligence and Security.) This annex pertains to intelligence requirements and procedures and OPSEC considerations.

- Intelligence–The S2 has the responsibility of collecting information to assist the commander in reaching logical decisions as to the best COA to pursue. Priority intelligence requirement (PIR) may include the location, type, and strength of the air defense threat; location, type, and strength of enemy air defense radars; known or suspected CBRN activity; the health threat; and issues which the commander considers to be PIR. In addition to PIR, the commander's critical information requirements (CCIR) are also considered.
- Intelligence reports–The S2 is responsible for disseminating all applicable estimates, analyses, periodic intelligence reports, and intelligence summaries generated within the battalion or received from higher headquarters. Information on submission of reports and suspense on intelligence products and reports should also be addressed in this appendix.
- Weather data–Weather conditions can disrupt ground evacuation efforts and planners must recognize their impact when coordinating AA operations.
- Reports–These include information acquired during the routine performance of duty by pilots, ambulance drivers, and medics.
- Counterintelligence.
  - Camouflage–When ordered or directed by the tactical commander, all units will initiate and continually strive to improve camouflage operations of positions, vehicles, and equipment. Noise and light discipline is emphasized at all times.
  - Communications security–These measures will be enforced at all times. The specific requirements and considerations are included.
  - Signs and countersigns–This paragraph lists the signs and countersigns to be used during hours of darkness. It also includes reporting requirements and procedures if the sign/countersign is lost or compromised.
  - Signal operating instructions.
  - Document security–This paragraph discusses the procedures for marking and safeguarding classified material, both work documents and completed documents. (Reporting requirements in the event of compromise are also included.)
- Captured or surrendered personnel, equipment, supplies, and documents–This appendix provides specific guidance on the handling of captured personnel, equipment, supplies, and

documents. The disposition of captured medical equipment and supplies is governed by the Geneva Conventions and is protected against intentional destruction.

- Security–This appendix discusses weapons security and checks, safeguarding of patient weapons, aircraft security, signal operating instructions (communications) security, TOC security, Sensitive Item Status Report, and escape and evasion.

C-16. Annex E. (Operations.) This annex establishes procedures for S3 operations within the medical evacuation battalion, and provides a basis for standardization of medical evacuation operations in a tactical environment. It is essential that these procedures be standardized to ensure common understanding, facilitate control and responsiveness, and enhance mission accomplishment. Information on readiness levels, threat levels, warning levels, camouflage, security, area damage control, operations, and leader checklists is also included.

- Operational situation report–Requirements for preparation, format, and submission of this report are discussed in this appendix.
- Operations security–This appendix provides the guidance and procedures for secure planning and conduct of combat operations.
  - Priority intelligence requirements and security classification.
  - Responsibilities–The commander is ultimately responsible for denying information to the enemy. The S3 is responsible to the commander for the overall planning and execution of operations. The S2 has the principle staff interest in assuming the required degree of OPSEC and has the primary staff responsibility for coordinating the efforts of all other staff elements in this regards. The OPSEC officer is responsible for the preparation of the PIR and providing classification guidance. Additionally, the OPSEC officer identifies the priorities for OPSEC analysis and develops OPSEC countermeasures. The S2 coordinates with the S3 in planning OPSEC.
  - Hostile intelligence threat–The different sources of intelligence (human intelligence, signal intelligence, and so forth) are discussed.
  - Operational security program–This includes physical security, information security, and signal security.
  - Document downgrading/declassification and classification authority.
- Terrorist threat and countermeasures–This appendix provides guidance on security measures to counter the terrorist threat, both on an individual and unit level.
- Minefield threat–This appendix provides guidance on the potential of minefields being established in the AO and delineates reporting requirements. (Specific minefield extraction techniques are contained in Annex L of this TSOP.)
- Operations security and countermeasures–This appendix discusses camouflage, light discipline, noise discipline, physical security, information security, and signal security.
- Unit location update–his appendix provides timely information on the location of main and forward CPs, location of subordinate unit CPs, location of helipads, and location of POL points.
- Flight operations–This appendix provides information concerning Army aviation LZs throughout the corps area and division areas; required reports; airdrop information; and mission debriefing.
- Communications-electronics–This appendix establishes communications policies, procedures, and responsibilities for the installation, operation, and maintenance of CE equipment.
  - Responsibilities of the battalion CE noncommissioned officer in charge.
  - Concept of operations.
  - Command and control.
  - Radio communications.
  - Radio teletypewriter communications.
  - Message/communications center service.
  - Message handling procedures.
  - Wire communications.

- Switchboard operations.
- Communications security and operations.
- Intelligence security.
- Meaconing, intrusion, jamming, and interference (MIJI) reporting and electronic communications countermeasures.
- Security violations–This prescribes procedures for reporting any event of action which may have jeopardized the security of communications.
- Destruction of material.
- Daily shift inventory.
- Physical security.
- Transmission security.
- Security areas–This discusses access procedures and rosters, access approval requirements, and prohibited items.
- Inventory of classified materials.
- Communications security officers and custodians–The appointment procedures, orders requirements, and duties of personnel are described.
- Safety–This discusses requirements for grounding, handling, and storing COMSEC equipment.
- Power units.

- Rear battle responsibilities–This appendix discusses rear battle responsibilities, task organization, and support for reaction forces.

C-17. Annex F. (Chemical, Biological, Radiological, and Nuclear Operations.) This annex prescribes the policy, guidance, and procedures for CBRN defensive operations.

- Responsibilities.
- Chemical, biological, radiological, and nuclear reporting requirements and procedures.
  - Contamination avoidance.
  - Protection–Protection pertains to those measures each Soldier must take before, during, and after a CBRN attack to survive and continue the mission.
  - Decontamination–This discusses equipment requirements, procedures, and types of decontamination (such as hasty). This paragraph also provides guidance on deploying clean ambulances into a contaminated area.
  - Mission-oriented protective posture–This appendix provides guidance on the garments required for the different MOPP levels and identification procedures for personnel in MOPP.
  - Radiation exposure guidance–Establishes OEG. Discussion in this appendix includes determining what constitutes a radiological hazard, prescribing acceptable limits of potential casualty producing doses of radiation, minimizing exposure, and protecting against electromagnetic pulses.
  - Masking and unmasking procedures.
  - Radiological monitoring and survey operations.

C-18. Annex G. (Logistics.) This annex establishes logistics procedures for subordinate units when operating in a field environment.

- Nonmedical supply and services–A discussion of applicability, responsibilities, policy, classes of supply, requisition and delivery procedures, hours of operation, and other supply-relevant topics and available services (such as laundry and bath) can be addressed in this section.
- Medical logistics–This paragraph discusses supply and resupply procedures for Class VIII; policies and procedures for the backhaul of emergency Class VIII on medical vehicles; and hours of operation for the supporting medical logistics facility.
- Food service–This appendix discusses responsibilities, hours of operation, Class I supplies, sanitation requirements, layout of field kitchen, fuel storage, maintenance, safety precautions,

and administration, such as headcounts, required reports, shift schedules, meals, ready-to-eat (MRE), and inspections/visits of subordinate unit kitchens.

- Transportation/movement requirements–This appendix may cover the following areas: deployability data, applicability; responsibilities; policies on speed, vehicle markings, transporting flammable materials, transporting ammunition and weapons, and so on; convoy procedures; safety; and accident reporting.
- Fire prevention and protection–Guidance on the use of the tent stoves, flammable materials, use of cigarettes, matches, and lighters, electrical wiring and appliances, safety of tents and occupants, spacing of tents, stoves and ranges, installation of British thermal unit (BTU) heaters, and fire fighting equipment are presented in this appendix.
- Field hygiene and sanitation–This appendix provides uniform guidance and procedures for the performance of functions related to field hygiene and sanitation. It includes policies, communicable disease control, field water supply, water containers and cans, water purification bags, food sanitation, latrines, liquid waste disposal, and garbage and rubbish disposal.
- Conventional ammunition down/upload procedures–This appendix delineates responsibilities and provides guidance and procedures for the requisition, storage, and distribution of ammunition and weapons, reporting requirements, and safety.
- Petroleum, oils, and lubricants accounting.
- Medical logistics support. The medical logistics concept of operations, requisition and distribution procedures, accountability, and reports are provided in this appendix.
- Maintenance–This appendix includes information on the maintenance requirements of the battalion and the location and hours of operation of maintenance units and collection points. Maintenance for medical equipment, vehicles, aircraft, and communications and other categories of equipment are discussed.

C-19. Annex H. (Safety.) This annex establishes minimum essential safety guidance for commanders and units. It includes risk assessment, accident reporting, safety measures, emergency procedures, vehicle safety, ground guide procedures, fire prevention and protection, antennas, climate, survival training, animal and arthropods hazards, personal protective measures, hearing conservation, carbon monoxide poison, helicopter safety, and refueling operations.

C-20. Annex I. (Civil-Military Operations.) This annex discusses the participation in CMO. Medical elements are often involved in CMO, humanitarian and civic assistance, and disaster relief operations. The activities which may be covered include providing—

- Direct support medical evacuation for the operation.
- Guidance on developing a medical evacuation system in a HN.
- Training to a HN's medical infrastructure.

C-21. Annex J. (Mass Casualty Situations.) This annex discusses the procedures for providing medical evacuation support to MASCAL situations, to include coordination for nonmedical transportation assets and the augmentation of these assets with medical personnel to provide en route patient care. It discusses evacuation capabilities of US forces.

C-22. Annex K. (Minefield Operations.) This annex discusses notification procedures, training requirements, and techniques for clearing pathways to casualties, marking mines, and extracting personnel and equipment.

C-23. Annex L. (Religious Support.) This annex discusses procedures for processing unit requests for religious support and coordination of religious support to casualties, declared MASCALs, mortuary affairs support, unit memorial ceremonies/services, EPW, detainees, religious support logistics and supplies, and reports and administration.

# Appendix D

# Example of the Medical Evacuation Plan and of the Operations Order

## SECTION I — EXAMPLE FORMAT OF THE MEDICAL EVACUATION PLAN

D-1. Figure D-1 provides an example of the medical evacuation plan.

**(Classification)**

**Headquarters**
**Place**
**Date, time, and zone**

**Time Zone Used Throughout the OPLAN:** *The time zone used throughout the OPLAN (including attachments) is the time zone applicable to the operation. Operations across several time zones use ZULU time. Place the classification and short title of the OPLAN at the top of the second and any subsequent pages of the base OPLAN.*

**Task Organization:** *List the number and coordinates of medical evacuation units here or in a trace or overlay. If you do not list units here, omit this heading.*

**1. SITUATION.** *(State the general medical evacuation factors affecting support of the operation. Include any information essential to understanding the current situation as it influences medical evacuation.)*

**a. Enemy forces.** *(Refer to the intelligence annex if it has been published or is to be published. List information about the composition, disposition, location, movements, estimated strengths, and identification of enemy forces. List enemy capabilities that could influence the medical evacuation mission.)*

**(1) Strength and disposition.** *(The size of the opposition force and its placement on the battlefield is important during the planning process for medical evacuation operations. When evacuation routes are selected, caution must be exercised to ensure medical evacuation assets are not compromised by going through enemy-held territory or by being ambushed by isolated pockets of resistance. Further, this information is vital in determining if medical evacuation assets [both ground and air] will require a security escort provided by movement and maneuvers forces before entering areas of the battlefield.)*

**(2) Combat efficiency.** *(Information on actual combat units or other forces, their training status, and their level of expertise and experience can be identified here. The level of AHS training and the development of a health care delivery system can also be included.)*

**(3) Capabilities.** *(This paragraph should discuss the potential capability to interfere with or disrupt medical evacuation operations.)*

**(4) Logistics situation.** *(This can include information on how well supplied the enemy/ opposition force is with food, clothing, or other vital logistics factors. It may also include the financial backing and availability of future support from outside individuals/groups/nations.)*

**(5) Weapons.** *(This should include a discussion of the enemy's weapons that present the greatest threat to air and ground evacuation personnel and vehicles. Ground assets are more likely to face a small arms threat, rocket propelled grenades (RPGs), and improvised explosive devices (IEDs) while performing their mission; air assets are vulnerable to surface-to-air weapons and when on the ground to the same threats as ground elements.)*

**b. Friendly forces.** *(List pertinent information concerning friendly forces that might influence the medical evacuation mission.)*

**(1) Strength and disposition.** *(This should include all forces [US, allied, coalition, and HN] and should be maintained on overlays. It may also include liaison officers, interpreter support, and coordination requirements.)*

**(2) Combat efficiency.** *(The health of the command has a significant impact on this factor. Additionally, when there are significant numbers of DNBI casualties, the medical evacuation workload will increase.)*

**(3) Present and projected operations.** *(Medical evacuation planners must be familiar with plans for current and projected operations. A risk management assessment of current and projected operations is conducted. Casualty collecting points and AXPs must be designated during the planning process. Medical evacuation planners must also be able to anticipate changing requirements to ensure continuous medical evacuation is available to the supported force.)*

**(4) Logistics situation.** *(Since medical evacuation vehicles conduct emergency resupply of forward deployed medical units, the medical evacuation planner must maintain visibility of the current medical logistics situation.)*

**(5) Sustainment area protection plan.** *(Medical units play an important role in the sustainment area protection plan and should be included in the planning process.)*

**(6) Weapons.** *(Medical units only have small arms for self-defense and defense of their patients. Medical vehicles and aircraft should not carry any automatic or crew-served weapons; to do so would result in the loss of the protections afforded under the Geneva Conventions for the unit, medical personnel, and patients under its care.)*

**(7) Health of the command.**

**(a) Acclimation of troops.** *(Medical evacuation crews require acclimation when introduced into an AO where the temperature range and elevation are different from their home station. Medical evacuation operations involve heavy lifting and may require extended periods of time to complete. To reduce the risk of injury to litters bearers and to facilitate the evacuation effort, medical and nonmedical personnel engaged in these operations should be acclimated to the AO and a work/rest schedule should be developed and implemented. Air ambulance crew rest requirements impact aircraft availability and must also be considered.)*

**(b) Presence of disease.** *(The presence of disease impacts medical evacuation operations in two areas. The presence of disease in the AO [usually at subclinical levels in the native population] contributes to the incidence of disease manifesting itself within the supported force. As DNBI rates increase, so do the requirements for medical evacuation. Medical evacuation personnel are also susceptible to the endemic and epidemic diseases within the AO and/or multinational force. High rates of DNBI for medical personnel will adversely impact the medical evacuation capability.)*

**(c) Status of immunizations and/or chemoprophylaxis.** *(Commanders must ensure appropriate measures are taken to protect their Soldiers from DNBI. The records of replacement personnel need to be screened to ensure all required immunizations have been received and appropriate chemoprophylaxis/barrier creams for the AO initiated/provided.)*

**(d) Status of nutrition.** *(The nutrition status of the troops involved impacts on the susceptibility to disease and environmental injuries, morale, and fatigue. The medical evacuation commander must also ensure that his evacuation crews carry sufficient supplies of MREs to sustain themselves and their patients should delays in evacuation be experienced [such as delays resulting from shifts in the battle or inclement weather].)*

**(e) Clothing and equipment.** *(Special requirements for clothing and equipment to support a particular operation should be obtained prior to the beginning of the operation [for example, cold weather clothing for extreme cold weather operations; additional blankets for patients being transported (either to add padding or warmth); or block and tackle equipment for medical evacuation operations conducted in mountainous terrain or in urban areas].)*

**(f) Fatigue.** *(Mandatory work/rest schedules and sleep plans for air crews and ground evacuation personnel must be developed and implemented.)*

**(g) Morale.**

**(h) Status of training.** *(This can include any specialized training required for the conduct of a specific operation [for example: helicopter crews require deck-landing qualifications to perform shore-to-ship evacuation with US Navy ships or orientation to the social, political, economic, religious, and ethnic issues of a HN or supported population, or training nonmedical personnel in the proper techniques for carrying litters].)*

**c. Environment.**

**(1) Terrain.** *(This paragraph should discuss any aspects of the terrain that will either hinder or enhance the execution of the evacuation mission. It should discuss both natural and man-made terrain, as medical evacuation urban areas can pose significant challenges not found on a natural battlefield. Medical personnel must be able to recover injured and wounded Soldiers from below ground and from upper levels of man-made structures. If UO are planned, armor ambulances provide added protection for medical personnel and their patients; augmentation should be requested if armor ambulances are not organic to the unit. If additional resources will be required to accomplish evacuation due to impassable or difficult terrain [such as in mountain operations where additional litter bearers and medical personnel are required], it should also be addressed here. Further, the type of terrain to be traversed, such as rugged mountain or jungle swamps may require a patient [who would otherwise be ambulatory] to be transported on a litter until easier terrain is encountered.)*

**(2) Weather.** *(This should include a discussion of current weather conditions and seasonal variants. Weather conditions impact both ground and air evacuation operations; however, the most significant impact may be on AAs as severely inclement weather can ground all aircraft. It should also discuss the impact that the weather has on the terrain [such as rivers being frozen in winter or tundra becoming impassable in spring]. The climate may pose problems with acclimation as well as place additional requirements to sustain personnel during evacuation on evacuation assets [litter evacuation in the mountains in extreme cold weather operations may require warming tents and the capability to sustain the patient during nighttime when evacuation is difficult].)*

**d. Civil considerations.** Dislocated civilian population, EPW, and detainees. *(Dislocated civilians fleeing an area of hostilities can clog roadways leading away from the conflict area. This congestion on road networks may make evacuation along these routes almost impossible. The establishment of camps to sustain these categories of personnel may interrupt the road network requiring detours and lengthened evacuation times. Injured, ill, or wounded EPW or detainees are evacuated using the same evacuation means but are segregated from US, allied, or coalition patients. Coordination for nonmedical guards for EPW/detainee* *patients being evacuated through medical channels must be accomplished with the echelon commander. A determination of eligible beneficiaries for evacuation should also be developed and disseminated.)*

2. **MISSION.** *(Statement of the medical evacuation mission.)*

3. **EXECUTION.**

a. **Concept of operations.** *(The concept of operations describes how the commander sees the actions of subordinate units fitting together to accomplish the mission. Commanders ensure that their concept of operations is consistent with their intent and that of the next two higher headquarters. The concept of operations describes any other details the commander considers appropriate to clarify the concept of operations and ensure unity of effort. If the integration and coordination are too lengthy for this paragraph, they are addressed in the appropriate annexes. When an operation involves two or more clearly distinct and separate phases, the concept of operations may be prepared in subparagraphs describing each phase.)*

b. **Coordinating instructions.** *(List only instructions applicable to two or more units and not covered in unit SOPs.)*

c. **Patient estimates.** *(Indicate rates and numbers by type unit/division/corps/EAC.)*

(1) **Number of patients anticipated.** *(The anticipated number of patients affects the number and type of medical evacuation resources required.)*

(2) **Distribution within the area of operations.** *(Distribution within the AO is an important consideration because varying types of terrain will generate different requirements for support. Operations conducted in mountainous terrain rely heavily on manual and litter evacuation techniques, are labor intensive, and require more time to complete. Differences in road composition, traffic density, and abundance/absence of paved roads will impact differently on the type of support required. Further, if the distance between Role 2 and Role 3 MTFs exceeds the capability of US Army AAs, the use of the USAF aircraft should be anticipated and preplanned. Coordination with the USAF is required.)*

(3) **Distribution in time during the operation (evacuation time).** *(As the battle progresses over time and space, medical evacuation assets may be reallocated to support the units in contact. The medical evacuation plan must be sufficiently flexible to enable the commander to shift resources as required.)*

(4) **Areas of patient density.** *(Areas with a high patient density may require augmentation of the medical evacuation assets supporting that location.)*

(5) **Possible mass casualties.** *(MASCAL situations should be anticipated when possible. Evaluation of the general threat and health threat in the AO can facilitate the medical evacuation planner in forecasting evacuation requirements. When dealing with MASCAL situations, the use of nonmedical transportation platforms should be included. Augmentation of these vehicles by medical personnel to provide en route care or CLS to provide first aid will assist in reducing the deterioration of the patient's medical condition until arrival at an MTF.)*

(6) **Lines of patient drift.** *(This indicates what routes injured Soldiers are most likely to take from the battlefield. It is usually the most direct route over the least demanding terrain, such as at the base of the hill rather than over the hill where climbing would be required.)*

(7) **Evacuation routes/corridors.** *(Evacuation routes should be preplanned, indicated on the medical evacuation overlay, and reconnaissance accomplished. Routes that provide lucrative targets, have significant obstacles to circumvent, or will be unduly congested due to fleeing refugees or other displaced persons should only be used if no other routes are available. Air corridors are provided by the supporting A2C2 element. Due the placement of hospitalization resources within the JOA, AA may be required to operate out of fixed sectors and coordination for air escorts must be coordinated.)*

(8) **Support requirements.** *(This paragraph discusses the type of direct support, general support, or operational control required for the operation and its command relationship. Specific guidance on trigger points and release points from a particular category of support should be included.)*

(9) **Medical evacuation procedures.**

(a) **Evacuation overlays.** *(Evacuation overlays must be developed to facilitate the evacuation effort. Both supporting and supported units must maintain and update [as required] overlays throughout the operation.)*

(b) **Communications.** *(Medical evacuation frequencies must be designated at the outset of the operation. Requirements for liaison officers, A2C2, and interpreters must be anticipated.)*

(c) **Casualty collecting points.** *(Location, staffing, and activation trigger [such as crossing a phase line] must be known to supported and supporting units.)*

(d) **Ambulance exchange points.** *(Location, staffing, and activation trigger [such as crossing a phase line] must be known to supported and supporting units).*

---

***Note*:** If an AXP is staffed with a treatment element, it is no longer considered to be an AXP but rather it becomes the forward element of a BAS or Role 2 MTF. The AXP is only a location on the ground where the patient is transferred from one evacuation platform to another. If it is staffed, it is usually staffed by one member of the evacuation platoon to direct returning ambulances to the area where the last ambulance departed from. The element designating the AXP is responsible for staffing it. Ambulance exchange points may be rendezvous points to be used only once during the operation.

***Note*:** Ambulance exchange points should not be used as resupply points for Class VIII. (Planning on pushing Class VIII to these points may result in the loss of the Class VIII materiel, as there may not be anyone at the point to receive the supplies.)

---

(e) **Ambulance shuttle system.** *(The ambulance [or litter] shuttle system is a management tool to facilitate the medical evacuation of forward areas.)*

(10) **Casualty evacuation (transportation).**

(a) **Medical augmentation.** *(Units and/or facilities capable of providing medical augmentation support should be identified and tasked as appropriate.)*

(b) **Manual evacuation.**

(c) **Litter evacuation.** *(A trauma specialist or CLS should be included as a member of the litter team when there is a significant distance to be covered.)*

(d) **Pack animals.** *(In some remote locations, the use of pack animals may be the most feasible form of transportation from the POI to a place where vehicles, aircraft, or watercraft become accessible. Whenever possible, medical personnel should accompany patients evacuated in this manner.)*

(e) **Nonmedical vehicle.** *(Nonmedical vehicles which could be preplanned for use for casualty transport should be identified. Further, guidance on the use of vehicles of opportunity should also be discussed. When possible nonmedical vehicles used for these purposes should be augmented with medical personnel or CLS.)*

(f) **Nonmedical aircraft.** *(Nonmedical aircraft which could be preplanned for use for casualty transport should be identified. Guidance on the use of aircraft of opportunity should also be discussed. When possible nonmedical vehicles used for these purposes should be augmented with medical personnel or CLS.)*

(g) **Nonmedical watercraft.** *(Nonmedical boats, ships, or other watercraft which could be preplanned for use for casualty transport should be identified. Further, guidance on the use of these platforms of opportunity should also be discussed. When possible nonmedical vehicles used for these purposes should be augmented with medical personnel or CLS.)*

(11) **Resources available.**

(a) **Mode of transportation.**

(b) **Personnel.** *(This includes a discussion on personnel strengths and the impact shortages have on the medical evacuation operations.)*

(c) **Supplies and equipment.** *(This includes both medical and nonmedical supplies and equipment. Litter and blanket exchanges should be covered here. If PMI are an issue, they should be discussed here.)*

(12) **Use of smoke and obscurants.** *(In most cases, medical units will not have a high priority for the use of smoke and obscurants; however, if the supported units have planned for its use, medical evacuation units can take advantage of the situation to clear the battlefield of casualties. Guidance on the use of colored smoke to identify the pickup location for patients can also be included.)*

d. **Warfighting functions.**

(1) **Medical planners must take the following warfighting functions into consideration.**

(a) **Intelligence.** (Known or suspected enemy situation. Terrain and weather. What kind of road network is in the AO? How will the predicted weather impact the traffic capability on and off the roads? Are there LZs along the axis of advance and in or around the urban area? How will the weather impact on the availability of aircraft?)

(b) **Movement and maneuver.** *(What is the combat scheme of maneuver and are there medical assets readily available to care for potential casualties? Particular attention should be paid to the employment of the armored ambulances.)*

(c) **Fire support.** *(Are there any buildings specifically targeted? If so, medical planners should ensure CCPs, evacuation routes, and treatment elements remain at a safe distance away from these buildings to prevent potential fratricide incidents.)*

(d) **Protection.** *(Preserving the force includes protecting personnel (combatant and noncombatant), physical assets, and information of the US and multinational partners.)*

(e) **Sustainment.** *(Typically, the BCT moves a forward logistics element toward the battle area to provide support. If this is the case, then how are medical assets integrated into the plan, and, more importantly, where are they positioned in the convoy and how are they secured?)*

(f) **Command and control.** *(The senior military on-site or the senior medical person at the scene, if available, assigns the evacuation precedence and determines if the patient should be evacuated by ground or air resources. What are the primary and secondary frequencies? This element is the central point for prioritizing assets and medical evacuation requests. It is imperative that this element is capable of continuous situational awareness through accurate battle tracking.) From information provided by the S2, medical planners must understand the tactics of the enemy and be prepared for casualties at all possible areas of contact. The medical planners should be able to gain insight into this matter during the war gaming process. The medical planners should provide input into the counterreaction portion of the action, reaction, counterreaction drill.)*

(2) **Additional planning information.**

(a) *Availability and numbers of ground and AA assets, evacuation routes and flight windows.*

(b) *Nonstandard ground and aircraft availability (for MASCAL situations) and how to trigger their involvement.*

(c) *Additional medical assets from higher levels such as treatment teams, ambulances, and combat configured loads of Class VIII.*

(d) *Existing and proposed locations of all available MTFs.*

(e) *Class VIII resupply plan.*

(f) *The need for litter bearer teams and special equipment for extracting casualties from buildings and rubble.*

e. **Casualty evacuation synchronization.**

(1) *Medical evacuation and CASEVAC assets must be incorporated into the evacuation plan. Ideally there should be a single dedicated controlling authority. This prevents duplication of effort and confusion and enables the battlefield to be cleared faster and more efficiently.*

(2) *In MASCAL situations CASEVAC assets once put into operation should be dedicated versus designated (lifts of opportunity). Potential problems that arise arising from relying on designated versus dedicated CASEVAC include excessive time spent acquiring the information needed to transport casualties. Medical regulating to higher roles of care facilities may have to be accomplished by blocks of beds rather than on a by name basis. If a facility is designated for certain specialty care [such as head trauma] the evacuator must know which facilities have been identified and where they are located.).*

(3) *Another consideration is aircraft configuration. Casualty evacuation may be a seats-out operation. By dedicating aircraft and crews, they have time to prepare for the mission. Casualty evacuation crews must be briefed on the details of their mission including LZs, PZs, air corridors, communications architecture, and supported units.*

(4) *Any CASEVAC plan should clearly articulate the following information.*

(a) *Who is responsible (such as the company 1SG) for coordinating the company CASEVAC mission?*

(b) *Combat lifesavers role.*

(c) *Class VIII and blood resupply procedures.*

(d) *Personnel to act as litter bearers.*

(e) *Locations of CCPs, forward treatment teams, BAS, and AXPs.*

(f) *Location and designation of MTFs.*

(g) *Medical regulating officer coordination.*

(h) *Medical treatment facility and MRO contact frequencies.*

4. **SERVICE SUPPORT.**

a. **Supply.** *(Refer to TSOP or another annex whenever practical.)*

(1) **General supply.** *(Provide special instructions applicable to medical evacuation units.)*

(2) *Consider supply levels for all classes of supply, in the event of mission requirements in an austere environment and at extended distances from the full compliment of logistics and sustainment resources.*

b. **Medical logistics.** *(Provide special procedures applicable to the operations.)*

(1) **Requirements.** *(For sustaining the ground and AA operations and personnel.)*

(2) **Distribution.** *(This should include the method of distribution and any limitations or restrictions that are applicable. Additionally, if special transportation requirements exist, they should also be noted.)*

(3) **Medical logistic activities.** *(This includes the location of the medical supply activity supporting the AO and the means of communicating requests for resupply.)*

c. **Supplies required for ground and AA operations and personnel.** *(Includes estimates of the number of patients anticipated to be evacuated.)*

d. **Health service support.** *(Include information and instructions for supported units. Prescribe the plan for ground and air medical evacuation.)*

(1) **Evacuation.** *(List CCPs and AXPs. Establish ambulances shuttles, routes, means, and schedules of evacuation and en route treatment policies for the use of nonmedical transportation assets; policies for evacuation by ground and air; and evacuation of CBRN-contaminated patients. Include information about medical evacuation request procedures and channels, and holding policies.)*

(2) **Treatment.**

(a) **Treatment of supported forces.**

(b) **Policies.** *(Should address treatment and hospitalization policies.)*

(c) **Units.** *(This includes information concerning the location, capabilities, and communications means of units providing support.)*

(d) **Other.** *(This can be information on other medical assets that may be used in emergencies.)*

e. **Personnel.**

(1) Current subordinate unit manning levels/critical MOS shortages (consider all manning influences—Task Organization, R&R flow, Boots On Ground data, and so forth).

(2) New personnel requirements resulting from the operation (language skills, additional skill identifier [ASI], and so forth).

(3) Casualty estimates (as developed).

(4) Forecasted replacement availability.

(5) Evacuation policy for the operation.

(6) Supporting HR organizations (location, command/support relationship, controlling element).

(7) Supporting medical elements

(8) Replacement fill priority (Coordinate w/S3/G3).

(9) Crew/key leader replacement.

(10) Projected postal flow/limitations.

(11) Change to established PASR reporting flow/times.

**(12)** Change to casualty reporting flow (CLT change, reporting changes).

**(13)** RSOI Reception HR impacts.

**(14)** R&R schedule/flow operational impact (during operations conducted during sustained operations).

**(15)** Civilian and JIM manning impacts on the operation focusing on strengths and personnel service requirements.

**(16)** Specific Army G1 personnel planning guidance impacts on the operation.

**(17)** Status of other personnel service or personnel support factors (postal, MWR, legal support, military pay support).

**f. Coordinating instructions.** *(List instructions applicable to two or more units and not routinely covered in unit SOPs. This is always the last subparagraph in paragraph 4. Place complex instruction in an annex.)*

**(1) Time or condition when a plan becomes effective.** *(Include the time or the conditions under which the plan is to be placed in effect.)*

**(2) Commander's critical information requirements.** *(List only in coordinating instructions and not in annexes.)*

**(3) Risk mitigation control measures.** *(These are measures unique to this operation and not included in unit SOPs. They must include MOPP, OEG, troop safety, vehicle recognition signals, and fratricide prevention measures.)*

**(4) Rules of engagement.** *(If required.)*

**(5) Boundaries.** *(Address measures established to protect units. Specify which tactical units are to provide protection.)*

**g. Special reports.** *(List reports requiring special emphasis that is not addressed elsewhere.)*

**5. COMMAND AND SIGNAL.**

**a. Command.** *(State the map coordinates for the CP locations and at least one future location for each CP. Identify the chain of command if not addressed in unit SOPs.)*

**b. Signal.** *(Include the headquarters location and movements, liaison arrangements, recognition and identification instructions, and general rules concerning the use of communications and other equipment, if necessary. Use an annex when appropriate.)*

**(Classification)**

**Figure D-1. Example of a medical evacuation plan**

## SECTION II — EXAMPLE FORMAT FOR AN OPERATIONS ORDER

D-2. Figure D-2 provides an example of the medical operations order.

**(Classification)**

Place the classification at the top and bottom of every page of the OPLAN.

**(Change from verbal orders, if any)**

The phrases "No change from verbal orders" or "No change from verbal orders except paragraph #" are required. This statement is only applicable if the commander issues a verbal order.

**Copy ## of ## copies**
**Issuing headquarters**
**Place of issue**

*Show the name of the town or place in capital letters, coordinates in parentheses, and the country in capital letters. You may encode both.*

**Date-time group of signature**

*(The effective time for implementing the plan or order is the same as the date-time group of signature unless the coordinating instructions state otherwise. Use time zone Z unless the order states otherwise. When orders apply to units in different time zones, use Z time. When an OPORD does not specify the actual time to begin an operation, state that time in terms of an event.)*

**OPERATION ORDER 00-001, Operation Liberation**
**References: OPORD 00-000**
**Time zone: Zulu**
**Task Organization**

Ground Ambulance Units participating in Operation Liberation
123 MED CO (GA) assigned to 1st MMB and provides emergency support to the 1st Marine Expeditionary Force (MEF)
234 MED CO (GA) assigned to 1st MMB
Ambulance Platoons assigned to:
C CO BSMC
C CO BSMC
C CO BSMC
C CO BSMC

Air Ambulance Units participating in Operation Liberation
C 2-1 MED CO (AA) CAB assigned to 1st Infantry Division
C 2-2 MED CO (AA) CAB assigned to 2nd Infantry Division
C 2-3 MED CO (AA) DS to the 1st MEF
C 2-4 MED CO (AA) CAB assigned to 1st Corps support

Air Force aeromedical evacuation assets
2d AF Combat Search and Rescue (CSAR), 1st AF Squad

1. **SITUATION.**

a. **Enemy forces.** *As addressed in OPORD 00-000, the ground-to-air threat is significant in the all enemy sectors. The air–to-air threat is minimal, in all sectors. All ground planned evacuation routes will be targets for ambush by enemy forces and sympathetic host nationals.*

b. **Friendly forces.** *A six (6) aircraft (HH-60) Air Force CSAR unit will be staged from the APOD. They will respond to downed aircrew evacuation recovery requests from the Air Force Combat Control Center.*

c. **Attachments and detachments.** *None.*

d. **Assumptions.** *None.*

2. **MISSION.** *On order provide tactical and operational ground and air intratheater medical evacuation or patient movement in support of Operation Liberation and continue coverage of all combined joint task force (CJTF) assets until relieved.*

3. **EXECUTION. Intent: State the commander's intent.**

a. **Concept of support operations.** *This is a three phase operation.*

**Phase 1—Will be prior to initiation of the attack.** *All US forces will be in the staging areas and in preattack positions. There will be no or very limited direct contact with enemy forces. C 2-1, C 2-2, and C 2-3 will not be required to* *provide evacuation support until D-3 days. Units in the staging area will be directed to call C 2-4 for evacuation at all times prior to D-3 days. All other evacuation units will concentrate on training and preparation for hostilities as directed by their commanders.*

**Phase 2—Will be the attack.** *C 2-1, C 2-2, C 2-3 will provide DS to their assigned, GS to the corps sustainment area and the backhaul mission.*

**Phase 3—Will be from the conclusion of the attack until stability operations commence.** *Evacuation assets will begin transferring to an area support mission during this phase.*

---

***Note.*** The concept of support operations takes into account both the tactical OPORD and the medical Annex to the OPORD. Information in both documents will shape the evacuation plan. Simultaneous planning should therefore occur.

---

**b. Tasks to maneuver units:** *Not applicable.*

**c. Tasks to movement and maneuvers units:**

**(1) Evacuation responsibilities.**

**(a) Phase 1.**

*C 2-1, C 2-2, C 2-3 will provide direct support to their division assets from D-3 on. C 2-1, C 2-2, C 2-3 will respond to any request for evacuation that is geographically close to their positions even if not in their division area.*

*C 2-4 will provide support to the entire staging area (AO Buffalo). C 2-4 will perform the ship-to - shore mission as directed by the 11th MEDCOM TPMC*

*The 123 MED CO (GA) will operate in the 1st MEF sector in a direct support role. 1st MMB has will cover the 1st and 2nd Division footprint.*

*The 234th MED CO (GA) will provide support to Camps Jones, Smith, and Franks and perform patient transfers from camps to the APOD.*

**(b) Phase 2.**

*C 2-1, C 2-2, C 2-3 will provide direct support to their division assets. C 2-1, C 2-2, C 2-3 will be responsible for moving patients from within the division area back to the staging AO Buffalo until FLOT is beyond phase line alpha.*

*After FLOT is beyond phase line alpha (at which time one way flight time for the H-60 back to AO Buffalo will exceed 2 hours), C 2-1 will designate 1 FSMT to stage at XYZ Airfield to perform the backhaul and the area support mission until relieved. United States AF aeromedical transports will be coordinated for by 11th MEDCOM flying from XYZ airfield.*

*Beyond phase line Alpha C 2-2 will provide one FSMT to support the greater casualties expected in C 2-2 sector. Coordinate internally for link-up.*

*C 2-4 will provide support to the entire staging area (AO Buffalo). C 2-4 will perform the ship-to-shore mission as directed by the 11th MEDCOM MRO. C 2-4 will designate 4 aircraft to go forward as required to backhaul patients to AO Buffalo.*

**(c) Phase 3.**

*C 2-1, C 2-2, C 2-3 will provide direct support to their Divisions' assets.*

*The 1st MEF will identify areas of high casualty density conducive to ground evacuation (urban areas) and divide the resources of the 123rd MED CO (GA) to cover these areas efficiently, if required.*

*The 234th MED CO (GA) will conduct convoy support and provide coverage for AO Buffalo.*

**d. Evacuation mission guidelines.**

**(1) Phase 1, 2, and 3.** *The senior military person advised by medical personnel on-scene will assign the evacuation precedence and determine the mode of evacuation (ground or air). Launch authority, a separate issue, based on risk will be according to the aviation SOP.*

**(a)** *Weather below acceptable minimums.*

**(b)** *No acceptable LZ in the immediate vicinity, except for hoist missions. Ground evacuation may be used for part of the evacuation.*

**(c)** *Chemical, biological, radiological and nuclear conditions at LZ would contaminate the aircraft. Ground evacuation may be used for part of the evacuation, followed by an air pickup after decontamination is complete.*

**(d)** *Patients are located at a LZ in direct enemy contact. Ground evacuation may be used for part of the evacuation, followed by an air pickup once clear of the enemy.*

**(e)** *Air evacuation assets are unavailable.*

**(f)** *Mission flight route requires an armed escort and one is not available.*

**(2) The following are additional guidelines for evacuation:**

**(a)** *Ground evacuation is the preferred method to clear casualties out of hostile urban areas. Air ambulance pick-up at a secured LZ outside of the urban area is preferred. For large urban areas, this may not be possible.*

**(b)** *If a mission requires the use of a hoist (such as thick forest or jungle) and the pick-up area has suspected enemy contact, ground evacuation is advised to a secure pick-up site.*

**(c)** *Prescheduled movement of ROUTINE or CONVENIENCE patients along secure routes of evacuation should be conducted using ground evacuation if the distance of evacuation is short.*

**(3) Coordinating instructions:**

**(a) Medical regulating.** *The 11th MEDCOM will oversee medical regulating during the operation.*

**(b) Tentative locations of MTFs.**

**Phase 1**

- *2nd CSH AB 1234 5678*

- *1st CSH AB 2345 6789*

**Phase 2**

- *2nd CSH AB 1234 5678*
- *1st CSH AB 2345 6789*

**Phase 3**

- *2nd CSH AB 1234 5678*
- *1st CSH AB 2345 6789*

(c) **Reporting.** *Evacuation units, through division G3 if assigned to a division, will submit a daily report of evacuation missions by 1200 the following day.*

(d) *If a MASCAL develops which overwhelms the resources of a single AA CO, contact GSAB for assistance.*

(e) **Casualty collection.** *Military treatment facilities can call 9 line missions to any medical evacuation asset for URGENT and PRIORITY missions however MTFs that need to clear beds will coordinate mass patient movements through CJTF G3 air for possible USAF support.*

- *1st CSH will call C 2-4 for evacuation requests during Phase 1.*
- *2nd CSH will call C 2-4 for evacuation requests during Phase 1.*
- *1st CSH will call C 2-1 for evacuation requests during Phase 2/3.*
- *2nd CSH will call C 2-2 for evacuation requests during Phase 2/3.*

(f) **Ground evacuation.** *Seasonal monsoons are expected to washout many of the main planned MSRs and cause flooding.*

(g) **Air evacuation.** *If air evacuation routes become so long as to seriously degrade the capabilities of an AA company, contact CJTF G3 Air for additional support and possible reallocation of evacuation assets. An AA CO is considered to be operating beyond its capability or inefficiently if:*

- *More than 50 per cent of missions require a one way trip time exceeding 2 hours.*
- *On more than 50 per cent of missions flown, routes of evacuation are flown through another AA company's area of direct support coverage.*

(h) **Medical considerations for evacuation.**

- US Army evacuation assets will be used to evacuate all injured personnel on the battlefield without prejudice.
- Human immunodeficiency virus is at epidemic proportions in Simula population. Take appropriate precautions during treatment and when dealing with biologicals.
- A significant number EPW and displaced civilians and detainees are predicted to be evacuated. Estimated to be a much as 75 per cent of total patient load.

(i) **United States Air Force aeromedical evacuation.**

- Combined joint task force G3 Air will coordinate for USAF intratheater aeromedical missions.

4. **SERVICE SUPPORT.**

a. **Casualty estimates.** Reference the Medical Annex for specific casualty estimates.

b. **Aviation maintenance support.** C 2-3 will receive AVUM/AVIM support from the 2-4 CAB.

c. Divisions G3 will report the following CCIR to CJTF G3 Air. Use the CJTF CCIR format.

- Any MASCAL.
- Any evacuation company with degraded capability for—
- Evacuation distance(defined above)
- Maintenance: below 70 percent operational readiness rate
- Personnel: number of crews less than available aircraft
- Any aircraft Class A accidents/events or shoot downs.

5. **COMMAND AND SIGNAL.**

a. Fragmentary orders (FRAGOs) to this order will be published to adjust the corps evacuation plan as necessary.

b. 11th MEDCOM MIRC Chat net name, Phone xxx-xxxx

- 1st Division Surgeon (G3), MIRC Chat net name
- 2-1 CAB MEDEVAC Cell, MIRC Chat net name
- 2nd Division Surgeon (G3), MIRC Chat net name
- 2-2 CAB MEDEVAC Cell, MIRC Chat net name
- 1st MEF, MEDEVAC Cell, MIRC Chat net name
- 1st Corps CAB MEDEVAC cell, MIRC Chat net name, phone xxx-xxxx

**ACKNOWLEDGE:**
**PATRICK, LTG**

**Figure D-2. Example of a medical operations order**

# Appendix E

# Use of Smoke and Obscurants in Medical Evacuation Operations

This appendix provides guidance for AMEDD personnel in the use of smoke and obscurants in medical evacuation operations.

## TACTICAL COMMANDER

E-1. Throughout the battlefield, forces acquire and engage targets based on visual, laser, and microwave technologies. Friendly and enemy units use smoke and obscurants across the battlefield as a combat multiplier. The use of smoke and obscurants to mask combat operations is dictated by the tactical commander. He normally provides the operational guidance for units or elements operating in an area requiring obscuration. Permission to employ smoke and obscurants solely to mask medical evacuation operations may not be approved. However, if the tactical commander's plan indicates that smoke operations are to be employed in a given AO, the medical planner should consider both the advantages and disadvantages posed by their employment.

## FACTORS TO CONSIDER

E-2. The medical planner should consider the factors which might impact the use of smoke and obscurants in medical evacuation operations. Factors to consider are the—

- Phase of the tactical operation in which smoke and obscurants will be employed.
- Effect on ground and air evacuation routes when operating in an obscured environment (such as limited hours of use, checkpoint or convoy requirements, or the elimination of NOE approaches).
- Potential for exploiting the use of the cover and concealment provided for clearing the battlefield of casualties, especially in defensive operations.
- Potential requirements for smoke generation to perform the medical evacuation mission which would not detract from the tactical capability and requirements.

E-3. Smoke can also be used to identify unit areas or LZs for which a medical evacuation request has been received. Further, smoke can indicate wind direction at a landing site for AA operations.

## EMPLOYMENT OF SMOKE AND OBSCURANTS

E-4. Smoke and obscurants are employed to protect friendly forces from attack during the offense or defense. Smoke and obscurants disrupt enemy combat operations throughout the depth of the battlefield and across the operational continuum. They—

- Disrupt the ability to communicate.
- Conceal friendly forces.
- Deceive the enemy.
- Identify and signal.
- Degrade the effects of directed-energy weapons.

E-5. The benefit to AHS forces is derived through the tactical commander's use of smoke to obscure friendly tactical maneuvers. This obscuration—

- Prohibits the enemy from knowing how many casualties have been inflicted.
- Aids the movement of medical units and equipment.

- Enhances the ability to resupply forward deployed AHS elements.
- Aids in the tactical deception plan.

# GENEVA CONVENTIONS AND THE USE OF SMOKE AND OBSCURANTS IN MEDICAL EVACUATION OPERATIONS

E-6. As discussed above, the 1949 Geneva Conventions for the Amelioration of the Condition of the Wounded and Sick in Armed Forces in the Field (GWS) provides protection of medical personnel and units from intentional attack so long as they carry out no duties harmful to the enemy (Article 21, GWS). In order to facilitate their identification so as to prevent their intentional attack, medical units, equipment, and personnel are authorized to display the distinctive emblem of the Red Cross (Article 41 and Article 42, GWS). Under tactical conditions, when requirements for concealment outweigh those for recognition, all distinctive emblems may be obscured or removed from medical equipment if ordered by competent military authority and authorized by Army regulations. Display of the distinctive emblem is not required to afford the right against intentional attack; attack of medical units, equipment, and personnel not displaying the distinctive emblem is prohibited if opposing forces realize that the forces about to be attacked are medical units performing humanitarian duties.

E-7. The use of smoke or obscurants in medical evacuation operations does not differ from the use of camouflage techniques and is not prohibited by the GWS. Its only effect will be to obscure the identity of units as they perform their humanitarian mission. Given the lethality of the modern battlefield, however, it would be difficult, if not impossible, to say that such obscuration of these units, equipment, and personnel would necessarily increase their risk from unintentional attack.

E-8. It is recognized that, with the advent of precision-guided munitions and electro-optical or laser target acquisition devices, there will be a substantial use of smoke and other obscurants on the modern battlefield as a result of normal combat operations. The legitimate use of obscurants by combatants to thwart the accuracy of precision guided munitions may increase the risk to units and equipment not employing obscurants. This may possibly place medical units and equipment at greater risk if they fail to employ them. Further, medical evacuation operations will have to be carried out on the battlefield as medical personnel find it, which will include obscurants employed for normal combat operations.

# USE OF SMOKE IN AEROMEDICAL EVACUATION AND HOIST RESCUE OPERATIONS

## EFFECTIVE USE

E-9. Smoke can be used effectively in aeromedical evacuation and overland hoist rescue operations to—

- Identify the landing site. Colored smoke is an excellent daytime marking method. The smoke generated from a smoke grenade is difficult to detect more than 2 to 3 miles away, but an aircraft in the area should have little difficulty in noting its location. As more than one unit may be operating in a given area, it is important that the unit requesting an aeromedical evacuation mission be able to signal the aircraft as to the correct landing site to use. Radio communications produce an electronic signature. The electronic signature created from a prolonged transmission to guide an AA to the landing site may not be an acceptable tactical risk.
- Ensure the LZ is controlled by friendly forces. When a unit employs colored smoke to mark a landing site, the aircrew should identify the color and confirm it with the ground personnel. The transmission time required for this procedure is limited, thereby reducing the electronic signature.
- Determine surface wind direction. The employment of smoke at the landing site also enables the aircrew to determine the wind direction.
- Provide cover and concealment. In some environmental conditions (such as desert operations), the phenomenon of inversion occurs. When this occurs, the smoke and obscurants used in

normal combat operations may provide an upper layer of smoke under which the AA can operate.

### Disadvantage

E-10. The use of smoke on aeromedical evacuation operations can be a disadvantage if incorrectly employed or if the smoke generated in the tactical operation interferes with the medical evacuation mission. Smoke can obscure the landing site and make NOE approaches unusable. Further, smoke on the battlefield can force aircraft to fly at higher than planned heights. This increases the risk of being acquired by the enemy.

### Overwater Hoist Operations

E-11. In overwater hoist rescue operations, the employment of smoke for marking the patient pickup area, for determining surface wind conditions, and for spatial orientation is essential. The smoke employed by the aircrew must not interfere with the conduct of the operation or mask the location of the individual to be rescued.

## EMPLOYMENT OF SMOKE IN GROUND MEDICAL EVACUATION OPERATIONS

E-12. The employment of smoke during ground evacuation operations must be in consonance with the tactical commander's plan. Smoke can mask medical evacuation operations on the battlefield, but must not interfere with the tactical mission. In all combat operations, but especially in UO, smoke can be employed to cover and conceal—

- Movement across open areas.
- Extraction of casualties from vehicles and/or buildings.
- Entry and exit into/out of structures.

## Appendix F

# Evacuation Capabilities of United States Forces

This appendix provides the evacuation capabilities of US forces and aircraft.

## EVACUATION CAPABILITIES OF UNITED STATES AIR FORCE AIRCRAFT

F-1. Evacuation capabilities of USAF aircraft are provided in Table F-1. As the majority of USAF aircraft are not dedicated medical evacuation platforms, when used for medical evacuation the crew must be augmented with medical personnel to provide in-flight care.

F-2. The evacuation capability of Civil Reserve Air Fleet aircraft is provided in Table F-2.

**Table F-1. Evacuation capabilities of United States Air Force aircraft**

| *Transport Aircraft* | *Litter* | *Ambulatory* | *Combination* |
|---|---|---|---|
| C-130 Hercules | 74 | 92 | Variety |
| C-141 Starlifter | 48 | 170 | Variety |
| C-5 Galaxy | 70 | | |
| C-17A | *36 | *54 | |
| KC-135 and KC-10 | 8 | 24 | |
| *Normal configuration | | | |

**Table F-2. Civil Reserve Air Fleet capabilities**

| *Transport Aircraft* | *Litter* | *Ambulatory* | *Combination* |
|---|---|---|---|
| Boeing B-767 | *111 | 22 | 87 Litter<br>22 Ambulatory |
| * When activated, these aircraft are preconfigured for medical evacuation. The crew must be augmented with medical personnel | | | |

## EVACUATION CAPABILITIES OF UNITED STATES ARMY VEHICLES AND AIRCRAFT

F-3. The evacuation capabilities of US Army vehicles and aircraft are provided in Table F-3.

**Table F-3. Evacuation capabilities of United States Army vehicles and aircraft**

| *Vehicle/Aircraft* | *Litter* | *Ambulatory* | *Combination* |
|---|---|---|---|
| **Ground Vehicles** | | | |
| M996 Truck, Ambulance | 2 | 6 | 1 Litter<br>3 Ambulatory |
| M997 Truck, Ambulance | 4 | 8 | 2 Litter<br>4 Ambulatory |

| | | | |
|---|---|---|---|
| M1010 Truck, Ambulance | 4 | 8 | |
| M792 Truck, Ambulance | 3 | 6 | 2 Litter<br>3 Ambulatory |
| M1133, Stryker MEV | 4 | 6 | |
| Bus, Motor, 44 Passenger | 18 | 37 | |
| M113 Carrier, Personnel | 4** | 10 | |
| M998 Truck, Cargo/Troop Carrier (Four Man) | 3 | 4 | |
| M998 Truck, Cargo/Troop Carrier (Two Man) | 5 | | |
| Truck Cargo, 2 1/2 Ton, 5 Ton | 12 | 16 | |
| M997 Heavy Expanded Mobility Tactical Truck | 9 | | |
| M871 Semitrailer, Cargo | 16 | | |
| M1085 Truck, Cargo, Medium Tactical Vehicle, Long Wheel Base, 5 Ton | 12 | 22 | |
| M1093 Truck, Cargo, Medium Tactical Vehicle Air Drop/Air Delivery, 5 Ton | 8 | 14 | |
| M1081 Truck, Cargo, Medium Tactical Vehicle Light Air Drop/Air Delivery, 2 1/2 Ton | 7 | 12 | |
| **Rotary-Wing Aircraft** | | | |
| ***UH-60 Blackhawk | 6 | 7 | *4 Litter<br>1 Ambulatory |
| ****UH-60A Blackhawk | 3 | 4 | 4 Litter<br>1 Ambulatory |
| UH-1H/V Iroquois | 6 | 9 | *3 Litter<br>4 Ambulatory |
| CH-47 Chinook | 24 | 31 | # |
| **Fixed-Wing Aircraft** | | | |
| C-12 Huron | | 8 | |

**LEGEND**

* Normal configuration
** Spall liner must be removed
*** without internal hoist installed, requires 6 man litter configuration kit
**** With internal hoist installed
# See Table F-5

## Railway Car Capabilities

F-4. Although railway cars are not available within the US Army inventory, it is important to know the approximate capacities of rail transport in the event they become available and/or required in civil support operations or through wartime HN support agreements. The approximate capabilities of railway cars are provided in Table F-4.

**Table F-4. Capabilities of railway cars**

| *Rail Transport* | *Litter* | *Ambulatory* | *Combination* |
|---|---|---|---|
| Pullman Car (US) | 32 | 48 | |
| Sleeping Car (NATO/HN) | 32 | 32 | |
| Ambulance, Railway Car (NATO/HN) | 24 | 30 | |
| Ambulance, Railway Car, Personnel | 21 | 21 | |
| Railbus | | | 40 Litter<br>16 Ambulatory |

# EVACUATION CAPABILITIES OF UNITED STATES NAVY SHIPS, WATERCRAFT, AND ROTARY-WING AIRCRAFT

F-5. The evacuation capabilities of USN ships, watercraft, and aircraft are provided in Table F-5. The entries for ambulatory and litter patients on ships are the same because all patients require a bunk.

**Table F-5. Evacuation capabilities of United States Navy ships, watercraft, and aircraft**

| *Ships/Watercraft/Aircraft* | *Litter* | *Ambulatory* |
|---|---|---|
| **Hospital Ships** | | |
| T-AH 19, US Naval Ship Mercy | 1000 | 1000 |
| T-AH 20, US Naval Ship Comfort | 1000 | 1000 |
| **Casualty Receiving And Treatment Ships** | | |
| Amphibious Assault Ship (LHD)(Multipurpose) | 604 | 604 |
| Amphibious Assault Ship (LHA)(General Purpose) | 367 | 367 |
| Amphibious Assault Ship (LPH) | 222 | 222 |
| Amphibious Transport Dock (LPD) | 14 | 14 |
| Dock Landing Ship (LSD) | 108 | 108 |
| **Limited Medical Capability Watercraft** | | |
| Amphibious Cargo Ship (LKA) | 12 | 12 |
| Amphibious Command Ship (LCC) | 24 | 24 |
| **Rotary-Wing Aircraft** | | |
| CH-46 Sea Knight | 15 | 25 |
| CH-53 Sea Stallion | 24 | *55 |
| V22 Osprey | 12 | 24 |

* With centerline seating installed

# Appendix G

# Selection of Patients for Aeromedical Evacuation and Patient Classification Codes and Precedence

Patient classification codes provide information to evacuators and treatment personnel in an abbreviated form. They can be used to complete administrative reporting requirements pertaining to the evacuation of patients.

**Paragraphs G-2 through G-5 implement STANAG 3204.**

## BRIEFING OF PATIENTS PRIOR TO AEROMEDICAL EVACUATION

G-1. When patients are evacuated by aircraft (routinely in Roles 2, 3, 4, and 5), they should be briefed on the following points—

- A number of ambulatory patients will be detailed to assist with the evacuation of litter patients in any emergency.
- Safety belts and litter straps are to be properly fastened according to orders given by the pilot.
- Patients are instructed on the proper position to assume in preparing for an emergency. Flight crews and CCATT ensure that seat safety harnesses have been tightened.
- Ambulatory patients, with the exception of those designated to assist litter patients, are the first to leave a downed aircraft.
- Immobilized litter patients are freed from litters and assisted in leaving the aircraft. Litters will not normally be removed from their fastenings in view of the limited time available to evacuate the aircraft.
- Mentally disturbed patients should be quieted so that the orderly removal of other patients will not be jeopardized.

### INTERNATIONAL STANDARDIZATION EVACUATION AGREEMENT

G-2. Table G-1 provides the patient classification codes defined in international standardization agreements.

### Table G-1. Patient classification codes

| Code | Classification |
|---|---|
| ***Class 1. Psychiatric Category*** | |
| 1A | Stretcher patients who are frankly disturbed and inaccessible, requiring close supervision and sedation, or the use of restraint equipment. |
| 1B | Patients who are not at the time of request grossly disturbed, but who may react badly to air travel, or who may commit acts likely to endanger themselves or the safety of the aircraft and its occupants. |
| 1C | Sitting patients who are cooperative and have proved themselves to be reliable under preflight observation. |
| ***Class 2. Stretcher Patients Other Than Psychiatric*** | |
| 2A | Patients, who in an emergency, would be unable to leave the aircraft without assistance. |
| 2B | Patients, who in an emergency, would be able to leave the aircraft unaided. |
| 2C | Patients who can walk onto an aircraft but on a long flight would benefit from a litter. |
| ***Class 3. Sitting Patients Other Than Psychiatric*** | |
| 3A | Sitting patients, including handicapped persons, who may need medical or nursing attention en route and who, in an emergency, would require assistance to escape. |
| 3B | Sitting patients who may need medical or nursing attention en route and who would be able to escape unassisted in an emergency. |
| ***Class 4. Sitting Patients Other Then Psychiatric*** | |
| 4 | Patients who will not need medical or nursing attention en route and are capable of traveling unescorted. Class 4 patients are normally required to make their own way from the deplaning airfield to their destination. |

## International Standardization Evacuation Precedence

G-3. Patients for AE will be given appropriate degrees of priority so that, if aircraft space is limited, the more urgent patients may be evacuated before those whose conditions are less serious. The degrees of priority are depicted in Table G-2. (The evacuation precedence used by the USAF is essentially the same as this listing. It contains a few word changes and introduces specific time limits. It does not contain Priority 4.)

### Table G-2. Patient priorities as designated in STANAG 3204

| | |
|---|---|
| PRIORITY 1/URGENT | Emergency patients for whom speedy evacuation is necessary to save life or limb, to prevent complication of serious illness, or to avoid serious permanent disability. |
| PRIORITY 2/PRIORITY | Patients who require specialized treatment not available locally and who are liable to suffer unnecessary pain or disability unless evacuated with the least possible delay. |
| PRIORITY 3/ROUTINE | Patients whose immediate treatment requirements are available locally but whose prognosis would definitely benefit by air evacuation on routine scheduled flights. |
| PRIORITY 4 | Patients for whom air evacuation is a matter of medical convenience rather than necessity. |

## Patient Classification

G-4. Table G-3 provides the patient classification codes used aboard USAF aircraft and that can be used in completing DD Form 601 (Patient Evacuation Manifest). These codes are expanded to include categories of patients and other personnel which may or may not apply on the battlefield (such as infants, relatives, or friends).

## UNITED STATES AIR FORCE EVACUATION PRECEDENCE

G-5. The evacuation precedence used by the USAF is dramatically different than that employed by the US Army medical evacuation system. These precedence should not be confused. Table G-4 (page G-5) provides the evacuation precedence and time frames used by the USAF.

**Table G-3. Patient classification codes**

| *Code* | *Classification* |
|---|---|
| ***Class 1.*** | ***Neuropsychiatric Patients*** |
| 1A | Severe psychiatric litter patients requiring the use of restraining apparatus, sedation, and close supervision at all times. |
| 1B | Psychiatric litter patients of intermediate severity requiring tranquilizing medication or sedation, not normally requiring the use of restraining apparatus, but who react badly to air travel or who may commit acts likely to endanger themselves or the safety of the aircraft. Restraining apparatus should be available for use. |
| 1C | Psychiatric walking patients of moderate severity who are cooperative and who have proved reliable under observation. |
| ***Class 2.*** | ***Litter Patients (Other Than Psychiatric)*** |
| 2A | Immobile litter patients unable to move about of their own volition under any circumstances. |
| 2B | Mobile litter patients able to move about of their own volition in an emergency. |
| ***Class 3.*** | ***Walking Patients (Other Than Psychiatric)*** |
| 3A | Nonpsychiatric and nonsubstance abuse patients who require medical treatment, assistance, or observation en route. |
| 3B | Recovered patients who are returning to their units and require no medical attention en route. |
| 3C | Ambulatory drug or alcohol substance abuse patients. |
| ***Class 4.*** | ***Infant Category*** |
| 4A | Infants under three years, occupying a seat or in a bassinet or a car seat secured in an ambulatory seat. |
| 4B | Recovered infants under three years, occupying a seat or in a bassinet or car seat secured in an ambulatory seat. |
| 4C | Infants in an incubator. |
| 4D | Infants under three years on a litter. |
| 4E | Outpatients under three years on a litter for comfort. |
| ***Class 5.*** | ***Outpatient Category*** |
| 5A | Ambulatory, nonpsychiatric, and nonsubstance abuse outpatients who are traveling for an outpatient visit and do not require a litter or medical assistance in flight. |
| 5B | Ambulatory drug or substance abuse outpatients going for treatment. |
| 5C | Psychiatric outpatients going for treatment. |
| 5D | Outpatients on a litter for comfort or safety. |
| 5E | Returning outpatients on a litter for comfort or safety. |
| 5F | Other returning outpatients. |
| ***Class 6.*** | ***Attendant Category*** |
| 6A | Medical attendants, either physician, nurse, or technician, who are assigned to give specialized medical treatment or nursing care to a particular patient. |
| 6B | Nonmedical attendants, either relatives or friends, who may assist with the patient's care and who may also require support. |

**Table G-4. Evacuation precedence used by the United States Air Force**

| *Evacuation Precedence* | *Definition And Time Frames* |
|---|---|
| URGENT PRECEDENCE | Applies only to the need for immediate life, limb or eyesight saving. These patients should be picked up and delivered to the designation facility as soon as possible. |
| PRIORITY PRECEDENCE | Applies to the need for prompt medical care not available locally. These patients should be picked up within 24 hours and delivered to the destination facility with the least possible delay. |
| ROUTINE PRECEDENCE | Applies to all other patients. Routine patients will be picked up and delivered on regularly scheduled flights. |

## Appendix H

# Troop Leading Procedures, Precombat Checks/Precombat Inspections, and Leader Checklists

Troop leading procedures (TLPs) are universal in their application. They are used by leaders at all levels regardless of branch or MOS. For purposes of this appendix the leader titles of ambulance platoon leader and ambulance platoon sergeant are used for familiarity.

## SECTION I — TROOP LEADING PROCEDURES

H-1. Leaders must conduct detailed planning and preparation before each mission in order to be successful. That is why the Army developed an eight-step process of TLP for leaders to follow.

H-2. Each step of the TLP process is designed to give leaders a framework on which to thoroughly develop their plans and orders. If these steps are not followed the resulting plans and orders may be missing information which may be critical to successfully execute the mission.

## THE EIGHT STEPS OF TROOP LEADING PROCEDURES

### STEP ONE: RECEIVE THE MISSION

H-3. Once staff planning is initiated the ambulance platoon leader gathers information concerning the operation from the company commander and disseminates to his subordinate leaders. Examples of this information include but are not limited to intelligence preparation of the battlefield (IPB) and METT-TC. If the company commander has not completed a full plan, he may still issue a warning order (WARNO) to the ambulance platoon leader.

- In keeping with the rules of time management, the ambulance platoon leader uses one-third of his overall time for planning the operation. The reverse planning sequence is recommended.
- The ambulance platoon leader's OPORD must also be understandable to subordinate personnel.
- Throughout the planning and development process the ambulance platoon leader may find it necessary to change or refine the plan as the tactical situation changes, as long as it follows the commander's intent.

### STEP TWO: ISSUE A WARNING ORDER

H-4. The ambulance platoon leader issues a WARNO based on information received from the company commander's WARNO or OPORD.

- The ambulance platoon leader can present the order in any format.
- As a minimum, the platoon should use the company or battalion tactical standing operating procedures (TACSOP). The best method is to use already developed and implemented platoon level TACSOP that cover preparation for offensive or defensive missions. These TACSOPs should be thoroughly trained and rehearsed.
- Whatever the mission, the platoon can begin work on its preparation by assessing equipment needs through the various military supply classes (such as Class I, III, IV, V, or VIII).
- Every Soldier in the platoon should have a specific task and purpose (identified in the platoon SOP) once the ambulance platoon leader has issued a WARNO.

## Step Three: Make a Tentative Plan

H-5. The ambulance platoon leader considers the following while preparing the tentative plan:

- Mission analysis, IPB, METT-TC considerations, higher mission and intent (two levels up).
- The ambulance platoon leader must know what systems are available for his use and how to employ them. He must fully understand his own equipment and capabilities as well as those of his enemy.
- Consider **Observation** and Fields of Fire, **Avenues** of Approach, **Key** Terrain, **Obstacles** and **Cover** and Concealment, (man-made and natural), (OAKOC), weather, morale, and flexibility to react to last-minute changes. As the commander updates the ambulance platoon leader, the ambulance platoon leader, in turn, continually updates the platoon with FRAGOs.
- Experience is a valuable asset for mental preparation. Knowing how Soldiers are going to react to situations, based on past experiences, can help any officer or NCO develop his plan.
- During the planning process, the leader should have a *vision* of the enemy and understand the threat using available intelligence. The better the ambulance platoon leader understands the enemy, the greater his opportunity for success.

## Step Four: Start Necessary Movement

H-6. Even though listed as Step Four, necessary movement can begin as soon as the mission is received.

- If movement is required, the ambulance platoon sergeant, with the assistance of squad leaders, can move the platoon to a new tactical assembly area (TAA). This allows the ambulance platoon leader more time to prepare his order and graphics.
- At the new TAA, preparations continue with precombat checks (PCCs). This is a good opportunity to conduct PCCs on squad equipment by using the platoon SOP checklists and to certify squad rehearsals. If the ambulance platoon sergeant is unable to conduct squad inspections, the platoon SOP should designate someone else, such as the senior squad leader, for the task.
- If platoon linkups are time driven, movements and time management become critical. Failure to plan or allow for ample movement time creates late linkups and results in late rehearsals, late PCCs, and various other events that will not be accomplished on time or to standard.
- If movements were conducted late or linkups were late, the mission timeline will suffer.

***Note.*** Time management is mission essential!

## Step Five: Reconnoiter (Conduct Reconnaissance)

H-7. Although listed fifth, this step may be implemented at any time and as often as necessary during TLP. When the risk of enemy contact is too great, or establishing additional orders for reconnaissance parties is time consuming leaders may have to rely on map reconnaissance and or photographic imagery to reconnoiter an AO. Ambulance platoon leaders must ensure that ground evacuation routes are reconnoitered and strip maps developed for security escort vehicle drivers and ambulance drivers.

- A leader's reconnaissance must be performed! Leaders who see more of the AO are better able to develop more detailed and useful plans.
- Information gained by the leader's reconnaissance allows more accurate terrain models to be built and greater detail can be provided when the ambulance platoon leader is briefing the OPORD.

## Step Six: Complete the Plan

H-8. Complete the plan—

- With reconnaissance completed, movement accomplished, and the initial plan established, the ambulance platoon leader should have most of the information that is needed to finish writing the OPORD.
- Any additional resources needed should be en route or already on-hand so the ambulance platoon leader, when briefing the OPORD, has positive control of additional equipment or manpower.
- Graphics must be finished before the OPORD is briefed to the platoon and quickly handed down to squad leader level. Logistics and sustainment graphics must include CBRN decontamination points, MTFs locations, and supported unit CCP and AXP.

## STEP SEVEN: ISSUE THE ORDER

H-9. Avoid rushing through the OPORD briefing as this may result in pertinent information being left out of the brief.

- When completed, distribute the OPORD with graphics to each squad and team leader. Also include a terrain model or sand table to enhance visualization of the terrain.
- Minimize distractions. Having subordinate leaders in the right mindset to receive the order is critical to allow them to collect and retain all of the information presented at the OPORD brief.
- If possible, present the OPORD with a representation of the terrain in the background to enhance visualization of the mission.
- After the OPORD has been presented, the ambulance platoon leader ensures that the information has been disseminated to every Soldier. Backbriefs from Soldiers and NCOs to the ambulance platoon leader should reveal any portion or portions of the OPORD that were missed or not fully understood. Everyone leaving the OPORD brief should have a clear understanding of the mission.
- Contingencies need special consideration. If, during the OPORD brief, a potential activity is discussed that has no clear plan, subordinates can help develop a contingency plan, unless the activity is already specified in the platoon SOP. Defer making a decision on topics discussed during the OPORD brief which require more thought and possibly guidance from higher headquarters. If at all possible, allow a follow-up period to develop clarity on the situation and the techniques needed to accomplish the mission.

## STEP EIGHT: REHEARSALS

H-10. Supervise rehearsals and plan to run as many rehearsals as time permits. After each rehearsal, conduct an after-action review (AAR). When conducting platoon rehearsals with attachments, ensure that the ambulance platoon leader/sergeant supervises the AAR with attachments. When conducting the AAR, ensure Soldier involvement. By this time all Soldiers understand the concept of the operation.

- When rehearsing the mission, ensure that the rehearsal lane coincides with what was briefed during the OPORD brief. Based on the latest intelligence, the rehearsal lane should match (as much as possible) what will be encountered during the actual mission. Do not allow for complacency. If there is information about the terrain and the obstacle that can be placed in the rehearsal lane to add realism, ensure that obstacle information is built into the lane. A company-level rehearsal is of limited value if the enemy obstacle does not resemble what was reconnoitered or templated.
- If time does not allow for a full rehearsal, backbriefs must be conducted. A successful understanding of the mission can be accomplished with key personnel by walking troops through the mission on the terrain model and asking questions. The backbrief is never a substitute for a rehearsal, but is an acceptable recourse when time is limited.
- Conduct final precombat inspections (PCI).
- If there are any last-minute concerns, they should be satisfied during this time.
- Conclude the brief.

H-11. Some steps of the TLPs require more development than others. Regardless of the mission, the ambulance platoon leader must not, for any reason, omit any step from the TLP.

## SECTION II — PRECOMBAT CHECKS AND PRECOMBAT INSPECTIONS

H-12. Precombat checks and PCIs are tools which leaders have to ensure that their Soldiers—

- Fully understand the mission and their role.
- Have the equipment that is necessary to accomplish the mission on hand and that it is serviceable.

H-13. Leaders must conduct PCCs and PCIs for every mission. Failure to perform these vital checks could result in unnecessary casualties, loss of equipment, and failure to accomplish the mission.

H-14. When conducting PCCs/PCIs, as deficiencies are identified corrections must be made immediately.

H-15. Who performs PCCs and PCIs are both intended to accomplish the same thing, which is to have a leader ensure that personnel and equipment are ready. The only difference is who is doing the inspecting.

- Commanders conduct PCIs. Ambulance platoon leaders, ambulance platoon sergeants, and squad leaders conduct PCCs.
- Why do we identify a commander's inspection separately? Equipment readiness is essential for mission success. Squad leaders and ambulance platoon leaders are certainly capable of ensuring equipment readiness, but the commander's first-hand knowledge of mission requirements cannot be substituted.

H-16. Units need to establish standard PCC/PCI checklists based on their units mission profile. They should be included in the TACSOP and issued to all leaders. They should include routine daily checks and checks that are mission specific. They do not need to be extremely detailed; a list of items to check and the leader's knowledge of the equipment and the appropriate manuals should suffice. The PCCs checklists provided below are examples that units may use to develop their own or they can be modified as appropriate to meet their own requirements.

## SECTION III — LEADER CHECKLISTS

H-17. This section provides sample checklists which may be used as is, or modified, to meet unit specific requirements. As with any checklist they are only effective if they are incorporated into a unit's TACSOPs and then used routinely.

# GENERAL CONSIDERATIONS

H-18. Leader receives the mission from the next higher headquarters.

___ Clarifies any questions.

___ Coordinates with next higher headquarters as required.

___ Conducts mission analysis.

___ Produces the estimate.

___ Revises mission statement.

___ Produces tentative time schedule.

___ Issues a WARNO.

H-19. Unit members perform readiness, maintenance, and functional checks under the supervision of unit leaders which includes:

___ Medical equipment sets are present and all perishable and dated items are current and on-hand.

___ Vehicles/aircraft/generators.

___ Night vision devices.

___ Communications equipment.

___ Weapons and ammunition.

___ Field sanitation equipment and supplies.

___ Special equipment (such as hoist and forest penetrator).

___ Common table of allowances equipment.

H-20. Leader makes a tentative plan.

___ Uses estimate of the situation to depict plan of support.

___ Develops courses of action (COA).

___ War games COA.

___ Determines best COA.

___ Completes his plan.

H-21. Leader issues OPORD.

H-22. Leader uses sand table or sketches to depict plan of support.

H-23. Leader affects coordination for the mission.

___ Support requirements.

___ Current intelligence (to include health threat) update.

___ Control measures.

___ Communications and signal information.

___ Time schedule.

H-24. Leader receives attachments/augmentation (personnel), if appropriate.

___ Attachments are oriented to unit.

___ Attachments are briefed on the mission.

H-25. Leader supervises AHS mission preparation.

___ Key leaders brief back unit leader.

___ Key personnel rehearsals are conducted.

H-26. Unit leaders:

___ Supervise.

___ Inspect.

___ Ensure adequate security.

___ Conduct brief backs.

___ Rehearse.

___ Continue coordination.

H-27. Unit plans for support of combat operations.

___ Analyze mission.

___ Patient acquisition and medical evacuation requirements.

___ Area support requirements.

___ Requirements for water, Class VIII, and other supply classes.

___ Requirements for CBRN defense.

___ Transportation requirements.

H-28. Unit leaders:

___ Execute a work/rest plan based on work priorities and statutory (crew rest) requirements.

___ Monitor current situation.

___ Issues appropriate FRAGOs based on intelligence or operational updates.

___ React to messages or orders from higher headquarters.

___ Execute any actions and coordination resulting from change.

H-29. Unit headquarters remains current on positions and missions of higher, adjacent, and subordinate units.

## GROUND AMBULANCES INSPECTED FOR

___ Authorized MES are on hand.

___ Medical equipment is complete and serviceable.

___ Oxygen is on hand and serviceable.

___ Authorized medications are on hand and current.

___ Packing list is available.

___ Strip maps and/or road maps are available (with overlays).

___ On-vehicle equipment is on hand.

___ Log book is present and current.

___ All drivers are licensed.

___ Situational awareness equipment (position locator) is on hand and serviceable.

___ Communications equipment is on hand and serviceable and set to correct frequency.

___ Medical unit identification markers (according to the Geneva Conventions) are displayed.

***Note.*** Markers are red on a white background only; camouflaged or subdued markers are not authorized.

## CHEMICAL, BIOLOGICAL, RADIOLOGICAL, AND NUCLEAR EQUIPMENT

___ Individual protective equipment is on hand and serviceable. (One set is issued; the other is maintained in support.)

___ Protective masks are issued and serviceable.

___ Nerve Agent Antidote Kit is available and distributed, if required.

___ Convulsant antidote for nerve agent is available and distributed, if required.

___ Decontamination apparatus is complete and serviceable.

___ Basic load of decontamination supplies is on hand.

___ Chemical agent alarms are on hand and serviceable.

___ M256A1 detector kits are issued.

___ Chemical, biological, radiological, and nuclear contamination marking kits are distributed.

___ Chemical agent monitors are on hand, if authorized.

___ Replacement filters for protective masks are on hand.

___ Biological and chemical warfare agents prophylaxis, immunizations, and barrier creams have been accomplished, if appropriate.

___ Nerve agent pretreatment packets are available.

___ Radiac sets are on hand.

## Miscellaneous Equipment

___ Inspect binoculars.

___ Inspect camouflage nets and support systems, if appropriate.

___ Inspect night vision devices.

___ Ensure batteries are on hand and serviceable.

___ Inspect tentage.

___ Inspect global positioning systems, if available.

## Personnel

___ Ensure Soldiers are in the correct uniform.

___ Ask questions to ensure that Soldiers have been briefed on mission and situation.

___ Implement appropriate MOPP level.

___ Check for drivers licenses.

___ Brief Soldiers on operations safety and environmental injuries.

___ Individual equipment is on hand and stowed properly.

___ Soldier has eaten and is briefed on future field feeding.

___ Identification cards and tags are on hand and serviceable.

___ Camouflage self and equipment, when directed. (If medical vehicles are camouflaged, the Geneva Convention emblem must be removed.) (Refer to STANAG 2931.)

___ Work/rest plan (to include crew rest) is implemented.

___ Water discipline plan is implemented, if appropriate.

## Communications Equipment

___ Radios are operational (communications check completed).

___ Telemedicine equipment is available and operational, if available.

___ Speech security equipment functions, if available.

___ Frequencies are set.

___ Matching units are operational.

___ Antennas are tied down properly.

___ Connectors are clean and serviceable.

___ Field telephones (TA-312) are on hand and serviceable, if appropriate.

___ Batteries are on hand and charged.

___ Man-pack radio sets are complete.

___ Switchboard is on hand and serviceable.

___ Communications wire (WD-1) is on hand and serviceable.

___ Antennas and remotes are present and serviceable.

___ Signal operating instructions are available and secured.

___ Call signs, frequencies, and challenge passwords have been disseminated.

___ Perform communications check again.

## VEHICLES

___ Loads are according to load plan; load plan is posted in the vehicle.

___ Hazardous cargo is properly identified and stored toward the rear of the vehicle for easy access and inspection.

___ Ammunition is issued and properly stored.

___ Vehicle fuel tank is topped off.

___ Package POL products and small arms lubricant are present.

___ Water cans are full.

___ Meals, read-to-eat are issued and stowed.

___ First-aid kits are present and complete.

___ Operator's manuals and lubrication orders are present for the vehicle, radios, and associated equipment.

___ Critical on-vehicle equipment and basic issue items are present.

___ Vehicle dispatch is complete.

___ Department of the Army (DA) Form 2404 (Equipment Inspection and Maintenance Worksheet) is complete.

___ No deadline deficiencies exist.

___ Before operation PMCS has been completed.

## INDIVIDUAL WEAPONS

___ Clean and functional.

___ Cleaning tools/kits, bolts, and ruptured cartridge extractors are present.

___ Range cards are on hand.

___ Ammunition is issued, accounted for, and secured.

___ Magazines are issued.

___ Blank adapter installed, if appropriate.

___ Function check performed.

## MEDICAL EVACUATION SUPPORT PLAN

___ Mission statement/commander's intent/task organization.

___ Intelligence preparation of the battlefield considerations.

___ Patient estimates by phase.

___ Identify expected areas of patient density.

___ Anticipate and plan for likely patient conditions.

___ Formulate/evaluate/implement/reevaluate COA.

___ Location of supported units.

___ Location of supporting MTFs/FSTs/hospitals.

___ Location of supporting evacuation units.

___ Augmentation of personnel and vehicles coordinated supporting MTFs.

___ Class VIII supply/resupply coordinated and available.

___ Location of supported unit CCPs and AXPs.

___ Evacuation routes primary and alternate.

___ Trafficability (enemy activity, hazards, condition, obstacles).

___ Security (gunship escort, MP escort, convoy support).

___ Strip maps/overlays/maps.

___ Landing zones designated for rotary-wing evacuation aircraft (primary and alternate)

___ Ensure that platoon/company/battalion/treatment facility litter teams are available to assist ambulance crews in loading patients.

___ Available litters and types (NATO standard, folding nonrigid, SKED®, and so forth).

___ Ground ambulances available (type/number/carrying capacity).

___ Air ambulances available (type/number/carrying capacity).

___ Supporting evacuation assets available (requested/ prepositioned).

___ Nonmedical vehicles for CASEVAC (type/number/carrying capacity).

___ Available aircraft (UH-60, CH-47).

___ Medical evacuation support plan for units without evacuation resources.

___ Treatment and evacuation of EPW and detainees.

___ Chemical, biological, radiological, nuclear considerations (contaminated evacuation vehicle plan, patient decontamination site, decontamination sites, contaminated aircraft).

___ Other.

## CASUALTY CARE CONSIDERATIONS

___ Self-aid/buddy aid/CLS/trauma specialist.

___ Combat lifesavers trained and CLS bags stocked and issued.

___ Emergency care specialists and flight medics are rehearsed, aid bags fully stocked.

___ Trauma specialist Class VIII resupply sets fully stocked and pushed forward to company command post.

___ Locations of platoon/company/battalion/treatment facility CCPs are known.

___ Other.

### MASS CASUALTY PLAN (RESOURCED/REHEARSED/COORDINATED)

___ Litter teams.

___ Litters (all).

___ Medical evacuation/CASEVAC vehicles (all).

___ Class VIII supplies (on hand).

___ Class VIII resupply (Prepackaged and prepositioned).

___ Company/battalion command net.

___ Brigade support medical company command net.

___ Personal effects, weapons, equipment, and property exchange.

___ Chemical, biological, radiological, and nuclear casualty/evacuation plan.

___ Location of decontamination site.

## LANDING ZONE OPERATIONS

### SAFETY

___ At all times take all the necessary safety precautions.

___ Platoon that is responsible for running the LZ provides all required safety gear. Included are—

  ___ Safety goggles.

  ___ Visual Signal, VS-17 panels/chemical lights.

  ___ Smoke (Appendix E).

___ Is the LZ secure.

### MARKING THE LZ

___ Use colored smoke to identify the LZ and wind direction.

___ Use an inverted "Y" to mark the LZ.

___ During daylight, camouflage the LZ so the VS-17 panels cannot be seen until evacuation aircraft are confirmed and inbound.

___ At night, emplace chemical lights only when an AA is confirmed and is inbound.

### LANDING REQUIREMENTS FOR LIGHT HELICOPTERS

___ Minimum real estate required is a cleared area 30 meters in diameter.

___ Area must have an approach and departure zone clear of obstructions and not over the established company area.

## Appendix I

# Safety on the Battlefield

The tactical environment provides ever-changing demands and unpredictable problems, often under stressful conditions. The interface of man, machine, and environment is constantly shifting. In this environment, mission accomplishment requires continuous leader involvement and flexible decision making. Not surprisingly, accidents and injuries increase during tactical operations. Safety in the tactical environment depends upon compliance with established standards. However, due to fluid conditions in the tactical environment, safe mission accomplishment relies heavily on complete integration of risk management into both planning and execution phases. Risk management assists commanders in anticipating and controlling hazards in the planning phase and in dealing with unexpected hazards as they arise in the execution phase.

## SECTION I — GENERAL

## COMMON ACCIDENTS

I-1. During all operations, it is critical commanders and leaders minimize risks through appropriate risk management procedures to protect the force and preserve the unit's war-fighting capability. Over half of all Army accidents during tactical operations are accredited to the following operations.

### VEHICLE OPERATIONS

I-2. Most accidents in this category are caused by excessive speed for conditions (weather, traffic). Other causes are recklessness, fatigue, unfamiliarity with roads, and untrained and inexperienced drivers. A lack of equipment/vehicle handling characteristics knowledge also contributes to accidents. Strict enforcement of standards is needed for all vehicle operations. The senior occupant must be responsible for this enforcement. Only trained, licensed personnel should be assigned to operate vehicles or equipment. Ground guides are mandatory during movement in bivouac and assembly areas, when backing, and during periods of reduced visibility. Operators must be familiar with proper operation and maintenance of commercial equipment.

### MATERIEL HANDLING

I-3. These accidents occur when an object is too large or too heavy to handle for those attempting to move it. As a result, someone may sustain a muscle or back injury, or may be crushed. Overconfidence in one's ability, a lack of planning, and fatigue are common factors in such accidents.

### MAINTENANCE

I-4. These accidents are often caused by failure to follow procedures, using the wrong tools, and/or fatigue.

# SEVERE ACCIDENTS

I-5. The following activities produce fewer accidents than those listed above; however, when they do occur, they often result in catastrophic damage or death. Commanders and leaders must pay special attention to unit missions which involve these activities or put Soldiers in situations or locations where they will come in contact with associated hazards and ensure they properly brief their personnel and mitigate risk to the fullest extent.

## AMMUNITION AND EXPLOSIVES HANDLING

I-6. Horseplay, mishandling, disassembly, unauthorized use, and improper storage of ammunition and explosives account for many personnel injury accidents. It is essential to enforce accountability and security procedures for unexpended ammunition and explosives and to comply with explosive storage safety standards in AR 385–64 and DA Pam 385–64.

## EXPLOSIVE SOUVENIRS

I-7. Educate Soldiers to dangers involved and the serious consequences of collecting unexploded ordnance on the battlefield or ranges. Post-tactical training shakedown inspections for this type of materiel are a must. Amnesty boxes are also useful. Platoon sergeants and squad leaders policing their Soldiers can prevent most of these accidents through proper training.

## FIELD EXPEDIENTS

I-8. Be suspicious of shortcuts. Although tactical operations frequently involve employment of field expedients, risks and benefits must be carefully weighed. In many cases, field expedients are the result of a weak supply system or inadequate planning.

## FIELD HEATERS AND STOVES

I-9. Operators of all types of heaters and stoves should be trained and licensed in advance. Equipment should be maintained and operated in accordance with operating instructions including use of proper fuel. Ensure combustible material is kept well away from heaters and stoves and fire fighting equipment is available for each heater and stove. Heaters or stoves with self-contained fuel supplies (small space heater) should not be refilled while the heater is on or still warm. Do not use heaters or stoves in tents or other confined spaces without use of proper ventilation, such as tent vent flaps, doors, or windows.

## PETROLEUM, OILS, LUBRICANTS STORAGE AND HANDLING

I-10. The POL handlers must know and practice safety rules and procedures. Frequent inspections ensure safe storage and transfer of POL products. Proper grounding procedures must be followed. Field manual 10–67–1 describes use of protective equipment to prevent personnel exposure.

## SOLDIER FATIGUE

I-11. When a Soldier's sleep time is dependent upon the tactical situation, debilitating fatigue can occur. Soldiers suffering from sleep loss experience various symptoms of fatigue including decreased coordination, narrowed attention span, and reduced performance to standard. Anticipate fatigue-related errors and take action to prevent them.

## TACTICAL SLEEP PLAN

I-12. Commanders need to develop and enforce sleep plans. Control sleeping areas to prevent Soldiers from being crushed by moving vehicles in and out of the area.

### Water Operations

I-13. Plan water operations carefully, as risks of drowning and equipment loss are high during water operations. Pair strong swimmers with weak ones to protect personnel. Secure equipment and float it across rather than requiring individuals to carry their equipment. Use safety lines and personal floatation devices.

### Weapons

I-14. Most accidents occur when cleaning or clearing individual weapons, entering or exiting vehicles, or running with loaded weapons. Provide guidance for weapons handling, loading, and clearing, and ensure it is strictly enforced. Do not load weapons that are not essential for the current mission.

### Weather-Related Casualties

I-15. Consider the effects of weather during planning. Unit effectiveness is quickly lost through weather-related casualties such as frostbite, heat stroke, lightning strikes, and falls. Instruct Soldiers in awareness, prevention, and first aid for weather-related injuries and when these conditions can be expected.

## VEHICLE CONVOY OPERATIONS

I-16. Convoy operations can be very dangerous if not properly planned. Control of convoy speed and proper separation between vehicles is critical to reducing risk of an accident. See FM 4-01.45 for in-depth information on tactical convoy operations.

### Convoy Commander Responsibilities

I-17. Convoy accidents are most commonly caused by a leader failing to perform his or her duties as the commander of a vehicle movement as required in FM 55-30. This includes failure to:

- Control the group's movement.
- Ensure vehicles maintain proper march speed.
- Properly mark unit vehicles.

I-18. Convoy commanders are responsible for ensuring the safe movement of the convoy. To do this, they must positively control the convoy's movement by using communications equipment among the vehicles. This includes ensuring each vehicle has a properly trained, equipped, and supervised crew leading from the front in the absence of radios and/or other means of managing the march. Control of the movement includes enforcing speed limits, march intervals, and seat belt usage. It also requires the ability to stop the march if an unexpected hazard is encountered along the route.

I-19. Preparing vehicles and Soldiers for movement is a leader responsibility. Inexperienced Soldiers, personnel turbulence, and ever increasing training requirements have caused some units to become complacent in managing risks associated with vehicle movements. According to the risk management process, leaders must identify hazards associated with the mission and develop, implement, and supervise control measures to mitigate those risks. These control measures include marking the vehicles in accordance with local SOPs, briefing crews on hazardous conditions expected along the route, precombat checks of personnel and equipment, and developing preaccident emergency contingency plans.

I-20. Unit SOPs should address leader responsibilities during movements of any number of vehicles and identify means to implement common controls.

### Driver Skills

I-21. Operators should be taught the specific skills needed for tactical vehicle operations, which, in addition to requirements outlined in AR 600-55, include—

- Pulling and backing trailers.
- Vehicle recovery operations.

- Loading and lashing of cargo.
- Methods of negotiating difficult terrain such as sand dunes, rice paddies, mountainous terrain, icy roads.
- Ground-guide procedures and signals.
- Methods and procedures for retrieving vehicles stuck in snow, mud, sand, or other restrictive terrain.
- Proper parking and use of the proper-sized chock blocks.

### Use of Safety Equipment

I-22. When the tactical situation allows, flashers should be turned on immediately if a vehicle is disabled or impedes traffic, and every effort made to move the disabled vehicle from the roadway. Highway warning kits should be provided for each vehicle in a convoy. When a vehicle is disabled, place the warning triangle a minimum of 100 meters to the rear of the vehicle and ensure personnel remain clear of the road and rear of the vehicle.

### Night Operations

I-23. Personnel who are required to operate motor vehicles while wearing NVG must be trained and tested on use and operation of such devices, and the training recorded on the individual's driver training records. Other factors include:

- Ground guides should be used when moving vehicles at night in areas where troops are present.
- Blackout driving should be prohibited on roads open to the public.

## FIRE PREVENTION

I-24. The risk of fire is high in areas where a large number of Soldiers are in tents. The following guidance will reduce the risk of fires:

- Establish a fire prevention and protection plan including procedures for inspecting and recharging fire extinguishers during tactical operations.
- Appoint a fire marshal for each bivouac area and train them in their duties. Train Soldiers in fire prevention techniques as well as emergency procedures in the event of a fire.
- Establish safe distances between tents to reduce risk of multiple losses from one fire.
- Provide available fire fighting equipment (portable extinguishers, sand, water buckets, and shovels) to contain small fires. Ensure personnel are trained on their use.
- Establish procedures for sounding fire alarms.
- Ensure no-smoking areas are established and enforced.
- Establish an inspection system ensuring compliance with fire prevention standards.
- Ensure flammable materials are stored according to appropriate directives and checklists.
- Ensure vehicle fire extinguishing/suppression systems are operational and crews are proficient in using the systems.
- Provide a designated fire plan, equipment, and trained personnel for POL storage, ammunition supply points, motor pools, hospitals, and hangars.

# SECTION II — ASSEMBLY AREA OPERATIONS

## SAFETY OFFICER/NONCOMMISSIONED OFFICER

I-25. The safety officer and NCO must establish, in coordination with the 1SG, platoon leaders, and platoon sergeants, procedures for ground guides and incorporation of these procedures into the unit SOP. Company commanders must weigh mission requirements and personnel available while in combat operations and use risk management analysis to validate procedures for implementation into the SOP.

Operating within the modular force structure may require assistance from support units to meet necessary personnel and vehicle requirements to safely ground guide and move aircraft.

## GROUND VEHICLES

I-26. Commanders and leaders must be aware of hazards from ground vehicles in AAs. Vehicles moving in the AA should always have a ground guide leading the vehicle until out of the AA. This prevents injury to personnel or damage to equipment. Use of a one-way path through the unit area will assist greatly in preventing accidents. Additionally units may be operating in the vicinity of armored vehicles and Soldiers must take additional precautions due to the inability of armored vehicle crews to see personnel close to their vehicles.

I-27. During initial entry into the AA, platoon ground guides lead convoy parties to their platoon locations with each vehicle dismounting one person to ground guide their vehicle. During this period of movement into the AA, there is a higher probability of accident or injury due to the amount of traffic and unfamiliarity of the area. Commanders must ensure movement into the AA is thoroughly briefed.

## AIRCRAFT

I-28. Aircraft in the AA present movement hazards and contribute to an additional hazard of blowing debris. Blowing debris is a hazard to personnel, aircraft, and equipment. The variation in environments in which a unit may operate determines the type and amount of debris. Desert environments, for instance, produce brownout conditions which are a hazard to personnel, equipment, aircraft, and crews. Preventive measures should be incorporated into the unit SOP mitigating this risk. An example of risk mitigation for aircraft operations, in and around the AA parking areas, is brownout mats. These mats reduce the brownout effect of operating in a desert environment. Commanders must ensure SOPs are implemented for each of the environments the unit may operate within. This will allow the unit to establish ground guide procedures for aircraft, whether while hovering into and out of the AA or during ground movement of the aircraft within the AA.

I-29. Ground guide procedures for flight into or out of the AA will be based on the distance between aircraft and proximity to hazards (natural or manmade). In environments where the AA allows for wide scatter of aircraft, the need for ground guides while hovering into and out of the parking area is negligible. In contrast, when the AA parking areas require close interval spacing of aircraft, the need for guides is critical to safety. Additionally, if the AA requires aircraft to park in close proximity to natural or manmade hazards, ground guides are required. The need for ground guides is more critical with operations under night or with NVG devices.

I-30. Upon initial entry into the AA, ground guides should meet incoming aircraft. The situation may dictate crews entering the AA know limited information on where they will park their aircraft. The ground guides guide the aircraft into their assigned parking spot. In desert environments, the ability of personnel to guide aircraft into the parking location is decreased. Brownout upon landing requires a thorough prebrief to the flight crew on the AA parking plan and as exact a location as possible in relation to a ground marking, such as an inverted Y, of each aircraft parking location. Certain situations may require aircraft to land directly on a parking spot. The company safety officer and NCO must be involved in the design and procedures planned for aircraft operations into and out of the AA.

## WING WALKING

I-31. Ground guides will also be utilized while moving aircraft with ground handling equipment. Ground guiding aircraft while moving with a vehicle requires additional personnel to ensure rotor blades and aircraft do not come in contact with another aircraft or object. Commanders must ensure procedures designed by the safety officer/NCO are integrated into the SOP. The plan should address guides necessary for various settings (such as define how many are necessary for close proximity parking and scattered parking not in close proximity to another aircraft or hazard). Ground movement of aircraft is normally associated with aviation maintenance operations. The maintenance officer should be involved in the design of ground guide/wing walking procedures for the unit SOP. A critical component of the aircraft ground movement plan is communication. Personnel conducting these operations must conduct a team prebrief and ensure that the team leader has a device that will signal all members to immediately halt the operation (survival vest whistle).

### GROUND GUIDES

I-32. Commanders and leaders must ensure ground guide operations, vehicle and aircraft, are incorporated into their company SOPs. During AA operations, ground guides perform a key role in preventing personnel injury, and vehicle and equipment damage. Ground guide operations are critical to maintaining combat readiness for the unit.

## PYROTECHNICS

I-33. Aviation units normally use pyrotechnics while in an AA. Units utilize pyrotechnics along likely enemy avenues of advance within their security sector as a warning of activity along these avenues. Additionally, aircrews utilize pyrotechnics as signaling devices in downed aircrew rescue operations. Commanders, unit leaders, and the ALSE officer must train personnel in proper use of pyrotechnics and its associated hazards. These items will be addressed in the unit SOP.

I-34. Pyrotechnics require protection against moisture, dampness, and high temperature. Pyrotechnic items must be given high priority for the best available protection because of their sensitivity. Wet pyrotechnic material is hazardous to store; consequently, any boxes showing signs of dampness will be removed from a storage site and inspected. If the pyrotechnic materiel is wet, it will be destroyed. Certain kinds of materiel deteriorate with age and have expiration dates on the containers. Loose pyrotechnic tracer composition, flare composition, and similar mixtures spilled from broken containers should be carefully taken up and covered completely with SAE 10 (EO–10) engine oil and removed for appropriate disposal.

I-35. Pyrotechnics are hazard division 1.3 which includes items that burn vigorously and cannot usually be extinguished in storage situations. Explosions are confined to pressure ruptures of containers and do not produce propagating shock waves or damaging blast overpressure beyond the magazine distance specified in DA Pam 385-64, Table 5–16. Tossing about of burning container materials, propellant, or other flaming debris may cause a severe hazard of spreading fire.

I-36. The fire fighting extinguishing agent for pyrotechnic materials, Class D (combustible metals [magnesium, potassium, and so forth]) hazard, should be a dry powder agent. Pyrotechnics are a mass fire hazard due to their chemical agents. The mass fire hazard can be more hazardous based on other likely types of ammunition stored with pyrotechnics. These fires may be fought if:

I-37. Explosives are not directly involved.

- White phosphorous munitions are involved, there may be an explosion (immerse/spray with water continuously).
- Pyrotechnics and magnesium incendiaries do not use carbon dioxide, Halon extinguishers, or water on or near munitions. Allow magnesium to cool unless upon flammable materiel. In this case, use a 2-inch layer of dry sand or powder on the floor and rake the burning material onto this layer and re-smother.

> **WARNING**
>
> **DO NOT DISCHARGE PYROTECHNICS IN THE VICINITY OF AIRCRAFT FLYING IN THE AREA.**

## SECTION III — PROTECTING THE SOLDIER

# ENVIRONMENTAL CONSIDERATIONS

I-38. As the military looks into the next century, and even today, military units should try to avoid unnecessary environmental damage, not only in training, but also across the spectrum of operational missions. A mission's success may be determined by political or socio-economic stability, both of which are affected by environmental factors and resources. The Army must be able to identify ways to protect the natural environment while executing full range of their missions:

- Considering the environment in planning and decision making in conjunction with other essential considerations of national policy.
- Protecting the environment of home stations and training areas as a means of retaining resources for mission purposes.
- Using environmental risk assessment and management principles to integrate environmental considerations into mission performance.
- Instilling an environmental ethic in Soldiers.
- Understanding linkages between environmental protection issues and their associated impact on safety, force protection, and AHS.

I-39. Soldiers must be prepared to respond across full spectrum operations in any component (offense, defense, stability or civil support operations), sometimes within the same operation. Deployed forces must be able to conform to environmental protection requirements of the theater commander without impairing combat effectiveness. For in-depth information on environmental considerations and Soldier protection, see FM 3-100.4.

I-40. When deployed, company commanders will often deal with base camps. Base camps, while not installations, are comparable to small towns and require many of the considerations applied to installations. A mayor (often the HHC commander) assists base camp commander with control of base operations. A base camp coordination agency will provide expertise and support to the commander, largely through its subordinate base camp assistance/assessment team. Environmental expertise is resident or aligned with this team and available to support the base camp commander and designated mayor of the base camp. They provide technical recommendations and maintain appropriate standards. More information about this phenomenon is provided in the Center for Army Lessons Learned Newsletter 99-9, Integrating Military Environmental Protection.

I-41. Army Regulation 200-1 and the Army's Commander's Guide to Environmental Management specify commanders' environmental responsibilities. To carry out these responsibilities, commanders:

- Comply with an installation's environmental policies and legally applicable and appropriate federal, state, and local laws and regulations or country-specific final governing standards if outside the continental United States.
- Demonstrate a positive and proactive commitment to environmental stewardship and protection.
- Provide environmental training required by law, regulation, or command policy.
- Ensure all personnel perform their duties in compliance with environmental laws and regulations, and respond properly to emergencies.
- Promote proactive environmental measures and pollution prevention.
- Supervise compliance with environmental laws and regulations during operational, training, and administrative activities.
- Include environmental considerations in mission planning, briefings, meetings, execution, and AARs.
- Understand the requirements of Army environmental programs.
- Identify and assess environmental risks of proposed programs and activities.
- Coordinate unit activities with higher headquarters environmental elements.
- Appoint and train an environmental compliance officer (ECO) coordinator for the unit.

- Ensure SOPs contain all environmental considerations and regulatory requirements appropriate for the level of command.
- Conduct environmental self-assessment or internal environmental compliance assessments.
- Understand linkages between environmental considerations and their associated impact on safety, force protection, and AHS.

# ENVIRONMENTAL DUTIES

I-42. The key to fulfilling environmental requirements successfully at unit level is the ECO. Army Regulation 200-1 directs all Army unit commanders "appoint and train ECOs at appropriate levels to ensure compliance actions take place." In company-sized units, this duty will generally translate into an extra duty. The ECO manages environmental issues at unit level and ensures environmental compliance. He also coordinates, through the respective chain of command, with the supporting installation environmental staff to clarify requirements and obtain assistance.

I-43. The ECO accomplishes environmental compliance requirements on behalf of the commander. He also coordinates with supporting installation environmental staff to clarify requirements and obtain assistance. While this position of responsibility is not a formal staff position, the ECO is critical to the commander's environmental program. The ECO:

- Advises the unit on environmental compliance during training, operations, and logistics functions.
- Serves as commander's eyes and ears for environmental matters.
- Coordinates between unit and higher/installation headquarters environmental staffs.
- Manages information concerning unit's environmental training and certification requirements.
- Performs unit environmental self-assessment inspections.
- Performs environmental risk assessments.

# UNIT PLANNING

I-44. Staffs integrate environmental protection into planning for larger units. Unit leaders integrate environmental protection into unit planning for battalion- and company-level units. Soldier safety should be addressed in SOPs, OPORDS, and training plans.

## STANDING OPERATING PROCEDURES

I-45. Unit leaders develop SOPs reflecting environmental protection considerations for routine tasks and activities. The SOPs provide information to Soldiers on how to accomplish routine tasks in an environmentally sound manner. The SOPs incorporate local requirements. As local requirements change, unit leaders update their SOPs. The SOPs also help define environmental protection requirements for all unit activities—facility operations, field operations, deployment, and combat. Unit leaders ensure SOPs comply with local requirements by coordinating with higher headquarters staff—usually the environmental office, the surgeon and his staff, preventive medicine personnel, and the SJA or engineer coordinator.

I-46. Unit leaders conduct environmental risk assessments when planning operations or activities. Risk assessment is a standard element of TLP. Unit leaders perform environmental risk assessments for activities not addressed in the SOP or when conditions differ significantly from those described in the SOP. A maintenance unit does not perform a risk assessment every time it performs a lubrication or service. Rather, the SOP describes the correct manner to perform these actions. Risk assessments apply to garrison operations as well as field operations.

## OPERATION ORDERS

I-47. Unit leaders address environmental protection in their plans and orders including: WARNOs, OPORDs, OPLANs, concept of operation plans (CONPLANs), and FRAGOs. The higher headquarters staff develops an environmental appendix/annex to its OPORD/OPLAN/CONPLAN. Subordinate unit leaders draw environmental information from the environmental appendix to the OPORD/OPLAN/

CONPLAN, or from Annex L in a joint operations planning and execution system document. Field Manual 5-0 directs the inclusion of Appendix 2 (Environmental Considerations) to Annex F (Engineer) of the OPLAN/OPORD/CONPLAN (battalion level and above) and specifies that lower-level unit leaders/staffs include environmental information in coordinating instructions and service and support paragraphs.

## Risk Management Plan

I-48. Risk management is an integral part of the military decision-making process. The risk management process provides the framework for making risk management a routine part of planning, preparing, and executing operational missions and everyday tasks. Assessing environmentally-related risks is part of the total risk management process.

I-49. Knowledge of environmental factors is a key to planning and decision making. With this knowledge, leaders quantify risks, detect problem areas, reduce risk of injury or death, reduce property damage, and ensure compliance with environmental laws and regulations. Unit leaders should conduct risk assessments before conducting any training, operations, or logistical activities.

### *Tactical Risk/Accident Risk*

I-50. When assessing the risk of hazards in operations, the commander and leaders must look at two types of risk.

- Tactical risk is risk concerned with hazards existing because of the presence of either the enemy or an adversary, thus involving the considerations of force protection. It applies to all levels of war and across the spectrum of operations. For example, during the Gulf War, the enemy's demolition of oil fields created a significant health and environmental hazard to the surrounding countryside and those units maneuvering through the area.
- Accident risk includes all operational risk considerations other than tactical risk. It includes risk to friendly forces and risk posed to civilians by an operation, as well as the impact of operations on the environment. It can include activities associated with hazards concerning friendly personnel, civilians, equipment readiness, and environmental conditions. Examples of environmentally related accident risk are improper disposal of hazardous waste, personnel not properly trained to clean up a spill, and units maneuvering in ecologically sensitive terrain. Preventive medicine considerations also fall into this area of risk.

***Note***: Tactical risk and accident risk may be diametrically opposed. The commander may accept a high level of environmentally-related accident risk on reducing the overall tactical risk. For example, a commander may decide to destroy an enemy's petroleum storage area to reduce its overall tactical risk.

## Training Plans

I-51. Formal planning for training culminates with publication of the training schedule. Informal planning and coordination (preexecution checks) continue until training is performed. During rehearsals, leaders ensure all safety and environmental considerations are met.

I-52. To conduct effective, meaningful training for Soldiers, leaders, and units, thorough preparation is essential. This means well prepared trainers, Soldiers, and support personnel are ready to participate, and their facilities, equipment, and materials are ready to use.

I-53. A unit executes training the same way it executes a combat mission. The chain of command is present, in charge, and responsible. During operations, leaders ensure environmental practices and preventive measures are being employed. Once Soldiers understand what is expected of them, these practices become merely another measure of unit proficiency and level of unit discipline.

# FIGHTER MANAGEMENT

I-54. Fighter management describes a process essential to effective and efficient battle staff operations. At company level, commanders do not have staffs. The battle staff becomes those personnel in the unit who operate the CP. Successful continuous operations at the company level are more demanding than higher level organizations. The unit requires a TACSOP allowing rest, especially for critical personnel. At company level, the commander and 1SG will find it very demanding to try to rest. Their rest plan must be a priority for the organization to be lead effectively and their units to be successful on the battlefield.

I-55. The cycle of recurring events within a CP focuses the CP officer in charge/noncommissioned officer in charge (OIC/NCOIC) on meeting information and action requirements. Company personnel are normally required to attend battalion/company recurring meetings. This has an impact on company CP operations and schedules. These recurring events include—

- Shift changes.
- Battle update briefings to the commander.
- Battalion targeting meetings.
- Battalion reports.
- Battalion battle updates without the commander.
- Battalion commanders' collaborative sessions.
- Company CP OIC/NCOIC shift change briefing/collaborative sessions.

I-56. The company CP OIC/NCOIC must achieve battle rhythm for updating and viewing information and understand how to use it to affect operations. A well-established battle rhythm aids the commander and leaders with CP organization, information management and display, decision making, and fighting the battle. Battle rhythm demands careful planning and design. Many competing demands must be deconflicted. Even subordinate platoons and sections affect a company battle rhythm based on their needs and unit procedures.

I-57. Command discipline is required to enforce sleep cycles and create an environment where sound sleep can be achieved. Battle rhythm and associated planning is a proactive, not reactive, exercise.

## SLEEP PLANS

I-58. Units must develop detailed rest plans and enforce them. Leaders must rest to maintain their effectiveness; however, some leaders attempt to get involved in every aspect of planning and execution. An integral part of the planning process is determining when senior leader presence is required. It is just as important to identify when a leader's presence is not required. The planning process should include the following supporting techniques:

- Include a sleep plan in the METT-TC analysis.
- Ensure leaders have confidence in second and third echelons of leadership and their ability to make routine decisions.
- Instill trust and confidence in the officers, junior NCOs, and specialists by effective training and SOPs.
- Consider contingencies and establish criteria for waking leaders.
- Post sleep plans in CPs, platoon, and section areas; everyone needs to know the plan.
- Synchronize sleep plans with higher headquarters and subordinate platoons/sections.

## CONTINUOUS OPERATIONS

I-59. Continuous operations are combat operations continuing around the clock at a high pace, requiring Soldiers to fight without relief for extended periods. Opportunities for sleep are scattered throughout the day and night.

### SUSTAINED OPERATIONS

I-60. Sustained operations are operations conducted 24 hours a day with little or no opportunity for sleep. Sustained operations are when the same Soldiers or small units engage in combat operations with no opportunity for the unit to stand down and little time for Soldiers to sleep. Aviation units routinely conduct sustained operations to an extent even greater than the infantry due to aviation's maneuver, maneuver support and sustainment roles. To combat the negative effects of sustained operations, commanders and leaders must enforce fighter management principles on their organizations. A fighter management policy should be addressed in the unit SOP, and higher headquarters policy should correlate with company policy.

### DEGRADATION OF COMBAT CAPABILITY

I-61. As sustained operations continue, all Soldiers begin to show effects of general fatigue and lack of sleep. Unless counteracted, unit performance of combat tasks decline. Recent studies indicate performance is degraded by 25 percent for each 24-hour period without sleep. After 96 hours, performance can be expected to be near zero. Determination to endure must be supplemented by countering the adverse effects to slow the rate of decline. It becomes more difficult to perform assigned tasks to the required standard. Leaders need to recognize signs of serious sleep deprivation in their subordinates. Studies show performance in all duty positions does not degrade the same. Performance in a duty position where there is a heavy load of mental tasks, such as aviation duties (determining, calculating, thinking, decision-making), degrades faster than the performance in a position whose tasks are mainly physical (firing, running, lifting, digging).

I-62. In addition to the degradation caused by fear, fatigue, and loss of sleep, there is a significant loss of effectiveness caused by operation in MOPP 4. When units are using full CBRN protective equipment, judgment is degraded, communications are less effective, and information flow between units is reduced.

## FATIGUE IN FLYING OPERATIONS

I-63. Aviation operations are inherently dangerous. Commanders and leaders must be aware of increased dangers present in combat aviation continuous and sustained operations. Additionally, leaders must be able to recognize the symptoms of fatigue and how to deal with fatigued aircrews. Field Manual 3-04.301 provides an in-depth review of aeromedical factors associated with aviation.

## OPERATIONAL TEMPO AND BATTLE RHYTHM

I-64. The aviation company should be staffed for 24-hour operations; however, it also conducts cyclical missions. Standard operating procedures establish methods of ensuring the right personnel are available for either cyclical or 24-hour operations. Regardless of methods used, practice during exercises must determine strengths and weaknesses of headquarters personnel for CP operations and training for additional personnel who may be utilized to staff the CP during continuous or sustained operations. Such knowledge allows leaders to focus on critical areas and personnel requiring additional training.

I-65. In planning, the battalion staff must consider battle rhythm requirements of subordinate companies and platoons. Depending on the situation, the battalion may schedule missions allowing company or platoon rotations to maintain their battle rhythm.

### ABSENCE OF BATTLE RHYTHM

I-66. Without procedures establishing battle rhythm, leaders and units reach a point of diminished returns. This typically occurs between 72 and 96 hours of operations. As leader fatigue sets in, information flow, planning process, execution, and sustainment suffer—often greatly. Symptoms of diminished battle rhythm include the following:

- Leader fatigue.
- Leaders not fully aware of critical decision points (DPs).
- Leaders not available at critical DPs.
- Disjointed time lines between various levels of command.

## Presence of Battle Rhythm

I-67. Battle rhythm allows units and leaders to function at a sustained level of efficiency for extended periods. Effective battle rhythm permits an acceptable level of leadership at all times. It can focus leadership at critical points in the fight or during particular events. Procedures and processes facilitating efficient decision making and parallel planning are critical to achieving battle rhythm. Every component of battle rhythm contributes uniquely to sustained operations.

## Training

I-68. It is difficult, if not impossible, to establish battle rhythm while simultaneously conducting operations. Preplanning makes it happen. Planning, preparing, and training before deployment lays a solid foundation for viable battle rhythm during operations. Commanders must ensure all company personnel are trained on TOC set-up and operations, and they understand the team concept required to conduct CP operations during combat operations for extended periods. Additionally, company commanders must coordinate with battalion staffs during training to deconflict staff operations and requirements which are counterproductive to the company battle rhythm.

## Battle Rhythm Elements

I-69. Battle rhythm is a multifaceted concept including the following elements:

- Sleep/rest plans.
- Trained second- and third-tier leadership in CPs.
- Synchronized upward multiechelon timelines.
- Parallel planning.
- Established processes and SOPs.

## Command Post Personnel Depth

I-70. Established processes and SOPs relieve many antagonistic effects of extended operations. Standard operation procedures establishing and maintaining battle rhythm by facilitating routine decisions and operations are a step in the right direction. Soldiers trained to act appropriately in the absence of leaders or orders can relieve commanders and leaders of many of the time-consuming tasks robbing them of essential rest. Examples of areas NCOs and junior officers can accomplish for the commander include—

- Battle summaries and updates during a fight.
- Intelligence updates before, during, and after a battle.
- Sustainment updates before, during, and after a battle.
- Updates to the next higher commander.
- Shift change briefings.

## Challenges to Battle Rhythm

I-71. Challenges to battle rhythm include enlisted, NCO, and junior officer duties; and synchronization of planning, execution, rehearsal timelines, equipment maintenance, and sleep plans.

## Noncommissioned Officer and Junior Officer Responsibilities

I-72. At the company level, all personnel are critical to the success of the operation and provide valuable contributions. It is imperative commanders make certain each Soldier understands the importance of even the most menial tasks, such as CP security and TOC setup and teardown. Given the amount of personnel in a company, 24-hour operations and force protection requirements, commanders must utilize all personnel available to successfully accomplish CP operations and continuous missions. This becomes even more critical when a company has platoons or sections staged forward in support of BCTs. The improper use of personnel produces the following results:

- Key leaders become exhausted.

- CP operations trained personnel become exhausted.
- The initiative of trained subordinates is stifled, and the incentive to train is diminished.

I-73. The following techniques ensure proper use of personnel:

- Appropriate tasks are assigned to junior NCOs and specialists.
- Effective training and SOPs instill trust in officers and confidence in junior NCOs and specialists.
- Effective command guidance to the company that it is a team effort and all personnel available contribute to the effort.

## Continuous Operations and Timelines Synchronization

I-74. Timelines for the operation at hand must allow for not only the next operation, but also extended continuous operations. Synchronized, multiechelon timelines assist units in achieving battle rhythm. If units do not address critical events at least one level up and down, disruption results. An example of an unsynchronized timeline is a battalion rehearsal that conflicts with platoon precombat inspections or other events in their internal timeline. Lower echelon units seldom recover from a poor timeline directed by a higher headquarters, as well as platoons and sections at company level. Company commanders must coordinate with their battalion staffs on development of SOPs that include planning, rehearsal, and execution timelines one level below battalion to prevent these conflicts.

## Standing Operating Procedures Utilization

I-75. SOPs must be practiced and reviewed during professional development and sergeants' time. The existence of an SOP will not resolve troop-leading challenges unless the SOP is practiced often and internalized by unit members. Checklists are critical, as many leaders will often find themselves rushed, physically fatigued, distracted, and deprived of sleep. Checklists ensure each step is considered even when leaders are exhausted.

# SECTION IV — ACCIDENT/INCIDENT REPORTING

I-76. Persons involved in, or aware of, an accident must immediately report it to the commander or supervisor directly responsible for the operation, materiel, or person(s) involved. Table I-1 and Table I-2 identify reporting requirements for aviation and ground accidents/incidents (peacetime and combat). See AR 385-40, DA Pam 385-40, and AR 385-95 for in-depth instructions on Army accidents. Company commanders ensure unit SOPs cover accident/incident reporting, and it is fully addressed by the company aviation safety officer.

**Table I-1. Aviation accident notification and reporting requirements and suspenses**

| *Peacetime* | | | | *Combat ** | |
|---|---|---|---|---|---|
| *Accident Class* | *Notification Telephonic Worksheet* | *DA Form 2397* | *Reporting AAA report* | *Notification Telephonic Worksheet* | *Reporting AAA report* |
| *A* | Immediate–to USASC (telephonic notification–no hardcopy notification required (DSN 558-2660/2539/3410 or Commercial (334) 255-2660/2539/3410 | (CAI/IAI) 90 calendar days | Aircraft ground ACDTS only-30 calendar days | Same as peacetime to USASC or Safety Rep, forward | (Only when CDR determines DA Form 2397 investigation/ report not feasible) submit as soon as conditions, situation permit-Do not exceed 30 calendar days. |
| *B* | Immediate–to | (CAI/IAI) 90 | Aircraft ground | Same as | (Only when CDR |

| Peacetime | | | | Combat * | |
|---|---|---|---|---|---|
| Accident Class | Notification Telephonic Worksheet | DA Form 2397 | Reporting AAA report | Notification Telephonic Worksheet | Reporting AAA report |
| | USASC (telephonic notification–no hardcopy notification required (DSN 558-2660/2539/3410 or Commercial (334) 255-2660/2539/3410 | calendar days | ACDTS only 30 calendar days | peacetime to USASC or Safety Rep, forward | determines DA Form 2397 investigation/ report not feasible) submit as soon as conditions, situation permit- Do not exceed 30 calendar days. |
| ***C*** | Immediate–to USASC (telephonic notification–no hardcopy notification required (DSN 558-2660/2539/ 3410 or Commercial (334) 255-2660/2539/3410 | N/A | 30 calendar days | Same as peacetime to USASC or Safety Rep, forward | Same as peacetime |
| ***D*** | N/A (Unless SOF issue involved/ suspected) | N/A | 10 calendar days | Same as peacetime | Same as peacetime |
| ***E*** | N/A (Unless SOF issue involved/ suspected) | N/A | 10 calendar days | Same as peacetime | Same as peacetime |
| ***F*** | N/A (Unless SOF issue involved/ suspected) | N/A | 10 calendar days | Same as peacetime | Same as peacetime |
| ***Submis-sion Methods*** | Class A-C telephonic (immediate) Class D, E, F-IF SOF | Mail | Typed or hand printed AAA reports by mail/FAX or courier/ message/ format/elec-tronic submission, Include attachments as required | Same as peacetime | Same as peacetime |
| Notes * Only when the senior tactical commander determines that the situation, conditions, and/or time does not permit normal peacetime investigating and reporting. | | | | | |

### Table I-2. Ground accidents notification and reporting requirements and suspenses (3)

| *Peacetime* | | | | *Combat (2)* | |
|---|---|---|---|---|---|
| *Accident* | *Telephonic Notification* | *AGAR* | *DA Form 285* | *Telephonic Notification* | *AGAR only*<br>*By any means possible* |
| *Class* | *Worksheet* | | | *Worksheet* | *(Message, Electronic, FAX, Phone, Hand Carry, Mail)* |
| *On-Duty* | | | | | |
| *A* | Immediately (1) | Not required | IAI/CAI-90 days | Immediately (1) | As time permits (Not to exceed 30 days) |
| *B* | Immediately (1) | Not required | IAI/CAI-90 days | Immediately (1) | As time permits (Not to exceed 30 days) |
| *C* | Not required | Within 30 days | Not required | Not required | As time permits (Not to exceed 30 days) |
| *D* | Not required | Within 30 days | Not required | Not required | As time permits (Not to exceed 30 days) |
| *Off-Duty* | | | | | |
| *A* | Immediately (1) | Within 30 days | Not required | Immediately (1) | As time permits (Not to exceed 30 days) |
| *B* | Immediately (1) | Within 30 days | Not required | Immediately (1) | As time permits (Not to exceed 30 days) |
| *C* | Not required | Within 30 days | Not required | Not required | As time permits (Not to exceed 30 days) |
| *D* | Not required | Within 30 days | Not required | Not required | As time permits (Not to exceed 30 days) |

Notes:

(1) USASC must be notified IMMEDIATELY by phone at DSN 558-2660/2539-3410 or Commercial (334) 255-2660/2539/3410 or notify USASC Safety Rep forward (during combat).

(2) ONLY when the senior tactical commander determines that the situation, conditions, and/or time does not permit normal peacetime investigating and reporting.

(3) Army civilian injury only accidents should be reported on appropriate Department of Labor Forms according to this regulation.

# Glossary

## SECTION I — ACRONYMS AND ABBREVIATIONS

| | |
|---|---|
| **A2C2** | Army airspace command and control |
| **AA** | air ambulance/active Army |
| **AAA** | abbreviated aviation accidents |
| **AAR** | after-action review |
| **AATF** | air assault task force |
| **ABCA** | American, British, Canadian, and Australian |
| **ACDTS** | accidents |
| **ACR** | armor cavalry regiment |
| **AD** | air defense |
| **AE** | aeromedical evacuation |
| **AECT** | aeromedical evacuation control team |
| **AELT** | aeromedical evacuation liaison team |
| **AF** | Air Force |
| **AGAR** | abbreviated ground accident report |
| **AGPU** | aviation ground power units |
| **AHS** | Army Health System |
| **ALCT** | airlift control team |
| **ALSE** | aviation life support equipment |
| **amb** | ambulance |
| **AMEDD** | Army Medical Department |
| **AO** | area of operations |
| **AOC** | air operations center |
| **AOE** | Army of Excellence |
| **APOD** | aerial port(s) of debarkation |
| **AR** | Army regulation |
| **ASI** | additional skill identifier |
| **ARSOF** | Army special operations forces |
| **ARNG** | Army National Guard |
| **ASCC** | Army service component command |
| **ASF** | aeromedical staging facility |
| **ASMC** | area support medical company |
| **ASOS** | Army support to other Services |
| **ASTS** | aeromedical staging squadron |
| **ATM** | advanced trauma management |
| **ATS** | air traffic service |

| | |
|---|---|
| **attn** | attention |
| **aug** | augmentation |
| **avail** | available |
| **AVIM** | aviation intermediate maintenance |
| **AVN** | aviation |
| **AVUM** | aviation unit maintenance |
| **AXP** | ambulance exchange point |
| **BAE** | brigade aviation element |
| **BAS** | battalion aid station |
| **BCT** | brigade combat team |
| **bde** | brigade |
| **BH** | behavioral health |
| **BIFV** | Bradley infantry fighting vehicle |
| **bn** | battalion |
| **BOA** | basis of allocation |
| **BSA** | brigade support area |
| **BSB** | brigade support battalion |
| **BSMC** | brigade support medical company |
| **BSS** | brigade surgeon section |
| **BTU** | British thermal unit |
| **C2** | command and control |
| **CA** | civil affairs |
| **CAB** | combat aviation brigade |
| **CAI** | centralized accident investigation |
| **CASEVAC** | casualty evacuation |
| **CASF** | contingency aeromedical staging facility |
| **CBRN** | chemical, biological, radiological, and nuclear |
| **CBRNE** | chemical, biological, radiological, nuclear, and high yield explosives |
| **CCATT** | critical care air transport team |
| **CCIR** | commander's critical information requirements |
| **CCP** | casualty collecting point |
| **CE** | communications-electronic |
| **CJA** | Command Judge Advocate |
| **CJTF** | combined joint task force |
| **CLS** | combat lifesaver |
| **CLT** | casualty liaision team |
| **CM** | campaign module |
| **CMCC** | corps movement control center |
| **CMD** | command |
| **CMO** | civil-military operations |

| | |
|---|---|
| **CO** | company |
| **COA** | course(s) of action |
| **COCOM** | combatant command |
| **COMSEC** | communications security |
| **CONPLAN** | concept of operation plan |
| **CONUS** | continental United States |
| **COSC** | combat and operational stress control |
| **COSR** | combat and operational stress reaction |
| **CP** | command post |
| **CRTS** | casualty receiving and treatment ships |
| **CSH** | combat support hospital |
| **CSAR** | combat search and rescue |
| **CTC** | Combat Training Center |
| **CZ** | combat zone |
| **DA** | Department of the Army |
| **DCO** | defense coordination officer |
| **DCSCOMPT** | Deputy Chief of Staff, Comptroller, US Army |
| **DCSIM** | Deputy Chief of Staff for Information Managment, US Army |
| **DCSLOG** | Deputy Chief of Staff for Logistics, US Army |
| **DCSPER** | Deputy Chief of Staff for Personnel, US Army |
| **DCSSPO** | Deputy Chief of Staff for Support Operation, US Army |
| **DD** | Department of Defense |
| **DHS** | Department of Homeland Security |
| **div** | division |
| **DMOC** | division medical operations center |
| **DNBI** | disease and nonbattle injuries |
| **DOD** | Department of Defense |
| **DODD** | Department of Defense Directive |
| **DP** | decision point |
| **DS** | direct support |
| **DSN** | defense switched network |
| **DSS** | division surgeon section |
| **EAB** | echelons above brigade |
| **EAC** | echelons above corps |
| **EAD** | echelons above division |
| **ECO** | environmental compliance officer |
| **EEE** | early entry element(s) |
| **EEH** | early entry hospital |
| **EEM** | early entry module |
| **EMEDS** | expeditionary medical support |

| | |
|---|---|
| **EMT** | emergency medical treatment |
| **EPS** | essential personnel services |
| **EPW** | enemy prisoner(s) of war |
| **1SG** | first sergeant |
| **F** | Fahrenheit |
| **FARP** | forward arming and refueling point |
| **FAX** | facsimile |
| **FEMA** | Federal Emergency Management Agency |
| **FHP** | force health protection |
| **FM** | field manual/frequency modulated |
| **FMC** | field medical card |
| **FMI** | field manual interim |
| **FRAGO** | fragmentary order |
| **FRSS** | forward resuscitation |
| **FSB** | forward support battalion |
| **FSMC** | forward support medical company |
| **FSMT** | forward support MEDEVAC team |
| **FST** | forward surgical team |
| **G3** | Assistant Chief of Staff (Operations and Plans) |
| **GA** | ground ambubance |
| **GAMC** | ground ambulance medical company |
| **GCC** | geographic combatant commander |
| **GRREG** | graves registration |
| **GS** | general support |
| **GSAB** | general support aviation battalion |
| **GWS** | Geneva Convention for the Amelioration of the Condition of the Wounded and Sick in Armed Forces in Field |
| **HBCT** | heavy brigade combat team |
| **HD** | homeland defense |
| **HHD** | headquarters and headquarters detachment |
| **HMMWV** | high mobility multipurpose wheeled vehicle |
| **HN** | host nation |
| **hosp** | hospital |
| **HQ** | headquarters |
| **HR** | human resources |
| **HS** | homeland security |
| **HSS** | health service support |
| **IAI** | installation-level accident investigation |
| **IED** | improvised explosive device |
| **IG** | Inspector General |
| **IM** | information management |

| | |
|---|---|
| **IPB** | intelligence preparation of the battlefield |
| **IPMC** | intratheater patient movement center |
| **ISB** | intermediate staging base |
| **IT** | information technology |
| **JFC** | joint force commander |
| **JHSS** | joint health service support |
| **JMD** | joint manning document |
| **JMTB** | Joint Military Transportation Board |
| **JOA** | joint operations area |
| **JP** | joint publications |
| **JPMRC** | joint patient movement requirements center |
| **JTF** | joint task force |
| **LHA** | amphibious helicopter assault carrier |
| **LHD** | amphibious helicopter assault carrier dock |
| **LIN** | line item number |
| **LOC** | lines of communication |
| **LOTS** | logistics over-the-shore |
| **LPD** | amphibious transport dock |
| **LPH** | amphibious assault helicopter carrier–aviation |
| **LSD** | dock landing ship |
| **LST** | tank landing ship |
| **LZ** | landing zone |
| **MACA** | military assistance to civil authorities |
| **MASCAL** | mass casualty |
| **MASF** | mobile aeromedical staging facility |
| **MBA** | main battle area |
| **MC3T** | medical command, control, and communication telemedicine |
| **MC4** | medical communications for combat casualty care |
| **MCB** | movement control battalion |
| **MCO** | major combat operation |
| **MCP** | main command post |
| **med** | medical |
| **MEDBDE** | medical brigade |
| **MEDCOM** | medical command |
| **MEDEVAC** | medical evacuation |
| **MEDLOG** | medical logistics |
| **MEF** | Marine expeditionary force |
| **MES** | medical equipment sets |
| **METT-TC** | mission, enemy, terrain and weather, troops and support available, time available, and civil considerations |

| | |
|---|---|
| **MEV** | medical evacuation vehicle |
| **MH** | mental health |
| **MHS** | military health system |
| **MIA** | missing in action |
| **MIJI** | meaconing, intrusion, jamming, and interference |
| **MMB** | multifunctional medical battalion |
| **MOC** | medical operations cell |
| **MOPP** | mission-oriented protective posture |
| **MOS** | military occupational specialty |
| **MP** | military police |
| **MRE** | meals, ready-to-eat |
| **MRI** | medical reengineering initiative |
| **MRO** | medical regulating office |
| **MSB** | main support battalion |
| **MSC** | Military Sealift Command |
| **MTF** | medical treatment facility |
| **MTOE** | modification table of organization and equipment |
| **MWD** | military working dogs |
| **MWR** | morale, welfare, and recreation |
| **N/A** | not applicable |
| **NATO** | North Atlantic Treaty Organization |
| **NAVAIDS** | navigational aids |
| **NBC** | nuclear, biological, and chemical |
| **NCO** | noncommisioned officer |
| **NOE** | nap-of-the-earth |
| **NP** | neuropsychiatric |
| **NSHS** | National Strategy for Homeland Security |
| **NVG** | night vision goggles |
| **OCP** | operational command post |
| **OEG** | operational exposure guidance |
| **OEH** | occupational and environmental health |
| **OMF** | originating medical facility |
| **OPCON** | operational control |
| **OPLAN** | operation plan |
| **OPORD** | operation order |
| **OPS** | operations |
| **OPSEC** | operations security |
| **OPTEMPO** | operational tempo |
| **OR** | operating room |
| **OVE** | on vehicle equipment |

| | |
|---|---|
| **PASR** | personnel accountability and strength reporting |
| **PAD** | patient administrator |
| **PCC** | precombat checks |
| **PCI** | precombat inspections |
| **PDS** | personnel daily summary |
| **PERS** | personnel |
| **PIM** | personnel information management |
| **PIR** | priority intelligence requipment |
| **PMCS** | preventive maintenance checks and services |
| **PMI** | patient movement items |
| **PMM** | preventive medicine measures |
| **PMRC** | patient movement requestment center |
| **POI** | point of injury |
| **POL** | petroleum, oils, and lubricants |
| **PR** | personnel recovery |
| **prep** | preparation |
| **PRM** | personnel readiness management |
| **PT** | platoon |
| **PSYOP** | psychological operations |
| **PZ** | pickup zone |
| **PVNTMED** | preventive medicine |
| **QSTAG** | Quadripartite Standardization Agreement |
| **R5** | reception, replacement, return to duty, rest and recuperation, redeployment |
| **R&R** | rest and recuperation |
| **RC** | reserve component |
| **rep** | representative |
| **RPG** | rocket propelled grenade |
| **RSTA** | reconnaissance and target acquisition |
| **RSOI** | reception, staging, onward movement, and integration |
| **RTD** | return to duty |
| **S&RO** | stability and reconstruction operations |
| **S1** | Adjutant (US Army) |
| **S2** | Intelligence Officer (US Army) |
| **S3** | Operations and Training Officer (US Army) |
| **S4** | Supply Officer (US Army) |
| **S6** | Information Management (US Army) |
| **SBCT** | Stryker Bridage Combat Team |
| **SEC** | section |
| **SF** | special forces |
| **SIMLM** | single integrated medical logistics manager |

| | |
|---|---|
| **SINCGARS** | Single Channel Ground/Airborne Radio System |
| **SJA** | Staff Judge Advocate |
| **SOA** | special operations aviation |
| **SOF** | special operations forces |
| **SOP** | standing operating procedures |
| **SPT** | support |
| **SRC** | standard requirements code |
| **STANAG** | standardization agreement |
| **SURG** | surgical |
| **SUST** | sustainment |
| **TAA** | tactical assembly area |
| **TACC** | tanker airlift control center |
| **TAC CP** | tactical command post |
| **TAC SOP** | tactical standing operating order |
| **TAES** | Theater Aeromedical Evacuation System |
| **TC** | training circular |
| **tm** | team |
| **TMIP** | Theater Medical Information Program |
| **tmt** | treatment |
| **tng** | training |
| **TO** | theater operations |
| **TOC** | tactical operations center |
| **TOE** | tables of organization and equipment |
| **TPFDL** | time-phased force deployment list |
| **TPMC** | theater patient movement center |
| **TPMRC** | theater patient movement requirements center |
| **TSOC** | Theater Special Operations Command |
| **TSOP** | tactical standing operating procedures |
| **UO** | urban operations |
| **US** | United States |
| **USAF** | United States Air Force |
| **USAMEDDC&S** | United States Army Medical Department Center and School |
| **USASC** | United States Army Safety Center |
| **USN** | United States Navy |
| **USNS** | United States Navy Ship |
| **USNORTHCOM** | United States Northern Command |
| **USTRANSCOM** | United States Transportation Command |
| **vet** | veterinary |
| **WARNO** | warning order |
| **WIA** | wounded in action |

| | |
|---|---|
| **WMD** | weapons of mass destruction |
| **XO** | executive officer |
| **ZULU** | universal time |

# References

## SOURCES USED

These are the sources quoted or paraphrased in this publication.


### NATO STANAGs

2040. *Stretchers, Bearing Brackets, and Attachment Supports*, 23 September 1982 (Latest Amendment, 6 October 2004)

2087. *Medical Employment of Air Transport in the Forward Area*, 29 June 1983 (Latest Amendment, 10 September 1999)

2128. *Medical and Dental Supply Procedures*, 15 November 1982 (Latest Amendment, 21 November 1991)

2132. *Documentation Relative to Medical Evacuation, Treatment, and Cause of Death of Patients*, 7 August 1974 (Latest Amendment, 15 September 1986)

2454. *Road Movements and Movement Control*, 27 January 2005

2931. *Orders for the Camouflage of the Red Cross and Red Crescent on Land in Tactical Operations*, 18 October 1984 (Latest Amendment, 3 April 1998)

3204. *Aeromedical Evacuation*, 26 September 1973 (Latest Amendment, 10 September 2001)

### ABCA QSTAGs

436. *Minimum Labeling Requirements for Medical Materiel* (Amendment 1), 8 October 1980

470. *Documentation Relative to Medical Evacuation, Treatment, and Cause of Death of Patients*, 23 February 1979 (Latest Amendment, 14 August 1989)

### PRESIDENTIAL DIRECTIVE

Homeland Security Presidential Directive-5, *Management of Domestic Incidents*, February 28, 2003

### DODD

DoD Directive 3025.15, *Military Assistance to Civil Authorities*, February 18, 1997

DoD Directive 5200.31E, *DoD Military Working Dog (MWD) Program*, March 29, 2006

### JOINT AND MULTISERVICE PUBLICATIONS

Joint Pub 1-02, *DOD Dictionary of Military and Associated Terms*, 23 March 1994

Joint Pub 4-02, *Doctrine for Health Service Support in Joint Operations*, 26 April 1995

Joint Pub 5-0, *Joint Operation Planning*, 26 December 2006

Joint Pub 5-00.1, *Joint Doctrine for Campaign Planning*, 25 January 2002

Joint Pub 5-00.2, *Joint Task Force (JTF) Planning Guidance and Procedures*, 13 January 1999

### ARMY PUBLICATIONS

AR 40-5, *Preventive Medicine*, 22 July 2005

AR 71-32, Force Development and Documentation-Consolidated Policies, 3 March 1997

AR 200-1, *Environmental Protection and Enhancement*, 21 February 1997

AR 385-40, *Accident Reporting and Records*, 1 November 1994

AR 385-64, *U.S. Army Explosives Safety Program*, 1 February 2000

AR 385-95, *Army Aviation Accident Prevention*, 12 December 1999

AR 600-55, *The Army Driver and Operator Standardization Program (Selection, Training, Testing, and Licensing)*, 31 December 1993

DA Pam 385-40, *Army Accident Investigating Reporting*, 1 November 1994

DA Pam 385-40, *Ammunition and Explosives Safety Standards*, 15 December 1999

FM 1-202, *Environmental Flight*, 23 February 1983

FM 1-564, *Shipboard Operations*, 29 June 1007

FM 3-0, *Operations*, 14 June 2001

FM 3-04.301, *Aeromedical Training for Flight Personnel*, 29 September 2000

FM 3-05, *Army Special Operations Forces*, 20 September 2006

FM 3-06.11, *Combined Arms Operations in Urban Terrain*, 28 February 2002

FM 3-07, *Stability Operations and Support Operations*, 20 February 2003

FM 3-11.5, *Multiservice Tactics, Techniques, and Procedures for Chemical, Biological, Radiological, and Nuclear Decontamination*, 4 April 2006

FM 3-34.2, *Combined-Arms Breaching Operations (Including Changes 1—3)*, 31 August 2000

FM 3-97.6, *Mountain Operations*, 28 November 2000

FM 3-100.4, *Environmental Considerations in Military Operations (Including Change 1)*, 15 June 2000

FM 4-01.45, *Multi-Service Tactics, Techniques, and Procedures for Convoy Operations*, 24 March 2005

FM 4-02, *Force Health Protection in a Global Environment*, 13 February 2003

FM 4-02.1, *Combat Health Logistics*, 28 September 2001

FM 4-02.4, *Medical Platoon Leader's Handbook—Tactics, Techniques, and Procedures (Including Change 1)*, 24 August 2001

FM 4-02.6, *The Medical Company—Tactics, Techniques, and Procedures (Including Change 1)*, 4 September 2004

FM 4-02.7, *Health Service Support in a Nuclear, Biological, and Chemical Environment—Tactics, Techniques, and Procedures*, 10 January 2002

FM 4-02.12, *Health Service Support in Corps and Echelons above Corps*, 2 February 2004

FM 4-02.17, *Preventive Medicine Services*, 28 August 2000

FM 4-02.18, *Veterinary Services—Tactics, Techniques, and Procedures*, 30 December 2004

FM 4-02.21, *Division and Brigade Surgeon's Handbook (Digitized)—Tactics, Techniques, and Procedures*, 15 November 2000

FM 4-02.24, *Area Support Medical Battalion—Tactics, Techniques, and Procedures*, 28 August 2000

FM 4-02.25, *Employment of Forward Surgical Teams—Tactics, Techniques, and Procedures*, 28 March 2003

FM 4-25.12, *Unit Field Sanitation Team*, 25 January 2002

FM 4-02.43, *Force Health Protection Support for Army Special Operations Forces*, 28 November 2006

FM 5-0, *Army Planning and Orders Production*, 20 January 2005

FM 8-42, *Combat Health Support in Stability Operations and Support Operations*, 27 October 1997

FM 8-55, *Planning for Health Service Support*, 9 September 1994

FM 8-250, *Preventive Medicine Specialist (Including Change 1)*, 27 January 1986
FM 10-67.1, *Concepts and Equipment of Petroleum Operations*, 2 April 1998
FM 21-10, *Field Hygiene and Sanitation*, 21 June 2000
FM 31-70, *Basic Cold Weather Manual*, 12 April 1968
FM 55-30, *Army Motor Transport Units and Operations (Including Change 1)*, 15 September 1999
FM 90-3, *Desert Operations*, 24 August 1993
FM 90-13, *River Crossing Operations*, 26 January 1998
ST 4-02.121, *Multifunctional Medical Battalion*, October 2005
TC 1-400, *Brigade Aviation Element Handbook*, 27 April 2006
TB MED 507, *Heat Stress Control and Heat Casualty Management*, 7 March 2003
CALL Newsletter 99-9, *Integrating Military Environmental Protection*, 1999
FMI 4-90.1, *Heavy Brigade Combat Team Logistics*, 15 March 2005
FMI 5-0.1, *The Operations Process*, 31 March 2006

## DOCUMENTS NEEDED

These documents must be available to the intended users of the publication.

DA Form 285 Series, *U.S. Army Accident Report*
DA Form 1156, *Casualty Feeder Card*
DA Form 2397 Series, *Technical Report of U.S. Army Aircraft Accidents*
DA Form 2404, *Equipment Inspection and Maintenance Worksheet*
DD Form 601, *Patient Evacuation Manifest*
DD Form 1380, *U.S. Field Medical Card*

# Index

References are to paragraph numbers unless otherwise stated.

## S



## T

*TC 21-24

TRAINING CIRCULAR
No. 21-24

HEADQUARTERS
DEPARTMENT OF THE ARMY
WASHINGTON, DC, 10 September 1997

# RAPPELLING

## CONTENTS



**DISTRIBUTION RESTRICTION:** Approved for public release; distribution is unlimited.

*This publication supersedes TC 21-24, 24 September 1991.

# PREFACE

This circular provides basic rappelling techniques to soldiers and leaders for the conduct of rappelling operations. It serves as the primary reference for both resident and nonresident instruction presented to cadets, officer candidates, and both commissioned and noncommissioned officers. This circular also discusses several advanced techniques dealing with infiltration and exfiltration.

Safety is always the most important consideration when conducting training. This circular provides guidelines for commanders to conduct operations safely. Commanders at all levels must analyze the complete training event to determine the degree of risk involved to men and or equipment (Appendix A). After determining the risks, risk reduction options or controls will be integrated into the training activity. These options or controls may range from a safety briefing or providing additional safety resources, to selecting other means of accomplishing the mission.

Rappelling, FRIES, and SPIES operations are inherently dangerous; therefore, the safety notes and considerations presented in this circular are the minimum acceptable standards.

The proponent of this publication (except for Chapter 5) is the US Army Infantry School. Submit changes for improving this publication on DA Form 2028 and forward it to the Commandant, US Army Infantry School, ATTN: ATSH-RB, Fort Benning, Georgia 31905-5593 (e-mail address: IKNERH@BENNING-EMH2.ARMY.MIL).

The proponent for Chapter 5, Fast-Rope Insertion/Extraction System, is the US Army Special Operations Command. Submit changes or comments for Chapter 5 on DA Form 2028 and forward it to the Commander, US Army Special Operations Command, ATTN: AOOP-TRS, Fort Bragg, North Carolina 28307-5000.

Unless this publication states otherwise, masculine nouns and pronouns do not refer exclusively to men.

CHAPTER 1

# TOWER RAPPELLING

*The introduction to rappelling is taught on a static tower (Figure 1-1). Using the building block approach to training, soldiers systematically progress to more demanding platforms, including taller static towers, and finally to helicopters. The static tower used may vary in size and height from 34 to 90 feet. The concept of learning the basic rappelling techniques before helicopter operations does not vary. (Units may conduct ground training and wall-side tower rappelling without a rappel master. In this case, the unit commander would be responsible for the safety of the rappelling training.)*

**Figure 1-1. Rappel tower.**

**Note:** Each static rappel tower should resemble the structure or aircraft being trained to. The Army's standard design for building static rappel towers can be obtained through the Plans and Support Branch of the post Engineers, Building 75, Fort Benning, Georgia 31905 (commercial 706 545-3307 or DSN 835-3307). Strict adherence to the design specifications is required for safety considerations. Rappel towers should be inspected annually by post safety or post engineers.

## Section I. PERSONNEL

This section discusses the personnel involved in training rappelling and their duties and responsibilities.

### 1-1. RAPPEL MASTER

The proponent for accreditation, evaluation, and information for Army rappel master courses is Directorate of Operations and Training, US Army Infantry School, ATTN: ATSH-OTT-F, Fort Benning, Georgia 31905-5593.

a. **Duties and Responsibilities.** The rappel master is responsible for the safety of rappellers. He ensures that all equipment (installation, unit, and personal property) is serviceable. He personally supervises the rappelling operation.

b. **Qualification.** Rappel master qualification is awarded only after the successful completion of a TRADOC accredited rappel master course, which includes the following subjects:

- Duties and responsibilities of a rappel master.
- Safety SOP, regulations, and references.
- Construction of a deployment bag.
- Conduct of an equipment rappel off the rappel tower.
- Conduct of a lock-in.
- Talking a rappeller through completion of a rappel.
- Conduct of ground training.
- Inspecting for proper hook-ups.
- Inspecting and maintaining equipment.
- Inspecting and maintaining snaplinks.
- Inspecting and maintaining rappelling gloves.
- Inspecting and maintaining rappel ropes.
- Identifying the rappel capabilities of aircraft used.
- Controlling rappels from UH-1H or UH-60 aircraft.
- Tying knots (square, bowline, half-hitch, Prusik), safety lines, and rappel seats (Swiss seat, Australian seat).
- Inspecting a rappel seat.
- Aircraft rigging for rappelling operations.
- Aircraft command and control.

c. **Proficiency Maintenance.** To remain current, rappel masters must execute their duties in a tactical or training exercise once every six months. If rappel masters do not execute their duties once every six months, they must take a refresher class taught by a current rappel master. The refresher class includes the subjects listed in the rappel master qualifications paragraph (1-1b).

## 1-2. RAPPEL SAFETY OFFICER

The RSO is an air assault-qualified or ranger-qualified SFC or above who serves as the OIC for the conduct of rappel operations.

a. The RSO is responsible overall for the safety of all rappellers and ensures that all safety precautions are followed.

b. The RSO briefs VIPs, visitors, and inspecting authorities on training, safety requirements, and layout of training areas.

## 1-3. RAPPEL LANE NCO

a. **Duties and Responsibilities.** Safety is the rappel lane NCO's number one priority. The rappel lane NCO—

- Ensures proper safety procedures are followed.
- Ensures proper hookup once directed to a rope station.
- Issues commands and maintains eye contact with the rappeller at all times.

b. **Qualifications.** The rappel lane NCO holds the rank of corporal or above, is ranger- or air assault-qualified, and is selected by the commander. Each tower rappel lane must have a qualified rappel lane NCO.

c. **Training.** The rappel lane NCO must also train on the following subjects:

- Responsibilities and safety requirements.
- Inspection and maintenance of equipment.
- Identification of satisfactory anchor points.
- Identification of safe and unsafe hookups.
- Establishment of a rappel point.
- Inspection of a rappel seat.
- Coaching techniques.
- Rappelling procedures.
- Emergency procedures.
- Belay control procedures.

d. **Participation.** The rappel lane NCO must participate in at least seven rappel operations: three as a rappeller; two as an assistant rappel lane NCO; and two performing the duties of a rappel lane NCO under the supervision of a qualified rappel master.

e. **Proficiency Maintenance.** If a rappel lane NCO has not conducted his duties within the last six months, he must complete the training listed in paragraph 1-3c under the supervision of a current rappel master.

## 1-4. RAPPELLER

Rappel qualification requirements apply to the individual rappeller. Participants in tower rappel training must complete the following requirements under the supervision of a rappel master. The unit commander ensures that personnel successfully complete these requirements before beginning aircraft rappel training.

- Identify all rappelling equipment.
- Demonstrate the construction and attachment of the rappel seat and the rappel rope to the seat.
- Identify unsafe attachments, equipment, rope connections, and seat construction.
- Define terms used in rappelling operations.
- Identify knots used in rappel operations.
- Understand and demonstrate rappel commands.
- Demonstrate rappelling positions.
- Demonstrate belaying procedures.
- Exhibit satisfactory performance from a rappel tower of at least 34 feet in height (two rappels with equipment and weapon, two without equipment and weapon). Two rappels are conducted from the free side of the tower (no wall).
- Demonstrate the ability to lock in.

### 1-5. BELAYER

Belay requirements are a subtask of basic rappel requirements. Soldiers must know how to belay before conducting rappelling training. The belayer—

- Assumes a position at the base of the lane about one pace away from the tower area.
- Ensures that the rappel ropes are even with the ground during tower rappels.
- Loosely holds the rappel rope with both hands so as not to interfere with the rappeller but still be able to stop the rappeller should he fall.
- Immediately stops the rappeller by pulling downward on the rappel ropes if the rappeller shouts "falling" or loses control of his brake hand during descent.
- Does not wear gloves to ensure a firm grip on the rappelling rope.
- Watches the rappeller at all times, and maintains constant voice or visual contact.
- Wears a helmet to prevent injuries from falling debris.

### 1-6. BELAY SAFETY

The belay safety must be ranger- or air assault-qualified. He ensures belay personnel are performing their duties properly. Rappel training requires one belay safety for each two rappel stations.

## Section II. PREOPERATIONS BRIEFINGS AND SAFETY PROCEDURES

The rappel master ensures participants have a basic understanding of requirements and safety procedures before conducting training.

## 1-7. SAFETY

The following personnel and equipment must be present during static tower training.

- Two military rappel ropes for each rappel station.
- One safety officer.
- One rappel master for each rappel site.
- One rappel lane NCO per rappel station.
- One medic with medical kit and backboard.
- One safety or medical evacuation vehicle with driver.
- One belayer for each rope station. Rappellers alternate stations.
- One belay safety for each two rappel stations (four ropes).

## 1-8. SAFETY BRIEFING

As in all training, a safety briefing precedes rappel operations. The rappel master briefs all personnel on safety to include the following instructions.

a. Each rappeller ensures loose clothing and equipment are secured.

b. Rappel seats are tied by the soldier and inspected by the rappel master before climbing the tower. Rappel seats are removed upon completion of every rappel, retied, and reinspected by a qualified rappel master or rappel lane NCO before subsequent rappels.

c. Rappellers climb the tower only when directed by the rappel master or rappel lane NCO.

d. Rappellers stay in the center of the tower until instructed to move to a rappel point.

e. No more than three personnel are behind each rappelling point.

f. If using a troop ladder, only three soldiers are on the ladder at one time. Soldiers do not climb the ladder until told to do so by a rappel master.

g. All rappel masters, rappel lane NCOs, instructors, and anyone else standing near the edge of the top of the tower must wear a restraining strap or safety rope. The strap or rope must be attached to an anchor point.

h. No one should lean or sit on the railing or banisters of the tower.

i. No one is allowed within 3 feet of the edge of the tower without being secured.

j. When attaching the rappel rope to the snaplink, rappellers pull the slack towards the anchor point. The rappel master or rappel lane NCO physically check each hookup.

k. All personnel weighing more than 200 pounds will conduct a standard hookup rappel to determine if they require a friction hookup. A friction hookup is created by placing an additional two ropes in the gate of the snaplink (for a total of six ropes in the snaplink).

l. Combat equipment is positioned on the rappeller so that it does not interfere with the brake hand. The weapon must be slung diagonally across the back with the muzzle pointing down, and on the opposite side of the brake hand.

m. Heavy duty gloves are required for all rappel training.

n. Kevlar helmets with chin straps fastened are worn during tower rappel training.

o. While on the tower, the rappeller maintains eye contact with the rappel master or rappel lane NCO and receives all commands from them.

p. The rappeller ensures that he has a belayer on his rope.

q. The belayer does not wear gloves and keeps both hands on the rope at all times. He also faces the rappeller at all times.

r. All tower rappelling is performed with a double strand of rope.

s. No running is allowed on the tower.

t. No smoking or eating is allowed near the tower.

u. All participants who are unable to rappel, lack confidence, or refuse to rappel are reported to the rappel master or OIC. These participants are immediately removed from the training area.

v. The RSO and rappel master must be aware of overconfidence and carelessness of some rappellers. The rappel master ensures all personnel are tower-qualified before beginning aircraft rappel training.

## 1-9. TOWER SAFETY AND PREPARATION

The rappel master is in charge of the tower. He conducts a visual and physical inspection of every item of equipment to include the structural lumber and timber, the ladder, the platform floor, and all anchor points.

a. The static tower will not be used during thunderstorms or excessively high winds. If ice is present, or if the platform is slick from rain, rappelling will be delayed until conditions are safe.

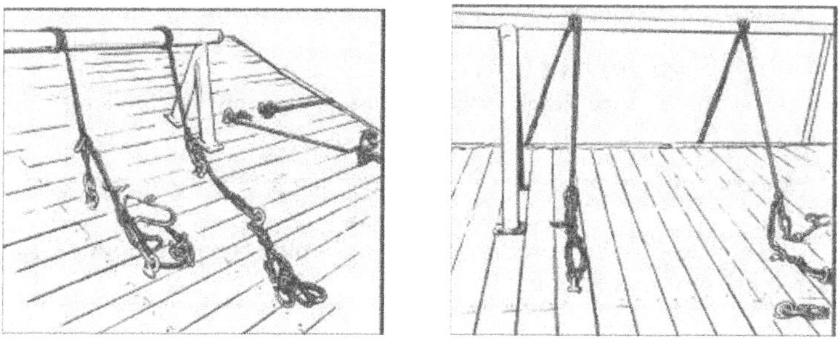

A B

**Figure 1-2. Example of rappel tower anchor points.**

b. All rope stations are rigged with two anchor points (Figure 1-2A, B, C). The first anchor point is a middle-of-the-rope knot, and the second is an end-of-the-rope anchor knot. The rappel master removes all the slack between the knots to create equal tension on the anchor points. He ensures that no less than 10 feet of rope is on the ground during static rappelling.

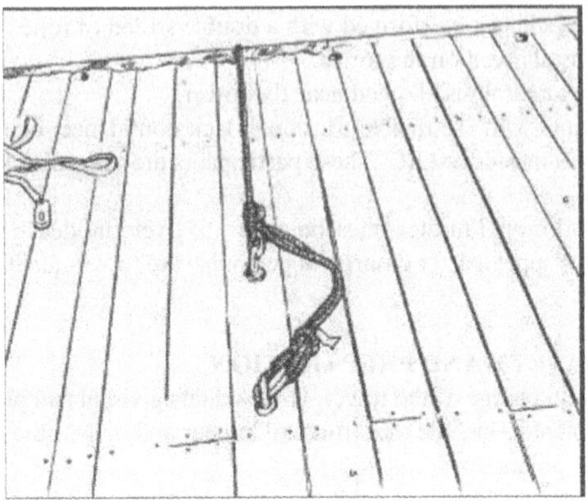

C

**Figure 1-2. Example of rappel tower anchor points (continued).**

## 1-10. RAPPELLER PREPARATION

Before conducting a rappel, each rappeller must prepare his individual clothing and equipment.

> **WARNING**
>
> **Failure to properly prepare rappellers could result in bodily injury or damage to equipment.**

a. Secure shirt tails, loose clothing, equipment, straps, and long hair.

b. Wear a helmet during rappelling. Properly fasten all straps and ensure the helmet is in serviceable condition.

c. Wear heavy leather workman's gloves.

d. Wear identification tags.

e. When rappelling with equipment, LBE or LBV should be unfastened in the front or fastened loosely behind the back of the rappeller. The rucksack should be worn high and tight on the back of the rappeller to allow the brake hand to reach the small of the back. Rucksack adjustment straps will be tied across the chest or tucked away.

f. Sling the weapon diagonally across the back with the muzzle down. Ensure the muzzle is on the guide hand side and the stock is towards the brake hand.

**Note:** Soldiers rappelling with equipment in excess of 50 pounds may want to consider using a friction hookup.

## Section III. RAPPELLING PROCEDURES

This section discusses procedures used in tower rappelling.

### 1-11. SEAT-HIP RAPPEL

When using the seat-hip rappel, friction is created by a snaplink that is inserted in a sling rope seat and fastened to the rappeller. This method provides a faster and more controlled descent than other methods. Wear gloves to prevent rope burns. An alternate technique is to insert a second snaplink into the first snaplink (attached to rappel seat) and run the rope through the second snaplink. This allows easier disengagement from the rappel rope without running the entire rope through the first snaplink. To disengage from the rappel rope using the alternate technique, release the tension from the rope by opening the gate of the first snaplink and removing the second snaplink (with the rope attached).

a. The rappel seat is constructed as follows (Figure 1-3A through 1-3T):

(1) Place the midpoint (center) of the length of the sling rope on the hip opposite the brake hand (the brake hand is the strong hand) (Figure 1-3A, B, C).

(2) Bring the sling rope around the waist above the hip bone. Tie a double overhand knot over the navel (Figure 1-3D, E, F, G, H).

(3) Let the two free ends of the sling rope fall to the ground in front (Figure 1-3I).

(4) Bring the two free ends of the sling rope down between the legs and up over the buttocks. Ensure that the two free ends do not cross (Figure 1-3J).

(5) Pass the ends of the ropes over the rope that is tied around the waist at the two points above the center of the two rear seat pockets (Figure 1-3K).

(6) Grab the free end of the rope that is on the left side of the body with the left hand, and the free end of the rope that is on the right side of the body with the right hand.

(7) Squat down and simultaneously pull on both running ends of the ropes and stand up. This will tighten the seat.

(8) Take the two running ends of the rope down and back over the waist rope from the inside. Bring the running ends back under the ropes that are going across the buttocks (Figure 1-3L).

(9) Tie the two running ends with a square knot and two overhand knots on the hip opposite the brake hand (Figure 1-3M, N).

(10) Place any excess rope in the trouser pocket near the square knot (Figure 1-3O, P).

(11) With the gate down and the hooked end of the snaplink against the navel, place the end of the snaplink through the single rope that is around the waist and the two ropes forming the double-hand knot (Figure 1-3Q).

(12) Rotate the snaplink a half turn so that the gate is facing up and will open away from the body (Figure 1-3R, S, T).

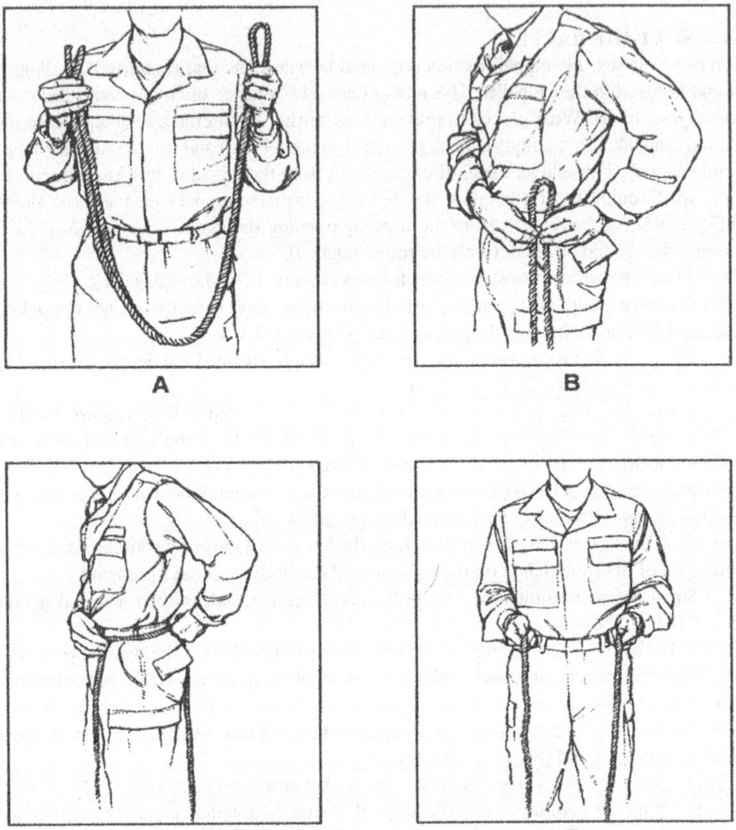


**Figure 1-3. Construction of rappel seat.**

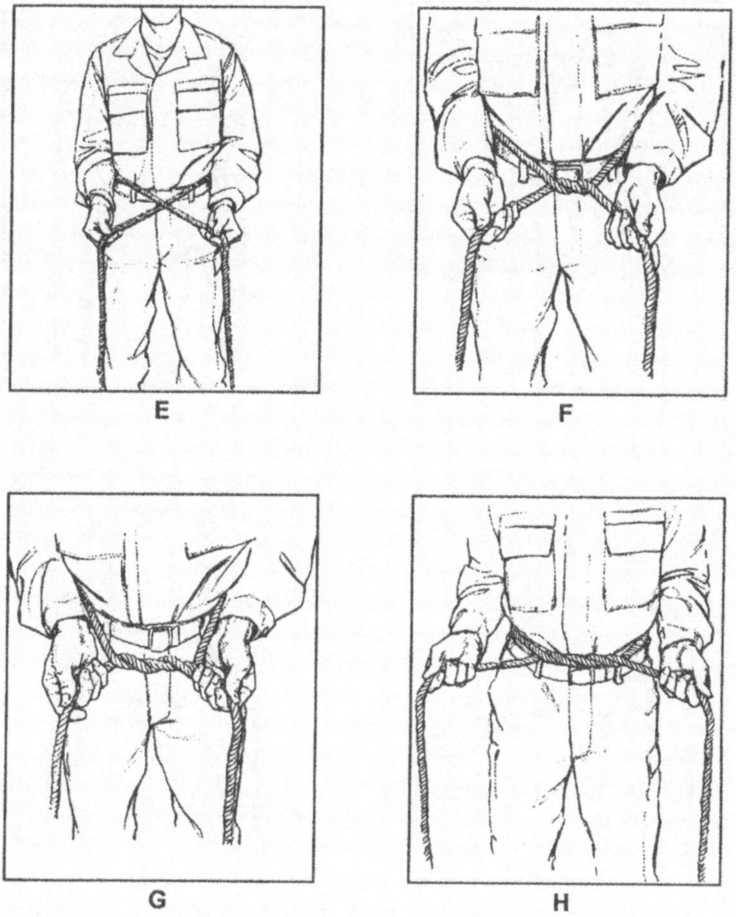


**Figure 1-3. Construction of rappel seat.**

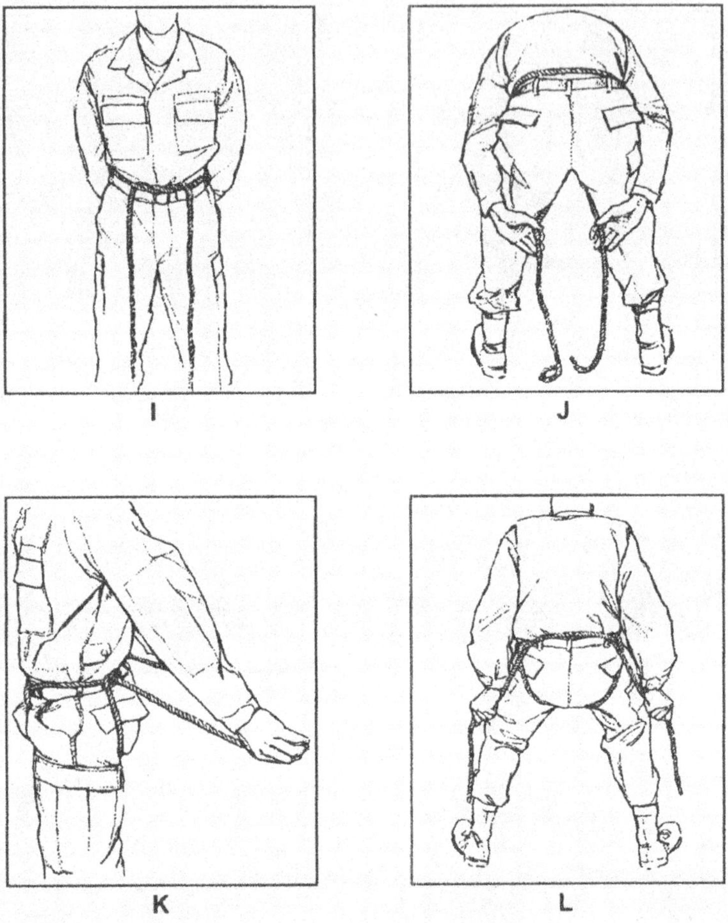


Figure 1-3. Construction of rappel seat.

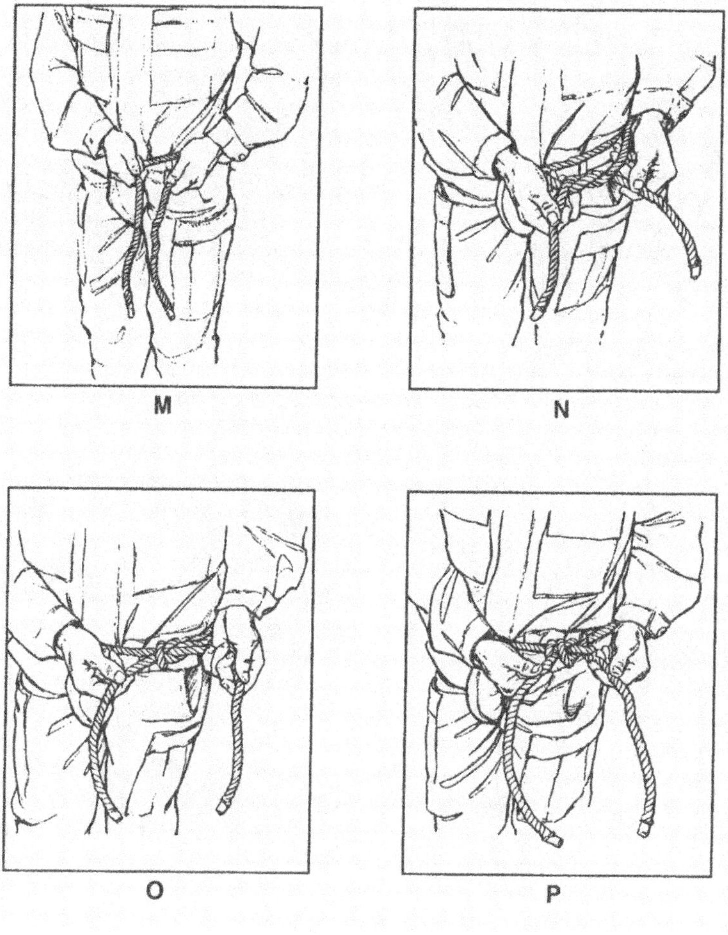


**Figure 1-3. Construction of rappel seat.**

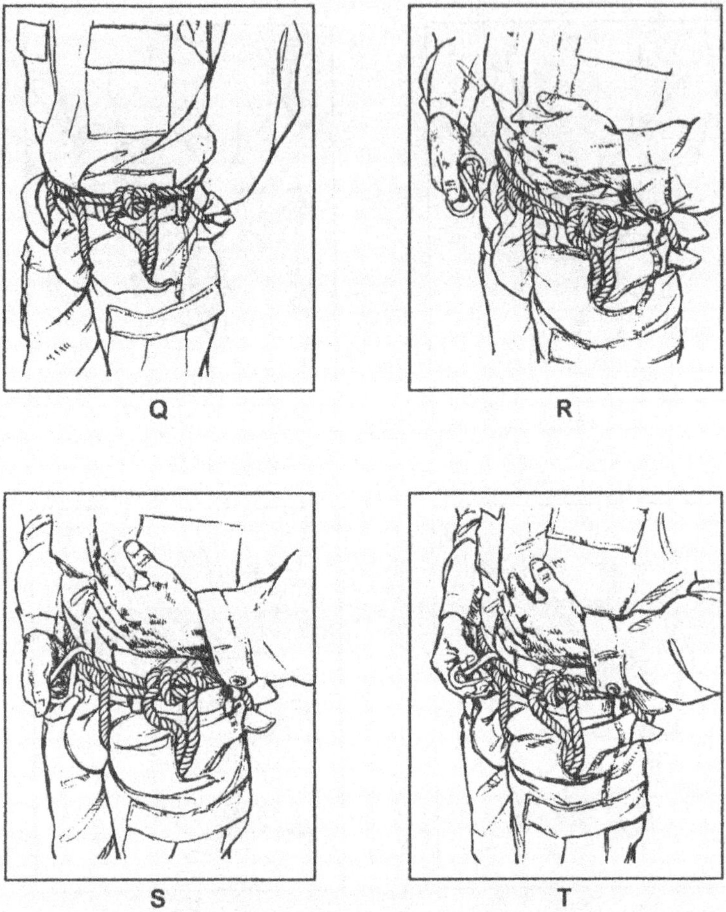


**Figure 1-3. Construction of rappel seat.**

b. To hook up using the seat-hip method, perform the following:

(1) Place the square knot with two overhand knots towards the anchor point for all seat-hip rappels.

(2) Grasp the two ropes with both hands and drop them through the gate of the snaplink. (At this point, two ropes should be running through the snaplink.)

(3) Using the hand closest to the anchor point, pull the slack towards the anchor point. Rotate the slack under and then over the top of the snaplink.

(4) Drop the two ropes a second time through the gate of the snaplink. (At this point, four ropes should be running through the snaplink.)

(5) Place the guide hand on the rope between the anchor point and the snaplink (palm facing up).

(6) Place the brake hand around the running end of the rope (palm facing down). Place the brake hand with the rope in the small of the back.

**1-12. AUSTRALIAN RAPPEL**

To hook up for the Australian rappel (Figure 1-4), perform the following:

a. Hook up to the Australian seat.

(1) Grasp the 9-foot sling rope at the midpoint (center) of its length, and double the rope.

(2) Place the doubled sling rope around the back and waist. Ensure that the rope is above the hipbone, but below the ribs.

(3) Tie a *square knot with two overhand knots* over the navel.

(4) Place any excess rope in the trouser pocket nearest the excess.

b. When the Australian seat is donned, face away from the anchor point and to the side of the rappel rope. (Stand to the same side of the rope as the brake hand. Determine left and right of the rope while facing the anchor point.) Place a snaplink onto the seat with the gate facing up, hinge closest to the body. Place it on the hip corresponding to the brake hand. The rappel master grasps the rappel rope and lays the rope into the snaplink. He then places one hand between the snaplink and the anchor point and draws slack toward the anchor point. He rotates the slack down, under, and over the rope and into the snaplink. The rappel master then slides the snaplink directly to the rappellers back.

c. Serve the running end of the rappel rope with the brake hand and prepare to rappel. During descent, brake by drawing the rappel ropes diagonally across the chest with the ropes running from near the waist to the pocket of the opposite shoulder.

**Figure 1-4. Australian rappel.**

## 1-13. CLIMBING PROCEDURES

Before climbing the ladder, the rappel master, safety OIC, or rappel lane NCO checks each rappeller's equipment.

a. The rappeller kicks the sand off his boots before climbing.

b. The rappeller grasps the outside of the ladder while climbing, not the rungs (when possible). If stairs are built for the tower, the rappeller grasps the railings as appropriate.

c. Just before climbing up the ladder, the rappeller sounds off, "(name) climbing," and then begins climbing up the ladder. Once at the top and clear of the ladder, the rappeller sounds off, "(name) clear."

d. Once off the ladder, the rappeller waits until the rappel master or lane NCO directs him to proceed to a rope station. At this time, the next rappeller in line may start to climb the ladder.

## 1-14. TOWER PROCEDURES

After the rappellers climb the tower, the following procedures are adhered to:

a. Once directed to a rope station, the rappel master or lane NCO ensures proper hookup for rappelling.

b. At this time, the rappeller sounds off with "on rappel" and the belayer sounds off with "on belay."

c. While maintaining his brake, the rappeller (on command from the rappel master or lane NCO) steps over the safety rail and faces the anchor point.

d. At this point, the rappel lane NCO sounds off with the following verbal commands and arm-and-hand signals.

(1) *Get Ready*—Rappel master or lane NCO extends both arms to the front with fists clenched and thumbs pointing upward (Figure 1-5). This alerts the rappeller. Each rappeller then looks over his brake hand shoulder to check for the belay man. The rappeller then looks at the rappel master. The rappel master or lane NCO makes his second check of the hookup, rappel seat, snaplink, and equipment.

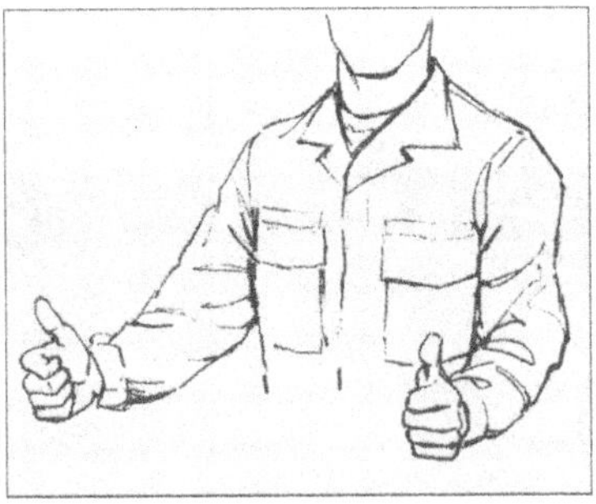

**Figure 1-5. Example of arm-and-hand signal GET READY.**

(2) *Position*—The lane NCO extends both arms to the front, elbows locked, forearms pointed downward, and fingers extended. He makes a circular motion with both forearms rotating in opposite directions (Figure 1-6). With the brake hand in the small of the back, the rappeller rotates 180 degrees out onto the wall or skid mock up and assumes an L-shaped position. The feet should be shoulder width apart, balls of feet on the wall or skid, knees locked, and body bent at the waist (Figure 1-7).

**Figure 1-6. Example of arm and hand signal POSITION.**

**Figure 1-7. L-shape position.**

(3) *Go*—The lane NCO extends his right arm with the elbow locked, fingers extended, thumbs around the index fingers, and points directly at the rappeller (Figure 1 8). This initiates the rappel. The rappeller flexes his knees and jumps vigorously backwards. At the same time, the rappeller throws his brake hand out at a 45-degree angle, letting the rope slide through both the brake hand and the guide hand. The rappeller looks over his brake hand shoulder at all times during descent.

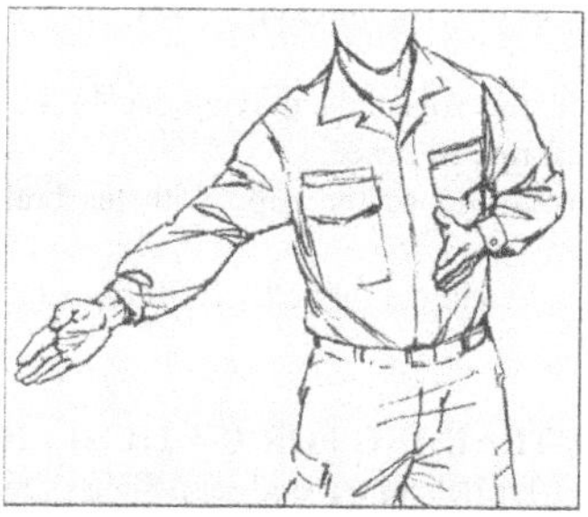

**Figure 1-8. Example of arm-and-hand signal GO**

e. The rappeller descends in a smooth, controlled manner.

f. The rappeller maintains eye contact with the ground at all times.

g. The rappeller maintains a modified L-shape position during descent with the feet shoulder-width apart, knees flexed, and buttocks parallel to the ground (Figure 1-9).

**Figure 1-9. L-shape position while rappelling.**

h. When carrying equipment or additional weight, a modified L-shape is used with the legs slightly lower than the buttocks to compensate for the added weight.

i. The rappeller's back is straight. He looks over the brake hand shoulder.

j. The guide hand is extended on the rope with the elbow extended and locked.

k. The rope slides freely through the guide hand, which is used to adjust equipment and to assist in balance during descent.

l. To brake, the rappeller places the brake hand (with rope in hand) in the small of the back and then grasps the rope firmly with the brake hand.

**Note:** Do not grip the rope firmly with the brake hand while the brake hand and brake arm are extended at the 45-degree angle. If this is done while rappelling, the brake hand and glove may become entangled in the snaplink causing injury to the hand and causing the rappeller to become hung up on the ropes.

m. Releasing tension on the rope and moving the brake hand out to the rear at a 45-degree angle regulates the rate of descent.

n. The rappeller never lets go of the rope with his brake hand until the rappel is completed.

o. After the rope is cleared and the rappeller is off rappel, he acts as the belayer for the next rappeller.

## 1-15. RAPPEL TOWER TRAINING FOR UH-1H HELICOPTER

Training on the rappel tower for helicopter skid rappelling prepares soldiers to rappel from a UH-1H helicopter (Figure 1-10). The rappeller is hooked up while he sits on the platform just above the helicopter skid. On the rappel master's command GET READY, the rappeller looks over the edge of the tower to ensure the running ends of the ropes are on the ground (Figure 1-10A). On the command SIT IN THE DOOR, the rappeller rotates his feet and legs off the platform and places them on the skid (Figure 1-10B). On the command POSITION, the rappeller turns around and assumes an L-shape position (Figure 1-10C). On the command GO, the rappeller bounds away from the helicopter skid and rappels to the ground. The rappel master is responsible for the proper procedures and safety.

A B C

**Figure 1-10. UH-1H helicopter skid rappel training.**

## 1-16. RAPPEL TOWER TRAINING FOR UH-60 BLACKHAWK HELICOPTER

Rappel tower training for the UH-60 Blackhawk is similar to that for the UH-1H with the exception that the UH-60 has no skids on which to stand. Therefore, the edge of the rappel tower is used as a pivot point to assume the L-shape position. All commands are the same except for SIT IN THE DOOR, which does not apply to the UH-60. The stances for each command are also different, as shown in Figure 1-11.

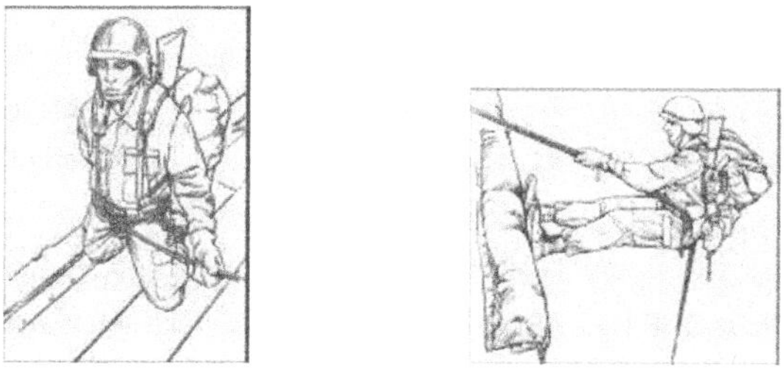

A B

**Figure 1-11. UH-60 helicopter rappel training.**

## 1-17. EMERGENCY LOCK-IN PROCEDURES

Mastering the lock-in procedure during tower training is critical before advancing to helicopter rappelling. Using the lock-in procedure allows personnel to hold in position for an extended period of time.

a. If the helicopter gains altitude above the length of the rappel ropes, the rappeller immediately brakes, locks in, and waits for the descent of the aircraft. Procedures for lock-in are as follows:

(1) Place the brake hand in the small of the back and brake to a complete stop.

(2) Release the guide hand from the ropes.

(3) Bring the guide hand around the back and grasp the running end of the two rappel ropes behind the brake hand.

(4) Using the guide hand, bring the two running ends of the rappel ropes around to the front.

(5) Secure these two running ends of the rappel ropes with the two anchor ends of the rappel ropes in the guide hand. This is now the new brake hand.

(6) Take the old brake hand out of the small of the back. Bring it around to the front and grasp the two ropes from the anchor point at a point just above the new brake hand. The old brake hand is now the new guide hand.

(7) Face the rappel master and wait for his command to lower to the ground.

(8) When the command is received from the rappel master to continue the descent, bring the brake hand to a 45-degree angle to the rear. When it is time to brake, bring the new brake hand around to the front diagonally across the chest.

b. If an engine fails or an aircraft emergency occurs during rappelling, the rappellers on the ropes descend as rapidly as possible and move from beneath the aircraft to the sides. Rappellers will maintain control of ropes, if possible.

## 1-18. COMMUNICATIONS

The rappeller at the top of a rappel point must be able to communicate with those at the bottom. Radios, hand signals, and rope signals are considered during a tactical rappel. For training situations, use the following commands:

| COMMAND | GIVEN BY | MEANING |
|---|---|---|
| LANE NUMBER, ON RAPPEL | Rappeller | I am ready to begin rappelling. |
| LANE NUMBER, ON BELAY | Belayer | I am on belay and you may begin your rappel. |
| LANE NUMBER, OFF RAPPEL | Rappeller | I have completed the rappel, cleared the rappel lane, and am off the rope. |
| LANE NUMBER, OFF BELAY | Belayer | I am off belay. |
| LANE NUMBER, FALLING | Rappeller | I am falling. Be alert below—belay man brake. |

## 1-18. DEMONSTRATION

After explaining the procedures to all rappellers, the rappel master should have an assistant demonstrate one complete cycle of rappelling from the static tower. This ensures that the rappellers can hear all the proper commands and see the actions and techniques used on the static tower.

CHAPTER 2

# GROUND RAPPELLING

*Ground rappelling is a technique that allows soldiers to negotiate mountains and cliffs safely and rapidly. Before rappellers participate in mountain and cliff rappelling they should have completed all the requirements for tower training. To ensure the rappel is safely conducted and meets the mission requirements, rappellers should consider several factors.*

## 2-1. PERSONNEL

The personnel responsibilities for mountain and cliff rappelling are the same as for tower rappelling. As a minimum the following personnel are needed for mountain and cliff rappelling.

a. **Mountaineering Safety Officer (MSO).** The MSO is a SFC or above who is ranger qualified or is a graduate of the Northern Warfare Training Center (NWTC). He has overall responsibility for the safety of all mountaineering participants and ensures that all safety precautions are followed.

b. **Rappel Master**. The rappel master is a qualified officer or NCO in charge of training on ground, tower, and aircraft rappel safety procedures.

c. **Medical Officer or Noncommissioned Officer.** Qualified rappellers and aidmen must know MEDEVAC procedures.

d. **Radiotelephone Operator.** The RATELO monitors the radio and reports to the RSO.

e. **Rappel Lane NCO.** The rappel lane NCO is a qualified officer or NCO in charge of rappelling on a wall or ground rappel lane (Chapter 1, paragraph 1-3).

## 2-2. SUSTAINMENT TRAINING

Before conducting ground rappel training, the unit or element conducts sustainment training.

a. Review the construction of a rappel seat, seats to be used, and hook-up procedures.

b. Conduct two rappels on the 34-foot tower wall (20-foot tower is acceptable): one with equipment, one without equipment. Conduct two rappels from the free side of a 34-foot or higher rappel tower.

c. Demonstrate the ability to lock in during one of the rappels from the free side of the tower.

## 2-3. SELECTION OF A RAPPEL POINT

The selection of the rappel point depends on factors such as mission, cover, route, anchor points, and edge composition (loose or jagged rocks). The anchor point should be above the rappeller's departure point. Primary and secondary anchors must be solid (natural anchors are preferred). The rappeller should be sure that the rope reaches the bottom or a place from which he can further rappel or climb. Also, the rappel point should be carefully tested and inspected to ensure the rope will run freely and that the area is clear of obstacles that could be dislodged. If a sling or runner is used for a rappel point, it should be tied twice from two separate loops. Suitable loading and off-loading platforms should be available.

## 2-4. ESTABLISHMENT OF A RAPPEL POINT

A rappel lane is the area and rope that a rappeller will use during his descent. It should have equal tension between all anchor points by establishing primary and secondary anchor points (Figure 2-1). If one anchor point fails, the rappel rope should not extend. All of the methods discussed below can be performed with a single or double rope. A double rope application should be used for safety.

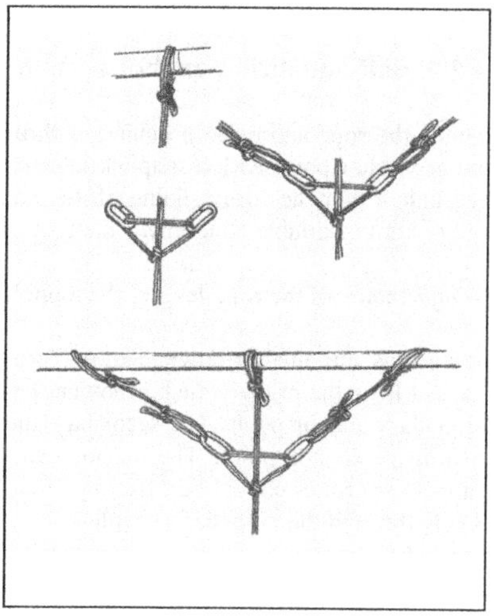

**Figure 2-1. Rappel point.**

a. If a rappel lane is *less* than half the rope length, the rappeller may apply one of the following techniques:

(1) Double the rope and tie a three-loop bowline around the primary anchor to include the primary anchor inside two loops and enough rope in the third loop to run to the secondary anchor (another three-loop bowline secured with an overhand knot).

(2) Double the rope and tie it around a secure anchor point with a round turn anchor bowline secured with an overhand knot (or any appropriate anchor knot).

(3) Double the rope and establish a self-equalizing anchor system (Figure 2-2) with a bowline-on-a-bight or figure-eight-on-a-bight knot. Tie off on the long standing and with a round turn anchor bowline.

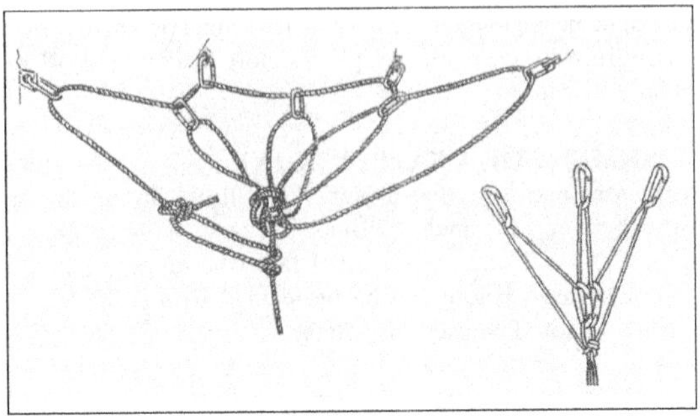

**Figure 2-2. Self-equalizing anchor system.**

(4) In an emergency, double the rope and place it behind or through a secure anchor point, or tie a runner around an anchor point with a snaplink inserted and place the rope through the snaplink. To preclude a rappeller from sliding off the end of the rappel lane, tie a double figure eight (square knot or double fisherman's knot) at the bottom end of the rope with both ends.

b. If a rappel lane is *greater* than half the rope length, the rappeller may apply one of the following techniques:

(1) Using two ropes, tie a round turn anchor bowline around a primary anchor point. Take the remaining rope (the tail from the primary anchor bowline) and tie another round turn anchor bowline to a secondary anchor point. The secondary anchor point should be in a direct line behind the primary anchor point. The anchor can be either natural or artificial. The ends of the rappel lane ropes should be offset by 15 centimeters (6 inches) so that the rope ends feed freely through the rappeller's snaplink.

(2) Using two ropes, establish a three-piton anchor system using a bowline-on-a-bight knot (or figure-eight-on-a-bight knot) and tie off on the long-standing end with a round turn anchor bowline.

(3) In an emergency, use two ropes and tie the two ends together with a joining knot. Place the joined ropes behind or through an anchor point, or tie a runner around an anchor point with a snaplink inserted and place the joined rope through the snaplink. Offset the joining knot to the left or right of the anchor. Tie off the bottom end of the rope with a joining knot to prevent a rappeller from sliding off the end of the rappel line.

c. Due to the length of the rappel, the rappel rope may not reach the anchor. If the rope is used to tie the knots, it may be too short to accomplish the rappel.

(1) When using a natural anchor—

(a) Tie a sling rope, runner, or another rope around the anchor with a round turn anchor bowline.

(b) Tie a fixed loop (figure eight or butterfly) in one end of the rappel rope, which is attached to the round turn around the anchor through the two snaplinks (opposing gates).

(2) When using an artificial anchor—

(a) Tie off a sling rope, runner, or another rope to form a loop.

(b) Put the loop through the snaplinks that are attached to the artificial anchor point.

(c) Bring the bottom of the loop up and connect it to the snaplinks that are between the artificial anchors (chocks, pitons, or bolts).

(d) Grasp the snaplinks that are between the chocks/pitons and pull them down and together.

(e) Tie a fixed loop (figure eight or butterfly) in the end of the rappel rope and connect this to the snaplinks that have been pulled together.

**Notes:** 1. Rerouted figure-eight knots can be used instead of bowlines.
2. Runners may be used from one or more anchor points.

d. Backfeed (stack) the rope to ensure it does not snarl when thrown. Take off one wrap at a time and let it fall to the ground, ensuring that no kinks, knots, or twists occur that might hinder the rope from feeding out. When the rope is backfed, anchor off one end of the rope. The two methods used to throw the rope are underhand and overhand. Use the overhand method when trees or shrubs are on or near the rappel point.

e. Set up a retrievable rappel point by applying one of the following techniques:

(1) When the rappel is *less* than half the total length of the rope, double the rope. Place the rope around the primary anchor with the bight formed by the midpoint. Join the tails of the rappel rope and throw the rope over the cliff. Tie a clove hitch knot around a snaplink just below the anchor point ensuring the locking bar inside the snaplink points away from the gate opening end and faces uphill. Snap the opposite standing portion into the snaplink. Upon reaching the bottom of the cliff, pull on the part of the rope to which the snaplink is secured to allow the rope to slide around the anchor point.

(2) When the length of the rappel is *greater* than half the length of the rope used, join two ropes around the anchor point (double fisherman's knot or square knot). Adjust the joining knot so that it is away from the anchor. Tie a clove hitch knot around a snaplink just below the anchor point ensuring the locking bar inside the snaplink points away from the gate opening end and faces uphill. Snap the opposite standing portion into the snaplink. Upon completion of the rappel, pull the rope to which the snaplink is secured to allow the rope to slide around the anchor point.

**WARNING**

**When setting up a retrievable rappel using only a primary point, take care in selecting the point. Ensure soldiers have a safety line when approaching the rappel point, with only the rappeller going near the edge.**

## 2-5. TYPES OF RAPPELS

Types of rappels include body rappel, hasty rappel, seat-hip rappel, seat-shoulder rappel, Australian rappel, and buddy-evacuation. (Seat-hip rappel and Australian rappel are described in Chapter 1, paragraphs 1-10 and 1-11.)

a. **Body Rappel.** Face the anchor point and straddle the rope. Pull the rope from behind and run it around either hip, diagonally across the chest, and back over the opposite shoulder (Figure 2-3). Then, run the rope to the brake hand, which is on the same side of the hip that the rope crosses (for example, the right hip to the left shoulder to the right hand). Lead with the brake hand down and face slightly sideways. The foot corresponding to the brake hand precedes the guide hand at all times. Keep the guide hand on the rope above to guide—not to brake. Lean out at a sharp angle to the rock. Keep the legs spread well apart and relatively straight for lateral stability and the back straight to reduce friction. Turn the BDU collar up to prevent rope burns on the neck. Wear gloves, and use other clothing to pad the shoulders and buttocks. To brake, lean back and face directly toward the rock area so the feet are horizontal to the ground.

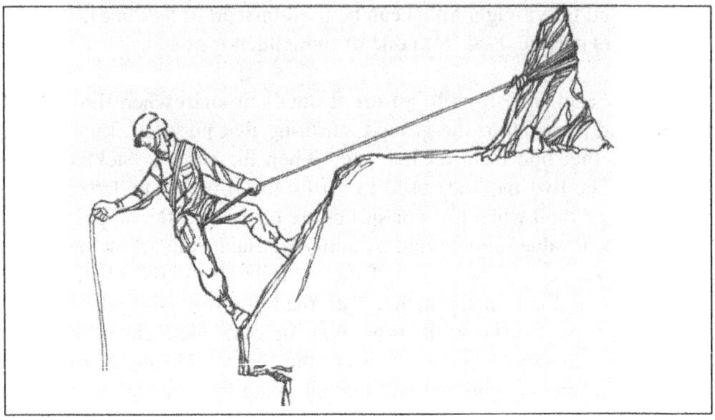

**Figure 2-3. Body rappel.**

b. **Hasty Rappel.** Facing slightly sideways to the anchor, place the ropes horizontally across the back (Figure 2-4). The hand nearest to the anchor is the guide hand, and the other is the brake hand. Wear gloves to prevent rope burns. To stop, bring the brake hand across in front of the body locking the rope. At the same time, turn to face up toward the anchor point. Use this rappel only on moderate rock pitches. The hasty rappel's main advantage is that it is easier and faster than the other methods, especially when the rope is wet.

**Figure 2-4. Hasty rappel.**

**Figure 2-5. Seat-shoulder rappel.**

c. **Seat-Shoulder Rappel.** To hook up for the seat-shoulder method, face the rappel point (Figure 2-5). Snap into the rope that passes up through the snaplink. Bring the rope over one shoulder and back to the opposite hand (left shoulder to right hand). Use the same technique in the descent as in the body rappel. This method is faster than the body rappel, less frictional, and more efficient rappellers with packs and during night operations.

d. **Buddy-Evacuation Rappel.** Use the buddy-evacuation rappel to evacuate an injured soldier from a cliff or steep terrain. Face the cliff and assume a modified L-shape body position to compensate for the weight of the victim on the back. The victim is top-rope belayed from above, which provides the victim with a point of attachment to a secured rope.

**Note:** To use this rappel, the victim must be conscious.

(1) The method for securing a victim to a rappeller's back is described below:

(a) To secure the victim to the carrier's back with a rope, the carrier ties a standard rappel seat (brake of choice, depending on the injury) and rests his hands on his knees while the victim straddles his back (Figure 2-6.)

(b) A 4.2-meter (14-foot) sling rope is used. A 45-cm (18-inch) tail of the sling is placed on the victim's left hip. (This method describes the procedure for a seat-hip rappel with right-hand brake.)

(c) The long remaining end of the sling rope is routed under the victim's buttocks and passed over the victim's and carrier's right hip. The rope is run diagonally, from right to left, across the carrier's chest, over his left shoulder, and back under the victim's left armpit.

(d) The rope is then run horizontally, from left to right, across the victim's back. The rope is passed under the victim's right armpit and over the carrier's right shoulder.

**Figure 2-6. Patient secured to carrier's back.**

(e) The rope is run diagonally, from right to left, across the carrier's chest and back across the carrier's and victim's left hip.

(f) The two rope ends should now meet. The two ends are tied together with a square knot and overhand knots.

(g) The knot is positioned on the victim's left hip. The carrier's shoulders should be padded to prevent cutting by the rope.

(2) An alternate method is to use two pistol belts hooked together and draped over the carrier's shoulders. The victim straddles the carrier, and the belay man secures the loose ends of the pistol belts under the victim's buttocks. Slack in the pistol belt's sling should be avoided, since the carrier is most comfortable when the victim rests high on his back. (See FM 8-35.)

(3) A large rucksack can be slit on the sides near the bottom so that the victim can step into it. The victim is belayed from the top with the carrier conducting a standard rappel. The carrier wears the rucksack with the victim inside.

**Note:** Rucksacks are not designed to support the weight of a soldier and his gear. Therefore, this technique is used only as a last resort.

(4) A casualty secured to a carrier, as described above, can be rappelled down a steep cliff using a seat-shoulder or seat-hip rappel. The casualty's and rappeller's shoulders should be padded where the sling rope and rappel lines cross if a seat-shoulder rappel is used. The buddy team should be belayed from above with a bowline tied around the victim's chest under his armpits. The belay rope must run over the rappeller's guide hand shoulder (Figure 2-7).

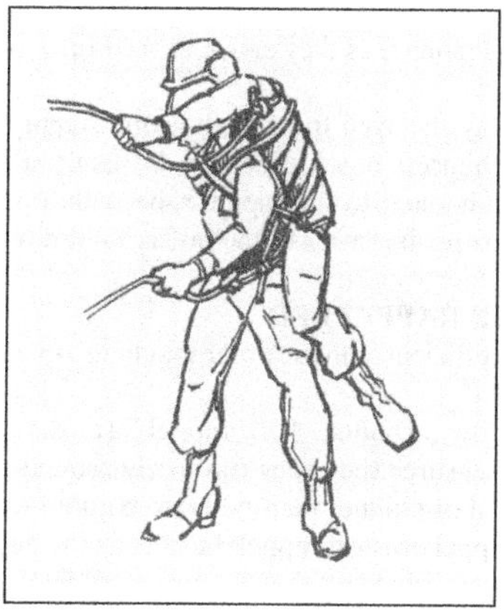

**Figure 2-7. One-man carry.**

## 2-6. RAPPELLING PROCEDURES

Proper rappelling procedures must be followed to ensure the safety of personnel.

a. The rappel lane NCO is tied off to a separate anchor point for safety, if possible. He—

(1) Ensures that the anchors are sound and the knots are properly tied.

(2) Ensures that loose rock and debris are cleared from the loading platform.

(3) Allows only one man on the loading platform at a time and ensures that the rappel point is orderly run.

(4) Ensures that each man is properly prepared for the particular rapel: gloves on sleeves down, helmet with chin strap fastened, gear prepared properly, and rappe seat and knots correct (if required). He also ensures that the rappeller is correctly hooked up to the rope and is aware of the proper braking position.

(5) Ensures that the proper signals or commands are used.

(6) Dispatches each man down the rope.

(7) Is the last man down the rope.

(8) Ensures that the rope does not run over sharp edges, and pads the rappelling surfaces as necessary to protect the rope.

(9) Ensures that the rope reaches the bottom or is at a place where additional rappels can be made.

b. The first rappeller down—

(1) Selects a smooth route for the rope that is clear of sharp rocks.

(2) Conducts a self-belay with a prusik sling tied from himself to the rappel rope.

(3) Clears the route and places loose rocks, which the rope may dislodge, far enough back on ledges to be out of the way.

(4) Ensures the rope reaches the bottom or is at a place from which additional rappels can be made.

(5) Ensures that the rope runs freely around the rappel point when pulled from below.

(6) Clears the rappel lane by straightening all twists and tangles from the ropes.

(7) Belays subsequent rappellers down the rope.

(8) Takes charge of personnel as they arrive at the bottom (off-loading platform).

**Note:** A rappeller is always belayed from the bottom except for the first man down. The first man belays himself down the rope by using a safety line attached to his rappel seat that is hooked to the rappel rope with a prusik knot. As the first man rappels down the rope, he "walks" the prusik knot down with him.

## 2-7. DUTIES OF THE RAPPELLER

To ensure the safe and efficient conduct of rappelling operations, each rappeller must know the duties of his job.

a. Each rappeller down shouts, "Off rappel" (if the tactical situation permits), untangles the ropes, and ensures the ropes run freely around their anchors. When silence is needed, a planned signal of pulling the rope is substituted for the voice signal. After the rope is cleared and the rappeller is off rappel, he acts as the belayer for the next rappeller.

b. All rappellers inspect ropes often when many soldiers are rappelling.

c. The last rappeller to descend constructs a retrievable rappel point and rappels down. Then he gently pulls the rope to prevent the rising rope end from entangling with the other rope. He stands clear of the falling rope and any rocks that it may dislodge.

**Note:** Soldiers wear gloves for all types of rappels for protection from rope burns. Also, bounding rappels are discouraged since this stresses the anchor and causes undue wear and friction on the rope.

d. Rappellers descend in a smooth, controlled manner. The body forms an L-shape with the feet shoulder-width apart, legs straight, and buttocks parallel to the ground. When carrying equipment or additional weight, a modified L-shape is used with the legs slightly lower than the buttocks to compensate for the added weight. The rappeller's back is straight. He looks over the brake shoulder. The guide hand is extended on the rope with the elbow extended and locked. The rope slides freely through the guide hand, which is used to adjust equipment and to assist balance during the descent. The rappeller grasps the rope firmly with the brake hand and places it in the small of his back. Releasing tension on the rope and moving the brake hand out to his rear at a 45-degree angle regulates the rate of descent. The rappeller never lets go of the rope with his brake hand until the rappel is completed.

e. During training, the lane number must be understood. In combat, a series of tugs on the rope may be substituted for the oral commands to maintain noise discipline. The number of tugs used to indicate each of the commands is IAW the unit SOP.

## 2-8. BELAYER

The belayer assumes a position at the base of the lane about one pace away from the rock area. He ensures that the rappel ropes are even with the ground during rock and tower rappels. The belayer loosely holds the rappel ropes with both hands so as not to interfere with the rappeller but still able to stop the rappeller should he fall. If the rappeller shouts "Falling" or loses control of his brake hand or descent, the belayer immediately stops the rappeller by pulling downward on the rappel ropes. To ensure a firm grip on the rappelling rope, the belayer does not wear gloves. (Because no friction exists between the belayer's hands and the rappelling rope, gloves are not required for safety. However, whether or not to wear gloves is decided by the OIC/NCO.) The belayer watches the rappeller at all times and maintains constant voice or visual contact with the rappeller. The belayer wears a helmet to prevent injuries from falling debris. All commands are spoken loudly and clearly.

CHAPTER 3

# HELICOPTER RAPPELLING

*Helicopter rappelling can provide a means of quick insertion with or without an LZ.*

## Section I. PERSONNEL

The personnel required for helicopter rappelling are the rappel master, the rappel safety officer, the pilot in command, the rappelers, and the belayers.

### 3-1. RAPPEL MASTER

A qualified rappel master is aboard each aircraft. Safety is the rappel masters number one priority. The rappel master—

- Ensures internal communication between the pilot and rappel master, and external communication between the aircraft and the ground.
- Inspects all equipment and uses only authorized, serviceable equipment.
- Inspects and tests all anchor points and knots before the mission starts.
- Ensures that all rappellers receive a safety briefing and the pilots and aircrew receive an air mission brief.
- Ensures that the rappelers are rappel qualified before conducting helicopter rappelling to include tactical rappelling.
- Maintains communications with the pilot at all times.

### 3-2. RAPPEL SAFETY OFFICER

The RSO is a SFC or above who is either air assault or ranger qualified. He has overall responsibility for the safety of all rappellers and ensures that all safety precautions are followed. He maintains communications at all times with the pilot and rappel master through FM radio. He alerts the rappel master and pilot of any unsafe acts.

### 3-3. PILOT IN COMMAND

The pilot in command (PIC) of the aircraft has the following responsibilities:

a. Ensures that the aircrew and all non-aircrew personnel are briefed and understand their responsibilities during rappelling operations, including aircraft safety and action in the event of an emergency.

b. Ensures that the donut ring anchoring device assembly and or aircraft anchor points have been inspected for completeness and functionality, and that they are installed properly.

c. Emphasizes procedural techniques for clearing, recovery, and or jettison of ropes.

d. Keeps the aircraft centered over the target with corrections from the rappel master as required.

### 3-4. RAPPELLER

In addition to the tower qualification requirements outlined in Chapter 1, the individual rappeller must complete advance training under the supervision of a qualified rappel master in order to participate in tactical helicopter rappelling operations. He must—

- Satisfactorily complete three rappels from a helicopter from a height of 60 feet (two rappels with combat equipment and weapon).
- Demonstrate confidence and proficiency in the techniques, procedures, and equipment used in rappelling from a helicopter.
- Know the rappelling equipment used in helicopter operations and any special equipment required for helicopter rappelling.

### 3-5. BELAYER

A belayer is assigned to each rope. He is responsible for walking the rope beneath the helicopter during the descent. (Walking the rope is defined as removing the slack from underneath the helicopter by walking backwards with the rope as the helicopter descends to land.) The belayer makes sure the ropes are not caught on the aircraft skids or tires to ensure a safe landing.

## Section II. TRAINING

Training for helicopter rappelling includes sustainment training and refresher training.

### 3-6. SUSTAINMENT TRAINING

Before conducting helicopter rappel training, the unit or element conducts sustainment training.

a. Review the construction of a rappel seat, equipment to be used, and hook-up procedures.

b. Conduct two rappels on the 34-foot or higher tower wall: one without equipment, one with equipment. Conduct two rappels from the open side of a 34-foot or higher rappel tower.

### 3-7. REFRESHER TRAINING

Refresher training is routinely conducted to maintain acquired skills. Soldiers who have not performed a helicopter rappel during the past six months will undergo refresher training consisting of three satisfactory rappels from a tower (one with weapon and equipment and one executing a lock in) before executing a helicopter rappel.

## Section III. PREOPERATIONAL BRIEFINGS AND SAFETY PROCEDURES

This section discusses safety, medical and communications requirements, and the procedures to follow during unusual conditions (adverse weather/terrain conditions, night operations). Personnel must use sound judgement to determine what action to take depending on the nature and severity of the conditions.

### 3-8. MEDICAL COVERAGE

A qualified and equipped medic will be present to respond to any mishap. Medical transportation must also be available. Absence of a medic, medical equipment, or transportation will terminate training. If the situation warrants and the installation cannot support a MEDEVAC mission, the rappel aircraft may be used as a last resort MEDEVAC vehicle.

## 3-9. COMMUNICATION REQUIREMENTS

During helicopter rappel training, the RSO will have radio communications with the aircraft. Voice communications are required before starting aircraft rappelling. Additionally, the RSO will inform the PIC to stop operations if an unsafe condition develops.

## 3-10. ADVERSE WEATHER/TERRAIN CONDITIONS

Rappel operations will not be conducted under the following conditions

- Ambient temperature is 30 degrees Fahrenheit or less.
- Winds in excess of 30 knots.
- Lightning strikes within 1 nautical mile of rappelling operations.
- Wind chill factors caused by the helicopter's rotor wash or extraction cruise air speeds, which could cause cold weather injuries.
- Water or ice on the rope inhibiting the ability of the rappellers to control their descent.
- The rope is exposed to the elements for a sufficient length of time to freeze, thereby reducing its tensile strength.
- Blowing particles produced by rotor wash causes the aircrew or the rappel master to lose visual contact with the ground.

## 3-11. NIGHT OPERATION REQUIREMENTS

The following requirements are necessary for night rappelling operations.

a. One chemlight will be attached to the end of the rope and one on each rappeller.

b. One chemlight will be secured to the attachment point of the rope.

c. Night vision goggles (NVG) will not be worn by rappellers during the descent. Aircrew members will wear NVG as required during night operations. NVG lighting criteria will be IAW Army regulations, specific aircraft aircrew training manuals, unit SOPs, or the tactical environment.

## 3-12. SAFETY BRIEFING

The following safety measures are enforced.

a. Loose clothing and equipment are secured.

b. Helmets are worn with chin straps fastened.

c. Rappellers wear identification tags and earplugs, carry identification cards, and role down their sleeves.

d. Weapons are slung diagonally across the back with the muzzle pointing down on the guide hand side.

e. All seats and rappelling equipment must be inspected by a rappel master before rappelling.

f. No running is allowed within 50 feet of the aircraft.

g. Personnel approach and depart the helicopter from the front and forward of the rear of the cargo doors. When approaching or departing the helicopter, personnel bend their bodies forward at the waist to ensure clearance of the rotor blades. At no time will personnel go near the rear of the aircraft.

h. Upon boarding the aircraft, the rappeller sits or kneels down, hooks up, and applies his brake hand to the small of his back.

i. While in the helicopter, the rappeller maintains eye-to-eye contact with the rappel master and receives commands from him.

j. The rappeller ensures that he has a belayer on his rope at all times when conducting training at a hover site.

k. During the descent, the rappeller maintains eye-to-ground contact.

l. If the rappeller sees his rope coming off the ground or sees that his belayer has lost control of his rope, he immediately brakes and executes a lock-in. He then waits for commands from the rappel master.

m. The rappeller brakes once every 30 feet during descent.

n. The belayer does not wear gloves. He keeps both hands on the rope and his eyes on the rappeller at all times.

o. All rappelling will be conducted using a double strand of rope.

## Section IV. DEPLOYMENT OF ROPES

Deployment of the ropes from a helicopter is a critical task. It can cause a planned rappelling operation to fail, or it can increase the time required to conduct the operation. This is due to the likelihood of the ropes becoming entangled. To prevent this, ropes must be deployed using a positive control technique. Two of the techniques that may be used are the *deployment bag* technique and the *log coil* technique.

### 3-13. DEPLOYMENT BAG TECHNIQUE

The rappeller places the deployment bag (D-bag) (standard) on a flat surface with the stow loop facing upward. If the D-bag still has a static line, the rappeller removes it by cutting the static line where it attaches to the bag. A 12-foot section of the static line can be used as a safety line in the helicopter.

a. Use a sandbag (about one E-tool spade full of sand) as a weight in the deployment system. Roll the sandbag into a rectangular shape, tie it, and then place it in a small plastic bag, rolled and taped. Place a retaining band over the middle of the weight, and place the weight on top of the D-bag. The weight should be about the same width as the D-bag.

b. Lay out and inspect the two ropes. The working ends (closest to the D-bag) should be even. Place a round turn with the two working ends of the ropes on the weight's retaining band. Working on top of the deployment bag, start forming figure eights. The stack should consist of 8 to 10 figure eights, one on top of another—do not exceed the width of the D-bag. Then, starting from either side, center a retaining band over the stack. Ensure that it is over all the figure eights in the stack. Repeat the process each time, placing one stack in front of the other. Continue until about 10 feet of rope remains. If one rope is shorter than the other, the end of the shorter rope should be about 10 feet from the last stack.

c. For a primary anchor point, measure down about 4 feet from the end of the shortest rope. Using both ropes, tie a bowline without a half hitch. This knot is the primary anchor point.

d. For a secondary anchor point, tie an end-of-line bowline with a half hitch toward the end of the ropes. The dressed knot should be 18 to 22 inches from the primary anchor point knot when using the UH-1H helicopter and 22 to 30 inches when using the UH-60 helicopter. If excess rope remains, S-fold and tape it between the two knots. Ensure that 3 to 6 feet of rope remain from the last stack of figure eights.

e. To place the ropes in the D-bag, remove the bag from under the stack of rope. Place the weight into the bottom of the D-bag, and place all the stacks of figure eights, in the order they were made, one on top of the other into the bag.

f. Two left, two right, and two rear loops are in the opening of the D-bag. ***Do not use the two rear loops***. Close the flap on the D-bag and push the loops through the aligning holes on the flap. Using an 8-inch piece of gutted 550-pound cord, tie the two loops together with a square knot and two hitches. Ensure that two 120-foot ropes are coming out of the center of the flap—not to one side. Repeat the process and tie the right loops together. Wrap the excess rope lengthwise around the bag.

**Note:** The sizes of the loops of the bowline should be no larger than an average size fist.

### 3-14. LOG COIL TECHNIQUE

The rappeller lays the running end of the double rope along the length of the coiling log (Figure 3-1). He then coils the double rope around both the running end of the rope and the coiling log. The rope must be coiled evenly and tightly.

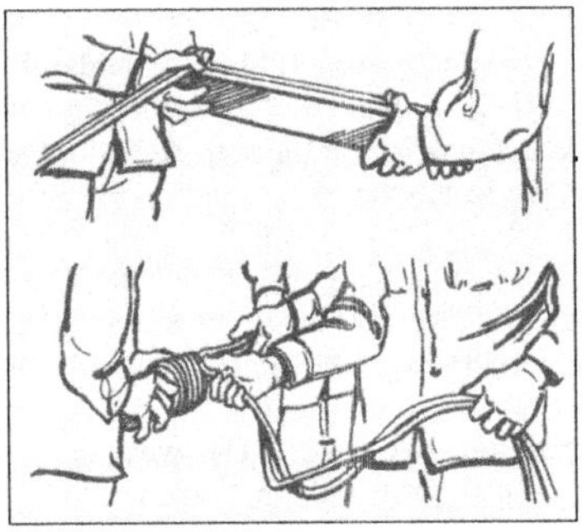

**Figure 3-1. Coiling Log**

## Section V. RAPPELLING OPERATIONS FOR UH-1H IROQUOIS HELICOPTER

The UH-1H helicopter provides a safe, stable aerial platform from which to conduct rappelling operations when landing is not feasible.

### 3-15. CHARACTERISTICS

The UH-1H is a single-engine, medium-speed, single-main rotor helicopter that can transport eight rappellers, one rappel master, and one three-man crew. It has a lift capacity of 2,300 pounds and is equipped with several floor-mounted tie-down fittings, seven of which are used during rappelling operations.

## 3-16. RIGGING OF UH-1H HELICOPTER FOR RAPPELLING

The rappel master performs the following to rig the UH-1H helicopter.

a. Removes all seats.

b. Locks the doors in the open position. If no locks are present, removes doors. Also, he removes small cargo doors.

c. Pads and tapes all sharp edges on the floor, door ledge, and all protrusions on the skids. Ensures each door ledge has a scuff pad to protect the rope from contacting the metal door ledge.

d. Secures the donut ring to the center of the floor of the helicopter. The donut ring has six snaphooks numbered clockwise, with 12 o'clock being toward the front of the helicopter. Positions the clamps of the donut ring toward the aft end of the helicopter. Ensures the front two snaplinks and rear two snaplinks are facing outside of the donut ring and hooks them to the tie-down ring with the snaphook facing down. Ensures the center two snaphooks face into the donut ring and are connected to the floor tie-down ring with the gate facing down (Figure 3-2). Attaches the free-floating safety ring to the center floor tie-down ring using a seventh snaphook.

**Note:** Using this technique has caused equipment failure in many aircraft. The cable clamp nuts must be inspected and certified as airworthy by the pilot or maintenance crew before flying the mission.

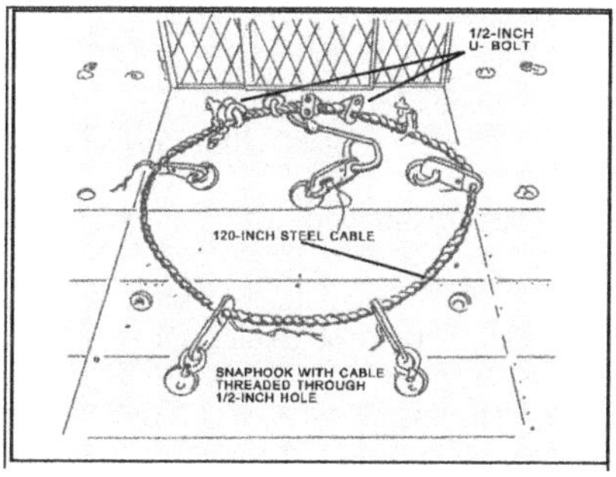


**Figure 3-2. Donut ring attached to the floor of the helicopter.**

## 3-17. CONSTRUCTION OF ANCHOR POINTS

In addition to the usual equipment requirements for rappelling operations, an anchor assembly is fabricated. This anchor assembly is commonly known as a *donut ring*. The secondary anchor point is a *floating safety ring*.

a. **Donut Ring.** This primary (or No. 1) anchor point for the rappelling ropes is constructed from a 1/2-inch steel cable with a steel wire core. The cable is 120 inches long, consists of 6 strands (18 wires per strand), and has a tensile strength of 21,000 pounds (Figure 3-2). The completed donut ring has a tensile strength of 3,000 pounds.

(Figure 3-3 shows the donut ring and rappel rope connection in a helicopter.)

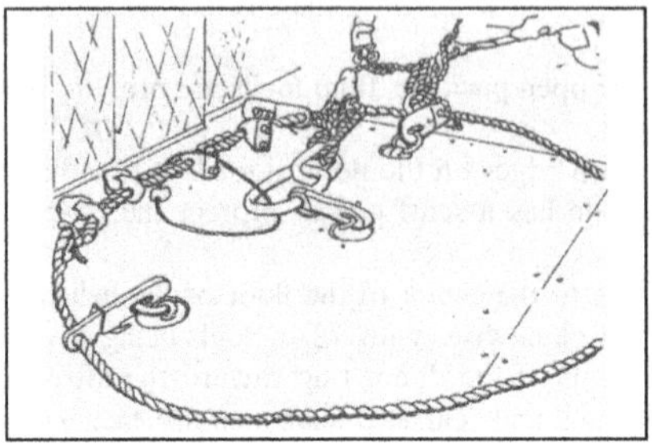

**Figure 3-3. Donut ring and rappel rope connection in a helicopter.**

(1) Thread six parachute static line snaphooks onto the 120-inch steel cable so that four snaphooks are facing out with the gates down, and the center two snaphooks are facing in with the gates down. Drill the end portion of each snaphook to make a 5/8-inch diameter hole, and thread the cable through the holes in the snaphooks.

(2) Overlap the ends of the 120-inch cable 20 inches to form a circle. Secure the ends with four 1/2-inch U-bolts placed at 2- to 3-inch intervals.

(3) Attach two U-bolts to each dead end of the cable so that the bolts engage the dead end.

(a) Before torquing the U-bolts, position a 12-inch length of chain or 1/8-inch diameter cable on the center of the overlapped 120-inch steel cable so that it remains in position between the two center U-bolts.

(b) Tighten each nut of the U-bolts with a torque wrench (if possible) to 40-foot pounds (480-inch pounds).

(c) After the U-bolt clamps have been attached and tightened, fasten a steel plate (drill to fit) over the open end of the U-bolt studs and spot weld in place to prevent loosening.

b. **Floating Safety Ring**. The floating safety ring is referred to as the secondary (or No. 2) anchor point for the rappelling ropes (Figure 3-4). The snaplink at the end of the rappelling rope is hooked to this connection. Either of the following two types of floating safety rings may be used:

(1) *First method.* Thread an elliptical rappelling ring through the free end of the keeper chain (cable). It is constructed of cold-rolled steel that is 1/4-inch in diameter, with inside dimensions of 2 1/4 inches (minor axis) and 4 inches (major axis). Thread a seventh parachute static line snaphook onto the ring before welding. Weld this ring together so that it withstands a force up to 3,000 pounds.

(2) *Second method.* Attach two snaplinks to the aircraft tie-down ring in the center of the donut ring. Insert the first snaplink through the free end of the keeper chain (cable) and the tie-down ring with the gate down. Insert the second snaplink through the free end of the keeper chain (cable) and the tie-down ring with the gate up. Tape the snaplink

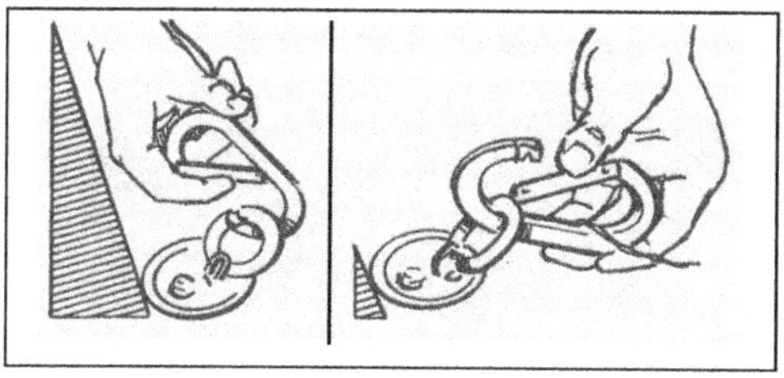

gates closed with masking tape. Then, tape the snaplinks together to ensure that the snaplink gates are on opposite sides of each other.

**Figure 3-4. Floating safety ring formed with two snaplinks.**

c. **Rappel Rope Anchor Points.** The rappelling rope is connected to the floating safety ring and the donut ring as follows:

(1) *Number 1 anchor point (donut ring).* Using a bight about 5 feet from the end of the standing part of the rope, make one turn through the snaplink forming a round turn. The bowline is the preferred method of attaching rappel ropes to snaplinks or anchor systems. The round turn with two half hitches is a reliable means of attaching ropes to anchor systems. Secure the round turn to the snaplink with two half hitches. Make the connection to the donut ring by attaching the snaplink gate upward. Ensure the gate faces upward with the opening away from the knot.

(2) *Number 2 anchor point (floating ring).* Attach the No. 2 snaplink the same as the first with some exceptions. Using a bight about 2 feet from the end of the standing part of the rope, connect the snaplink to the rope the same as the first connection. Tape the end of the standing part of the rope and the knots with masking tape (or green engineer tape) to secure them in place. Make the connection to the floating safety ring the same as the connection to the donut ring (Figure 3-5). Four ropes can be connected to the floating safety ring using two snaplinks (Figure 3-6).

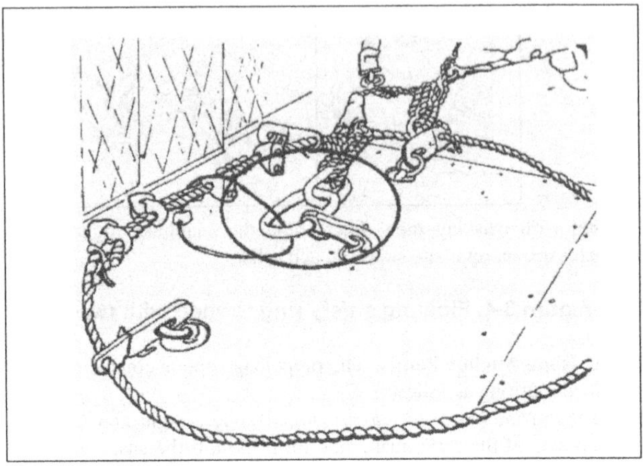

**Figure 3-5. Rappel rope connection using two snaplinks for the floating safety ring.**

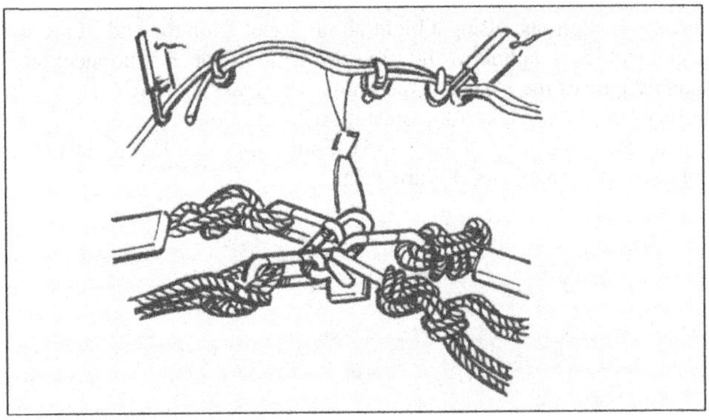

**Figure 3-6. Four rappel ropes connected to the floating safety ring (two snaplinks).**

## 3-18. SEATING ARRANGEMENTS AND LOADING TECHNIQUES

Rappelling operations are executed quickly and safely to ensure the success of the operation in a combat or training environment. All rappellers must know the seating arrangements and the procedures for loading (Figure 3-7). They must be rehearsed under the supervision of the rappel master before conducting the operation.

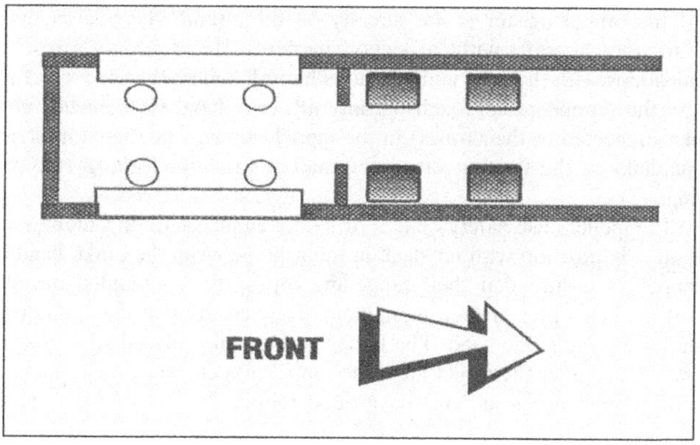


**Figure 3-7. Seating arrangement.**

a. The first four rappellers are hooked up to the donut ring and use safety belts while in flight. The remaining rappellers use seat belts or safety straps. For safety emphasis during flight, each rappeller's brake hand is in position with no slack in the rope between the brake hand and the donut ring. The rappeller ensures that his rappelling rope is not tangled on any part of the interior of the helicopter or his equipment. He ensures that the coiling log or D-bag is located in such a position that he drops the rope with ease, using his guide hand. The coiling log or D-bag is secured with half of it under the rappeller's inside leg and the other half on top of his outside leg.

b. The rappellers should board the aircraft and sit along the leading edge of each door. They hook up with the rappel seat square knot on the inboard side of the aircraft and the brake hand on the outboard side. The rappel master sits in the center to aid visual inspection of personnel and equipment.

c. For a flight time of less than five minutes and for training and operations where speed is a factor, the rappel master directs the rappellers to be seated in the door facing outward with their feet on the skid. In a tactical situation and before the helicopter comes to a hover, the rappellers assume an upright position facing inside the helicopter.

d. The aircraft is cleared as fast as possible. This reduces the time that the aircraft is in the high-hover attitude to a minimum.

## 3-19. RAPPELLING PROCEDURES

The rappel master ensures all personnel follow the rappelling procedures for the safe and efficient execution of the operation. These procedures are rehearsed under the supervision of the rappel master.

a. Rappellers (eight maximum) are identified, and rappel masters tell them which door of the aircraft to exit. Rappellers approach the aircraft once it is firmly on the ground, and the crew chief signals when it is safe to approach. The approach to the aircraft should be about 45 degrees to the nose of the aircraft allowing the pilot to see the rappeller.

b. If the rappel master is not already on the aircraft, he boards first and secures himself to the aircraft with a safety harness. He dons a headset to maintain communications with the pilot and positions himself behind the crew chief's chair.

c. As the rappellers approach the aircraft, they hand their anchor attaching points (snaplinks connected to their ropes) to the rappel master. The rappel master connects one set of snaplinks to the floating safety ring and one set of snaplinks (the primary) to the donut ring.

d. All rappellers use safety belts. For safety emphasis during flight, each rappeller's brake hand is in position with no slack in the rope between the brake hand and the donut ring. Rappellers ensure that their rappelling ropes are not tangled on any part of the interior of the helicopter or their equipment. They ensure that they can drop their D-bag or log coil using the guide hand. The D-bag or log coil is placed so it does not fall out of the aircraft. Once over the target area, the rappel master issues the rappel commands and ensures the rappel ropes are on the ground (about 20 feet of rope should be on the ground).

e. On the command GO, the rappeller flexes at the knees and vigorously pushes away from the skid gear, allowing the rope to pass through his brake hand and guide hand. The descent is accomplished smoothly at a rate of about 8 feet per second, avoiding jerky stops. The rappeller initiates his braking action slowly when he is about halfway to the ground. The safety officer or NCO ensures that at least a five-second delay is maintained between each rappeller. Another technique used to deploy rappellers is to issue the command GO to the rappellers diagonally opposite each other. In the early phase of performance, the rappeller's feet are together (particularly if over woods or jungle) and his guide hand is used (not the brake hand) to remove any entanglements.

f. Upon reaching the ground, the rappeller clears the rappel rope through the snaplink (or rappel ring) until the rope is free. If other rappellers are to follow on the same rope, the rappeller on the ground separates or untwists the ropes and becomes the belayer for each subsequent rappeller. If ropes are to be released from the aircraft following the last rappeller, untwisting and separating the ropes is not necessary. A two-second interval is maintained between exit groups (two men exiting at the same time).

g. The rappel master releases the rope from the donut ring after he confirms (by visual inspection) that the rappeller is off rappel. He then drops the rope away from the helicopter.

## 3-20. RAPPELLING COMMANDS

Helicopter rappelling is conducted in a noise-filled environment. Each rappeller is trained to know the rappel commands and to understand the actions required of him during the execution of each command. Due to the noise created by the helicopter, all rappel commands should be accompanied by arm-and-hand signals. All commands are issued by the rappel master and confirmed orally by the rappellers.

a. The helicopter rappelling commands are as follows:

(1) GET READY. The rappeller checks his combat equipment, looks toward the donut ring, and pulls the rope to check the anchor point connection. He then checks his

snaplink to ensure the rope is properly seated in the snaplink on his rappel seat.

(2) THROW ROPE. The rappeller drops his deployment bag out and away from the helicopter with his guide hand. He ensures that the rope does *not* fall between the side of the helicopter and the skid. The rappel master ensures the rope is touching the ground and is free of tangles and knots. The rappeller then replies, "Rope okay."

(3) SIT IN THE DOOR. The rappeller swings his legs to the outside of the helicopter and takes up a sitting position. (This command applies only to the UH-1H helicopter.)

(4) POSITION. The rappeller pivots 180 degrees on the skid. His feet are shoulder-width apart, the balls of his feet are on the skid, his knees are locked, his body is bent forward in an L-shape position, and his brake hand is on his buttock. The rappel master makes a visual inspection of the snaplink and rappel ring.

(5) GO. The rappel master points at the rappeller to exit. One rappeller from each side rappels at the same time. The rappel master gives the command GO to the rappellers that are diagonal to each other—for example, the front left rappeller and the rear right rappeller. The rappeller flexes his knees and jumps backward, letting the rope run through the brake hand and guide hand.

b. The arm-and-hand signals that are used with rappelling commands are as follows:

(1) GET READY (Figure 3-8). The rappel master extends both arms to the front with fists clenched and thumbs pointing upward.

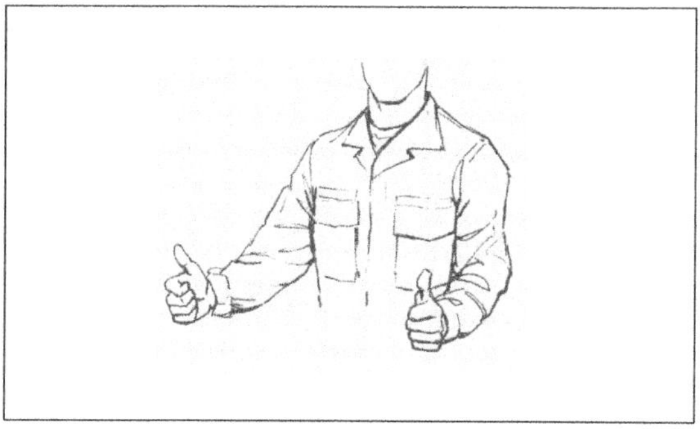

**Figure 3-8. Arm-and-hand signal for GET READY.**

(2) SIT IN THE DOOR (Figure 3-9). The rappel master extends both arms to the front with elbows locked, fingers extended, thumbs running along the index fingers, and palms facing downward. He bends slightly at the waist so that the arms are below the waist. Then, he moves both arms in a crisscross waving motion, alternating left over right and right over left.

**Figure 3-9. Arm-and-hand signal for SIT IN THE DOOR.**

(3) THROW ROPES (Figure 3-10). The rappel master extends both arms to the front with elbows locked, fist clenched and index fingers extended. He points at respective rappellers, brings the forearms to an upright position and then down to the elbow-locked position.

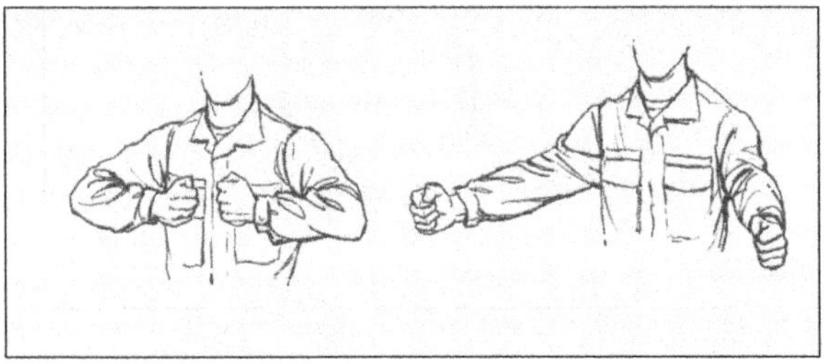

**Figure 3-10. Arm-and-hand signal for THROW ROPES.**

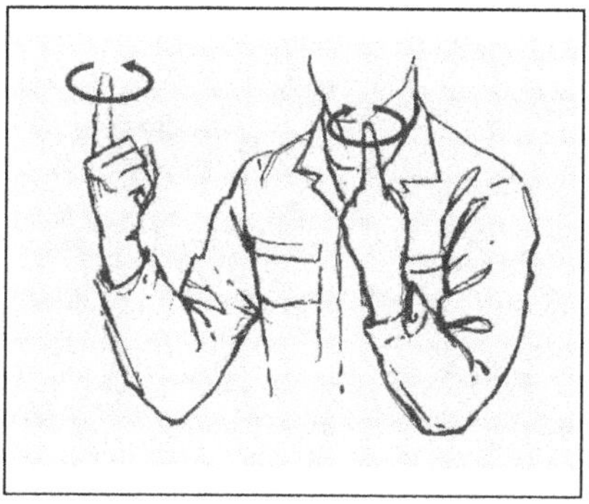

**Figure 3-11. Primary arm-and-hand signal for POSITION.**

(4) POSITION. The two arm-and-hand signals for POSITION are:

(a) Primary signal (Figure 3-11). The rappel master extends both arms to the front, elbows bent, forearms pointed upward, fists clinched, and index fingers pointing upward. He makes a circular motion with both forearms rotating in opposite directions.

(b) Alternate signal (Figure 3-12). The rappel master extends both arms to the front with elbows locked, fists clenched, and index fingers extended. He bends at the waist so that his arms are below his waist and makes a circular motion with arms rotating in opposite directions.

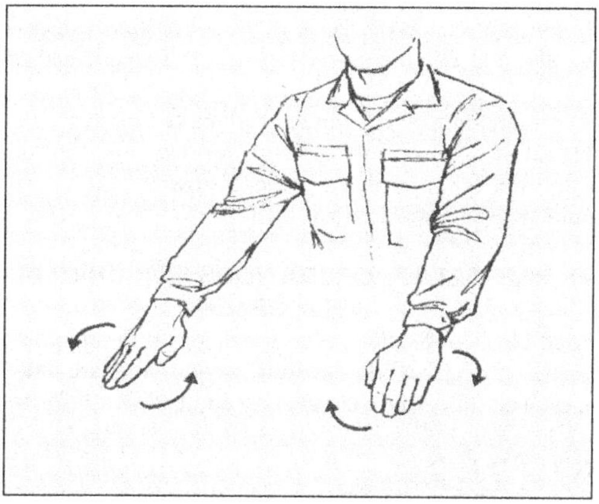

**Figure 3-12. Alternate arm-and-hand signal for POSITION.**

(5) GO (Figure 3-13). The rappel master extends an arm with elbow locked, fingers extended, and thumb along the index finger, and he points directly at the rappeller.

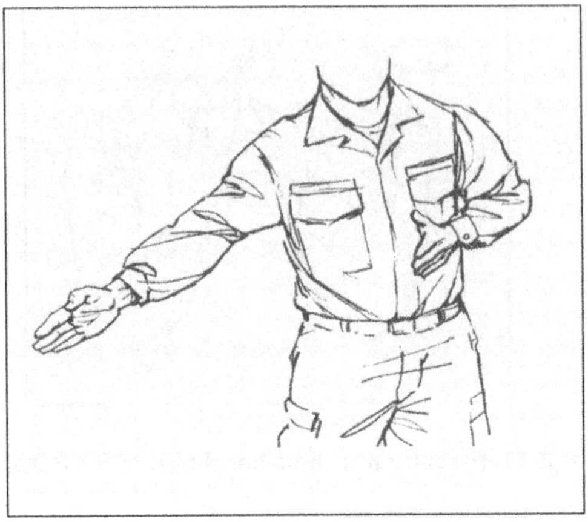

Figure 3-13. Arm-and-hand signal for GO.

### 3-21. INSPECTION AND SAFETY CONSIDERATIONS

The rappel master and pilot (or his representative) conduct a joint inspection of the aircraft to ensure the safety of all personnel and serviceability of equipment.

a. Cargo doors are locked in the open position or cleared for closing, depending on the mission.

b. All loose objects in the cargo compartment are removed or secured forward.

c. Sharp edges or protrusions on the cargo floor and door ledges are taped.

d. The donut ring and floating safety ring are serviceable and properly attached to tie-down fittings.

e. A headset and intercom jack for the rappel master are available and operational.

f. Serviceable safety harnesses are available for the rappel master and crew chief.

g. The recovery rope is installed and properly stowed.

h. Unused floor rings are taped down.

## Section VI. RAPPELLING OPERATIONS FOR UH-60 BLACKHAWK HELICOPTER

The techniques used by the rappeller when rappelling from different aircraft are similar. However, positioning, seating, and the tie-down anchor point are different. Each rappeller is well trained on each aircraft before conducting a rappel.

### 3-22. CHARACTERISTICS

The UH-60 is a twin-engine, medium-speed, single-main rotor helicopter that can transport 10 rappellers, 1 rappel master, and 1 four-man crew when the center row of troop seats is removed. It is equipped with four 4,000-pound (load limit) cabin ceiling tie-down fittings or rings that are located in the ceiling of the troop/cargo compartment. These fittings are used as the primary rappelling rope anchoring points. The UH-60 is also equipped with eight 3,500-pound (load limit) cargo restraint net rings. Four of the eight rings are located in the ceiling of the troop/cargo compartment. These rings are used as the secondary rappelling rope anchoring points.

## 3-23. RIGGING OF UH-60 FOR RAPPELLING

The rappel master rigs the UH-60 helicopter by performing the following:

a. Locks both cargo doors in the open position.

**Note:** For arctic or other cold-weather operations, or during flights of long duration, the cargo doors are closed and locked until the time specified for opening them.

b. Removes the center row of troop seats.

c. Tapes any sharp edges or protrusions on the cargo floor and door ledges that may come in contact with the rappeller or the rappelling rope.

d. Stows loose equipment forward in the cargo compartment.

e. Extends the rappel master's intercom cord to the rear over the aft utility drain line and tapes the cord to the overhead troop seat support tube.

f. Installs the floor restraint provisions for rappellers No. 1 through 6.

g. Rigs and connects rappelling ropes to the aircraft's primary and secondary anchoring points.

(1) *Primary anchor points.* Ties a bowline with a half hitch about 4 feet from the standing end of the rope. Attaches the two primary snaplinks to the respective cabin tie-down fitting ring with gates facing in the opposite directions (Figures 3-14 and 3-16)

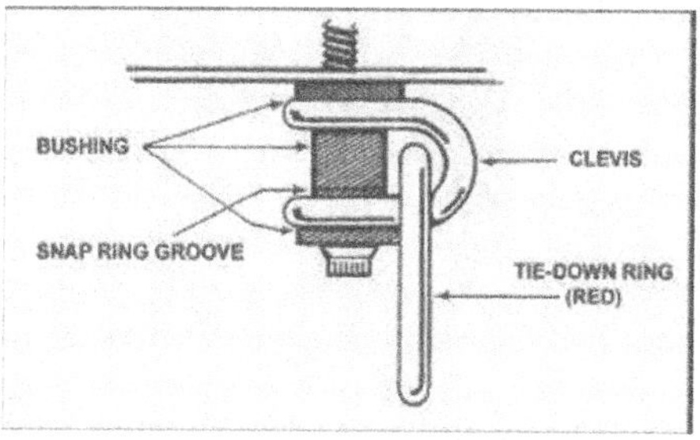


**Figure 3-14. Primary snaplink attaching point.**

(2) *Secondary anchor points.* Ties a bowline with a half hitch about 1 1/2 feet from the standing end of the rope. Attaches the secondary snaplink to the adjacent overhead cargo restraint net ring (Figures 3-15 and 3-16).

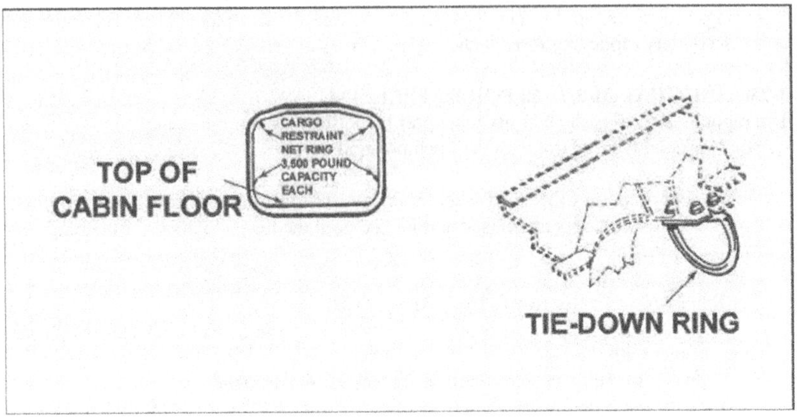


**Figure 3-15. Secondary snaplink attaching point.**

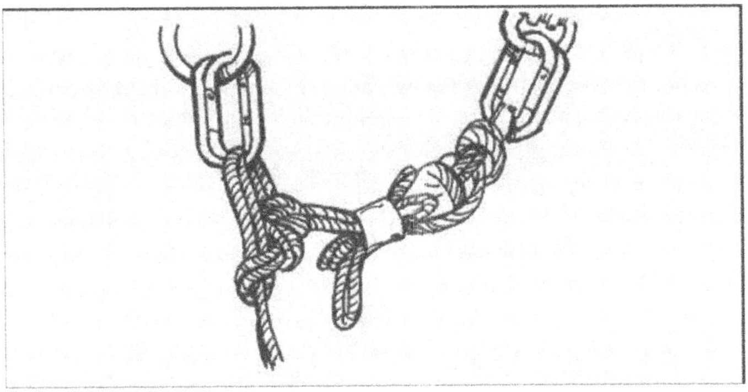

**Figure 3-16. Primary and secondary snaplink attaching points.**

h. Removes and secures the cargo hook access door and deploys the cargo hook in the DOWN position.

i. Installs the recovery rope for endangered rappellers.

j. Tapes the unused floor rings.

**Note:** There must be a minimum of 22 inches and a maximum of 30 inches between the anchor knots.

### 3-24. SEATING ARRANGEMENTS AND LOADING TECHNIQUES

Rappelling operations are executed quickly and safely to ensure the success of the operation in a combat or training environment. All rappellers must know the seating arrangements and the procedures for loading. They must be rehearsed under the supervision of the rappel master before conducting the operation.

a. A maximum of 10 rappellers, with and without combat equipment, and 1 static rappel master is seated and restrained aboard the UH-60 helicopter when the center row of troop seats are removed (Figure 3-17).

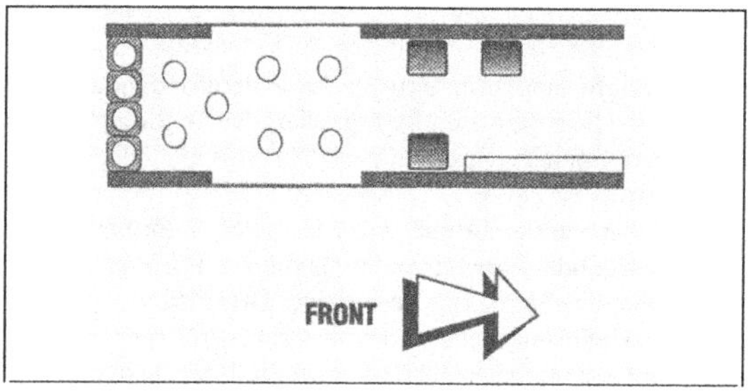


**Figure 3-17. Rappelling personnel seating arrangement.**

(1) Rappellers No. 1 through 6 and the rappel master are seated on the cargo floor. The rappellers are restrained using 3 1/2 feet of 1-inch tubular-nylon webbing (or equivalent) that is run through the cargo tie-down rings attached to the floor of the aircraft. The webbing is attached to the rappeller by means of a snaplink that is hooked to a Swiss seat. The rappel master is secured by his safety harness.

(2) Rappellers No. 7 through 10 are seated on the troop seats across the aft end of the cargo compartment. They are secured by the seat belts.

b. Rappellers are organized into 10-man sticks. They approach the aircraft from the left or right side in reverse order. They are divided into two groups and approach and enter the aircraft at the same time, depending on the situation. Rappellers No. 9, 7, 5, 3, and 1 enter the aircraft through the left door; they are seated and restrained. Rappellers No. 10, 8, 6, 4, and 2 enter the aircraft through the right door; they are seated and restrained. The static rappel master is restrained in the center of the cargo compartment using a C-3A troop-type safety belt or a safety harness during take-off and landing.

**DANGER**

**BECAUSE THE LOWEST ARC OF THE ROTOR BLADE OCCURS AT THE DIRECT FRONT OF THE AIRCRAFT, APPROACHING THE AIRCRAFT AT THIS POINT COULD RESULT IN PERSONNEL INJURY OR DEATH.**

## 3-25. RAPPELLING PROCEDURES

The rappel master ensures all personnel follow the rappelling procedures for the safe and efficient execution of the operation. These procedures are rehearsed under the supervision of the rappel master.

a. The rappel master is equipped with a headset and maintains communications with the pilot at all times. He wears a safety harness and stations himself in the center of the cargo floor to maintain control of the rappellers, their ropes, and all anchor attaching points.

**Note:** If all rappellers are rappelling into the same area, individual rappel ropes are not needed. The ropes used by rappellers No. 1 through 4 are used by the succeeding rappellers.

b. After all rappellers are in the aircraft, rappellers No. 1 through 4 hand the rappel master their primary and secondary anchor point snaplinks. The rappel master hooks the primary anchor point snaplink into the respective cabin tie-down fittings and the secondary anchor point snaplink into the respective cargo restraint net rings. After the primary and secondary anchor point snaplinks are secured, the first four rappellers hook onto their respective rappel rope. The rappel master checks the first four rappellers for the correct hookup.

c. After the first four rappellers exit and clear their ropes, the rappel master hands the next four rappellers their rappel ropes to hook up. Left-handed rappellers take extra precaution to ensure the correct hookup. Problems can be avoided by placing left-handed rappellers in the positions where they initially have the rope to their nonbrake side.

**WARNING**

**Extreme care must be taken in hooking up and positioning personnel that require repositioning of their rope to their proper brake-hand side. This could result in bodily injury or damage to Army property.**

d. After checking the rappellers for the correct hookup, the rappel master moves the next four rappellers into position on the floor of the aircraft. After the four rappellers exit and clear their ropes, the rappel master hands the last two rappellers their rappel ropes to hook up.

e. After checking the rappellers for correct hookup, the rappel master moves them into position on the floor of the aircraft. After the last rappeller is off the rope, the rappel master releases the ropes.

f. If the rappel master rappels, the crew chief is instructed to release or retrieve the ropes.

## 3-26. RAPPELLING COMMANDS

The UH-60 rappelling is conducted similar to the UH-1H rappelling with the exception that no skids are on the UH-60 on which to stand. Therefore, the edge of the floor, along the door of the helicopter, is used as a pivot point to assume the L-shaped rappelling position.

a. GET READY. This alerts the rappeller. The rappeller and rappel master perform final checks of the hookup, rappel seat, snaplink, and equipment.

b. THROW ROPES. The rappeller looks down below the aircraft to ensure that no one is under the aircraft. Keeping his brake hand in the small of his back, he tosses the deployment bag with the ropes *out and away* from the helicopter with his guide hand. The rappeller observes that the ropes are touching the ground and are not knotted or entangled.

c. POSITION. From a kneeling position with the brake hand in the small of the back, the rappeller rotates 90 degrees so that he is facing the inside of the aircraft and the rappel master. The rappeller then places his heels on the edge of the floor of the helicopter doorway and leans out into an L-shaped position. His feet are shoulder-width apart, the balls of his feet are on the edge of the helicopter doorway, his knees are locked, and his body is bent at the waist toward the helicopter.

d. GO. The rappel master initiates the rappel with this command. The rappeller flexes his knees and jumps backwards. At the same time, the rappeller throws his brake hand out at a 45-degree angle, letting the running ends of the ropes slide through both the brake hand and guide hand.

### 3-27. INSPECTION AND SAFETY CONSIDERATIONS

The rappel master and pilot (or his representative) conduct a joint inspection of the aircraft to ensure the safety of all personnel and serviceability of equipment.

a. Cargo doors are locked in the open position or cleared for closing, depending on the mission.

b. All loose objects in the cargo compartment are removed or secured forward.

c. Sharp edges or protrusions on the cargo floor and door ledges that may come in contact with the rappeller or his rappelling rope are taped.

**Note:** Do not tape the door latches or handles; this can interfere with door operations.

d. Primary and secondary rappelling anchor points are serviceable and securely attached to the aircraft structure.

e. A headset/helmet and intercom jack for the rappel master are available and operational, and the intercom extension cord is secured overhead.

f. Serviceable safety harnesses are available for the rappel master and crew chief.

g. The cabin ceiling tie-downs must have safety wire installed to ensure they do not come undone or unraveled, and the bolt head must be stamped H.

## Section VII. RAPPELLING OPERATIONS FOR MH-53 HELICOPTER

The MH-53 helicopter is a highly versatile aircraft. It is used for a variety of missions, usually in conjunction with special operations.

### 3-28. CHARACTERISTICS

The MH-53 is a two-engine, single-rotor, heavy-lift helicopter. It has a crew of six and can be refueled while in flight. It has a precision navigational and communications package, which makes it excellent for conducting deep infiltration and exfiltration missions. The MH-53 helicopter can fly at night in all weather and terrain conditions. It can follow down to 100 feet and operate from unprepared sites. It is fitted with three gun stations for a mix of 7.62-mm miniguns and caliber .50 machine guns.

### 3-29. RIGGING OF MH-53 HELICOPTER FOR RAPPELLING

When rappelling from the crew entrance door, rappellers connect the ropes with locking snaplinks to the two 10,000-pound tie-down rings under the left scanner's window. When rappelling from the ramp, rappellers connect the ropes with locking snaplinks to the two 10,000-pound tie-down fittings on the left and right sides of the aircraft at station 522. Any sharp edges that could damage the ropes should be padded or taped. A length of fire hose may also be used over the portion of the rope that comes in contacts with the door or ramp edge.

### 3-30. SEATING ARRANGEMENTS AND LOADING TECHNIQUES

The MH-53 is a large, versatile helicopter. The seating arrangements and loading techniques are numerous. The using unit develops a seating and loading SOP that ensures the safety of all personnel and permits smooth, efficient execution of the mission.

### 3-31. RAPPELLING PROCEDURES

The rappel master or his assistant are at the two exit points (ramp and door). The rappel master maintains communications with the helicopter commander and relays all commands and time warnings. The commander issues time warnings at 20-minute, 10-minute, 5-minute, and 1-minute intervals. During limited visibility, the rappel master may use NVDs to observe the safety/belayer. Chemical lights are attached to the top or bottom of the rappel rope. Arm-and-hand signals should be IAW The Air Assault School Handbook and FC 57-2.

**Note:** Night vision goggles cause limited depth perception and a tunnel vision effect.

a. The rappel master is secured by a harness and ensures the proper hook up of each rappeller. Once hooked up, the rappellers release their safety belts. At the command of the rappel master, the rappellers position themselves to ease immediate deployment.

**WARNING**

**Before conducting deployments off the ramp, the deploying team members must be briefed on the importance of maintaining separation between members. About 24 to 27 inches are desired. This separation helps maintain aircraft center of gravity limits.**

b. The rappel master deploys ropes only after the helicopter is in a stable hover over the target area and he has given the command ROPES. He ensures the ropes are on the ground before giving the command GO. After the last rappeller is off the rope, the rappel master retrieves or cuts the ropes away.

## 3-32. RAPPELLING COMMANDS

The helicopter rappelling commands for the MH-53 are as follows:

a. GET READY. This command is given as the helicopter approaches the rappel site. It alerts the rappellers of the approach to the site and signals them for a final equipment check.

b. POSITION. This command clears the first rappellers into position for deployment.

c. THROW ROPE. This command is given once the helicopter reaches a stable hover.

d. GO. The rappellers exit.

## 3-33. INSPECTION AND SAFETY CONSIDERATIONS

The rappel master briefs all personnel participating in the operation. He conducts hands-on inspection to ensure the safety of all personnel and serviceability of equipment.

a. Tie-down fittings are serviceable.

b. All sharp or protruding edges that may come in contact with the rappelling ropes are padded or taped.

c. All ropes are retrieved or cut away before forward movement of the helicopter.

d. Only three rappellers are deployed at a time—two from the ramp and one from the personnel door.

# CHAPTER 4
# SPECIAL PATROL INFILTRATION/EXFILTRATION SYSTEM

*The special patrol infiltration/exfiltration system (SPIES) provides an excellent form of exfiltration over short distances. It is not recommended for infiltration because team members are exposed the entire time. Due to the nature of SPIES operations, a thorough briefing is required for all participants before the operation. Personnel being extracted must train extensively in the SPIES extraction before infiltration. For the other personnel involved, a complete preoperations briefing is held before the operation. This is crucial when additional assets are involved other than the extraction helicopter such as gunships, aerial observers, or artillery support.*

## Section I. SPIES MASTER DUTIES AND QUALIFICATIONS

The SPIES master holds the rank of sergeant or above (may be waived by the commanding officer). The commanding officer ensures that this qualification is entered on the soldier's training record.

### 4-1. QUALIFICATIONS

To qualify to be a SPIES master, the sergeant must participate in at least three SPIES operations (observe twice and execute SPIES master duties once under observation). For example, he hooks up the helicopter, assists in the preparation of an operation, and conducts a successful operation under the supervision of a qualified SPIES master. A SPIES master is conscientious and thorough, and he knows all aspects of a SPIES operation. He can give an effective pilot's brief, use the aircraft communications equipment, and understand aviation terminology.

### 4-2. DUTIES

The SPIES master is responsible for the safe conduct of the SPIES operation.

a. **Preflight Duties.**

(1) Inventories and inspects all SPIES equipment.

(2) Briefs the pilot and other concerned personnel about details of the operation, especially the extraction and dismounting procedures.

(3) Ensures that an ICS helmet and gunner's belt are available for the SPIES master's use. Uses a sling rope if no belt is available. Connects and checks the operation of the ICS to be used. (ICS communications is established between the SPIES master and pilots on all SPIES operations.)

(4) Attaches the SPIES rope to the helicopter IAW the guidance in this chapter.

(5) Ensures that nothing is adrift in the aircraft that may fall on a team member.

(6) Checks the location of the emergency axe. Places the axe where readily available, yet secure enough so as not to endanger the men on the SPIES rope. (Inspects the axe to ensure that it is sharp.)

b. **Extraction Duties.**

(1) On arrival at the team's estimated position, assists the pilot in determining the exact location of the team members.

(2) As the aircraft approaches the team's location, aids the pilot (using the clock system) in placing the aircraft directly above the team.

(3) Requests permission from the pilot to drop the SPIES rope when the aircraft is hovering above the team.

(4) Drops the rope, taking care to avoid striking team members on the ground.

(5) Notifies the pilot when the rope is down, and reports all altitude corrections to ensure that team members reach all SPIES attachment points.

(6) Watches for the "thumbs-up" signal from all team members.

(7) On receipt of the "thumbs-up" signal, advises the pilot that the team is ready for extraction and requests a vertical lift-off.

(8) Advises the pilot of the team's position, the location of any potential obstacles, and the avoidance of horizontal movement.

(9) If a team member becomes entangled with an obstacle during the extraction, notifies the pilot and requests that the vertical lift be stopped. If the situation is critical, prepares to cut the SPIES rope (the anchor point or cargo straps) after the team members are secured to the obstacle or on the ground.

(10) When positive that all obstructions are clear, advises the pilot to obtain a safe altitude (about 350 feet AGL for training purposes or as the situation dictates in combat) or to transition into forward flight.

(11) At frequent intervals during the flight, advises the pilot on the safety status of all team members. Maintains a constant visual watch on the team and checks security of the SPIES attachments often.

c. **Dismounting Duties.**

(1) On arrival at the dismounting area, informs the pilot as to the approximate distance of the lower rope end from the ground.

(2) Once the pilot starts the vertical descent, continually informs him as to the approximate distance of the lower rope end from the ground.

(3) Informs the pilot of any horizontal drift that occurs and any obstructions near the SPIES rope. Also informs the pilot of any oscillation that may occur.

(4) Informs the pilot when the rope is about 25 feet above the ground and again when it is 10 feet above the ground. Ensures that the rate of descent is slow enough to enable the team members to land and get out from under team members safely.

(5) Reports when the first man initially touches down, when the last team member starts to safely move away from under the helicopter, and when all team members are disconnected.

(6) On order of the pilot, either retrieves the SPIES rope into the helicopter or disconnects the SPIES rope and drops it to the ground. When using the UH-1H/UH-60 helicopter, the only way to retrieve the SPIES rope while in the air is by having an arranged recovery rope attached with a 16-foot sling rope. In some cases, the SPIES master joins two 12-foot-long sling ropes to haul the SPIES rope aboard and attaches the rope about 5 or 6 feet below the cargo hook or cargo strap hookup point. The type of knot used to connect the sling (or recovery) rope to the SPIES rope is self-tightening in nature (for example, the prusik knot). The SPIES master fastens the standing end of the sling rope to the deck tie-down or uses a snaplink. Although it is important to keep the line out of the way, the primary consideration is its length. The rope must be long enough for any oscillation in the SPIES during flight.

## Section II. PREOPERATIONS BRIEFINGS AND PROCEDURES

Before conducting a SPIES training mission, the participants must have a basic understanding of the requirements. The SPIES master conducts briefings to ensure the soldier extracted and the pilot know the procedures. As SOPs are developed and units train together, the SPIES master simply refers to the SOP. He always gives a safety briefing and conducts an equipment inspection.

**Note:** SPIES has not been approved for Army-wide use. Approval must be obtained from division level or higher commander before the conduct of any training.

### 4-3. SAFETY

As in all training, a safety briefing must precede operations using the SPIES. The briefing should consist of, but not be restricted to, a review of the following:

- Characteristics of equipment associated with the SPIES.
- Equipment inspection.
- Proper donning of the harness.
- Methods of extraction and insertion used.
- Emergency signals.

### 4-4. COMMUNICATIONS

Due to the noise level associated in all helicopter operations, radios are used for communication. They are also used before the arrival of the helicopter. Precise arm-and-hand signals are established in case of radio failure or poor communications due to static or noise overriding the audio output of the radio. During the first part of the operations, the SPIES master either observes (daytime) or knows that a definite procedure is taking place (night or jungle) while the teams are hooking up to the SPIES rope for extraction. Headsets and voice suppressors replace the handset for better radio procedures. This allows the RATELO on the ground to use both hands while the helicopter hovers for a faster and safer hookup. If radio communications are hampered, special procedures are used along with hand or light signals.

### 4-5. EXTRACTION PROCEDURES

After the team has been located, the SPIES master assists the pilot in directing the helicopter to the proper distance over the team.

a. The team leader positions himself so that he can move and approach the rope as it is dropped by the SPIES master. Once the rope is clear of any obstacles, the team leader signals the team to their assigned positions along the 10 hookup points. Using the primary or harness snaplink, each team member hooks to the D-ring on his side of the line. This is the primary hookup. Once the team is hooked up, the team leader hooks into the alternate or second hookup point using the safety line and snaplink. He then faces forward along the line so that he is heading in the direction he is traveling when the aircraft starts its ascent.

b. The SPIES rope is held up and routed over the shoulder closest to the rope. With the other hand, the team members give thumbs-up signals to allow both the team leader and the SPIES master to see that they are ready to go. The team leader physically inspects (if time and situation permits) or hooks himself in on the lowest point along with the RATELO to ensure the running end is clear of all obstacles. He gives the thumbs-up signal to the SPIES master. This thumbs-up signal (at night, an arranged light signal) continues until the team is lifted off the extraction surface. The team leader also signals (arms held straight to the sides) when the team is at safe altitude (approximate 10 feet above the tallest obstacle at the extraction site). The helicopter starts a transition in a horizontal direction on its return flight.

## 4-6. EMERGENCY PROCEDURES

During the flight, from extraction until the team is safely and quickly detached from the SPIES rope, each team member should be aware of any problems that arise from above or below. The man above checks the man below. At the first sign of danger or if an emergency arises, the team leader or a team member places his free hand on his head. Upon observing anyone on the SPIES rope with his hand on his head, the SPIES master tells the pilot to make an emergency landing in the nearest and safest area.

## 4-7. DISMOUNTING PROCEDURES

During the familiarization training phase, the SPIES master ensures all members know they should immediately head in the direction of the nose (12 o'clock) of the aircraft. This allows the pilot to see that the team is out from under the aircraft. When an emergency situation arises, the pilot can make a better appraisal of the situation if he can see all the members of the team. When the helicopter is making a scheduled landing, the team ensures that the SPIES rope does not interfere with the aircraft and that the aircraft does not land on the rope.

## 4-8. INSPECTION OF EQUIPMENT

The SPIES is inspected by a certified SPIES master at six-month intervals and whenever the SPIES master questions serviceability. Outdated, spliced, abraded, or cut rope is removed from service. However, loose or broken stitching in excess of three stitches is not repaired.

> **CAUTION**
>
> **Acid contamination, cuts, or fraying of harness or sling webbing constitute nonrepairable damage.**

a. The certified SPIES master inspects the harness and suspension sling webbing for signs of contamination from oil, grease, acid, or rust. He also checks for cuts, twists, fading, excessive wear, fusing (indicated by unusual hardening or softening of webbing fibers), fraying, burns, abrasions, and loose or broken stitching. The damaged harness or suspension sling is removed from service and returned to supply for appropriate disposition.

b. The SPIES master inspects all hardware for signs of corrosion, pitting, ease of operation, security of attachment, bends, dents, nicks, burrs, and sharp edges.

Replacement of hardware (except chest strap adapter) that requires unstitching of webbing renders the harness unserviceable.

c. The SPIES master replaces the V-ring by cutting the strap above the stitching. He folds and stitches a new end section of the leg strap. If the leg strap is damaged, the SPIES master returns the harness or suspension sling to supply for appropriate disposition.

d. The SPIES master checks ropes, harnesses, and suspension slings for expiration of service or total life, which is 7 years of service (opening manufacturer's package) or 15 years from date of manufacture, whichever occurs first. The SPIES master also ensures the rope is free of splices. He ensures the eye loop at the end of the SPIES rope is not broken, frayed, or loose.

e. The SPIES master inspects the rope surface for cuts, excessive abrasions, and snags. Cuts on the rope are excessive when there are four or more cut strands in any 5-inch length. The two-to-one braided rope has 12 pairs of strands (24 individual strands) around the circumference. Abrasion is extensive when torn yarns are equivalent to that of four strands of any 5-inch length. Rope that has been subjected to heavy loads may display glazed areas where it has worked against hard surfaces. This condition may be caused by paint or fused fibers. After long use, the rope may become fuzzy on the surface (although this should be minimized with the surface coating). However, the effect on the rope's strength is negligible.

f. The SPIES master inspects the rope for signs of contamination by acid, alkaline compounds, salt water, fire extinguishing solutions, and petroleum-based solvents. Although the ropes gradually change color, such changes do not indicate decrease in strength unless the change is due to contact with strong chemicals. Changes in color caused by chemicals are usually spotty; changes caused by use are uniform throughout the length of the rope.

## 4-9. RIGGING OF UH-1H HELICOPTER FOR SPIES OPERATION

Personnel who are not familiar with SPIES are encouraged to observe or take part in the rigging of the helicopter. This builds confidence in the equipment and assists in a more comprehensive training of new SPIES masters.

a. **Equipment.**

- One 120-foot SPIES rope with deployment bag.
- Two 11-foot, 3- or 4-loop cargo slings or two 9-foot, 3- or 4-loop cargo slings.
- Two Type IV connector links (four if aircraft does not have cargo hooks).
- Heavy duty tape (100-mph tape).
- One 12-foot sling rope.
- Five oval snaplinks (nine if the aircraft does not have cargo hooks).

**NOTE:** If the aircraft does not have cargo hooks, use four 11-foot or four 9-foot, 3- or 4-loop cargo slings.

b. **Installation.**

(1) The primary attachment point for the SPIES rope is the cargo hook. The end of the SPIES rope having a polyurethane encapsulated eye is attached to the cargo hook.

The two 9- or 11-foot long cargo suspension slings are joined to form one continuous sling, using a Type IV link. The sling is stretched out on the helicopter deck. One end is taken under the helicopter and through the eye of the SPIES rope. It is then connected on the other end of the sling using a Type IV link assembly. The sling must pass between the helicopter skids and the fuselage.

(2) Once the the SPIES rope and cargo straps are in place, the straps running across the deck of the helicopter are secured in place by at least four and as many as eight snaplinks. The snaplinks are evenly spaced across the deck and alternated from one side of the strap to the other and top and bottom, so that the first snaplink is to the rear of the strap and going around the bottom two straps. The next snaplink is in the front of the cargo strap and around the top two sections of the strap. This process continues until at least four points are established.

(3) If eight snaplinks are available, then each tie-down has two reversed. If no hook is available or is not working properly, the SPIES can be used safely by doubling the cargo slings and Type IV links. Thus, two cargo straps are side by side for a total of four slings and four Type IV links. Padding is used around the edge of the fuselage to protect the sling from damage.

(4) Care must be taken when using the UH-1H helicopter because of the ways in which it can be outfitted. Some helicopters have a step attached at the doorway. This is an added obstruction not only during installation but also during the operation. Others may have rocket pods or machine guns mounted. Not all of the UH-1 helicopters are hooked up the same way each time.

## 4-10. RIGGING OF UH-60 HELICOPTER FOR SPIES OPERATION

The SPIES master inspects all materials required to rig the helicopter.

a. **Equipment.**

- One 120-foot SPIES rope with deployment bag.
- Two 11-foot, 3- or 4-loop cargo slings or two 9-foot, 3- or 4-loop cargo slings.
- Two Type IV connector links (four if aircraft does not have cargo hooks).
- Heavy duty tape (100-mph tape).
- One 12-foot sling rope.
- Five oval snaplinks (nine if the aircraft does not have cargo hooks).

**Note:** If the aircraft does not have cargo hooks, use four 11-foot or four 9-foot, 3- or 4-loop cargo slings.

b. **Installation.**

(1) The primary attachment point for the SPIES rope is the cargo hook. The end of the SPIES rope has a polyurethane encapsulated eye that is attached to the cargo hook. The two 9- or 11-foot cargo suspension slings are joined to form one continuous sling using a Type IV link. The sling is stretched out on the helicopter deck. One end is taken under the helicopter and through the eye of the SPIES rope. It is then connected on the other end of the sling using a Type IV link assembly. The sling passes between the helicopter skids and the fuselage. Padding is used around the edge of the cargo hatch to protect the sling from damage.

(2) Once the SPIES rope and cargo straps are in place, the straps running across the deck of the helicopter are secured in place by at least four and as many as eight snaplinks. The snaplinks are spaced evenly across the deck and alternated from one side of the strap to the other and top and bottom. Thus, the first snaplink can be to the rear of the strap, wrapping around the bottom two straps. The next snaplink is in the front of the cargo strap and around the top two sections of the strap. This process continues until at least four points are established.

(3) If eight snaplinks are available, each tie-down has two snaplinks connecting the same spot, and the swing gates are reversed. If no hook is available or is not working properly, the SPIES can be used safely by doubling the cargo slings and Type IV links. Two cargo straps are side by side with a total of four slings and four Type IV links.

### 4-11. RIGGING OF CH-46/CH-47 HELICOPTER FOR SPIES OPERATION

The SPIES master inspects all the materials required to rig the helicopter.

a. **Equipment.**

- Two 11-foot, 3-loop slings.
- Two 9-foot, 3-loop cargo slings.
- Four Type IV connectors.
- One 13-foot sling rope.

b. **Installation.**

(1) The SPIES rope is attached using two 9- or 11-foot cargo suspension slings and four Type IV links. The cargo slings are passed through the encapsulated eye of the SPIES rope and attached to the outboard cargo tie-down rings on the aircraft floor. Two tie-down rings are used for each sling. Padding is used around the edge of the cargo hatch to protect slings from damage.

(2) Not all of the tie-down rings will be in the exact same position on the helicopters. This is one of the main considerations in deviating from the installation procedures. When possible, the cargo straps are placed to form two U shapes. One strap is placed forward of the cargo hole in the center of the aircraft floor and one aft or toward the rear of the helicopter. Installed properly, the cargo straps hold the SPIES rope comfortably in the center of and slightly below the opening of the cargo hatch.

(3) The use of snaplinks attached close to all four faulty tie-down points not only ensures a backup in case of a faulty tie-down ring but also reduces the amount of movement in the cargo suspension straps. A total of eight snaplinks are used for added security at each point with the swing gates reversed.

## Section III. LAND AND WATER EXTRACTION

If the situation, mission, or terrain suggests the possibility of a SPIES extraction, the leader includes the SPIES harness in each individual's equipment list. If the mission or insertion precludes the wearing of the harness, it will be carried on the top and just inside the flap of the pack being used. Once the extraction helicopter has been requested, the harness is retrieved and donned.

## 4-12. LAND EXTRACTION PROCEDURES

The SPIES is used only when the team requires immediate extraction or is unable to move to a clear (open) position suitable for helicopter landing.

a. The extraction helicopter proceeds to the area and establishes radio or visual contact with the team. The backup helicopter equipped with the SPIES remains aloft and away from the area. The helicopter maintains visual contact with the LZ and monitors radio communications.

b. After the team has indicated readiness for pickup and the tactical situation has stabilized, the extraction helicopter moves to the LZ by the safest route. When the helicopter is above the team's location, the SPIES master drops the SPIES rope on order of the pilot after the aircraft has obtained a stable hover at slightly above treetop height.

c. The team hooks up the same as in the familiarization procedures. Team members sling individual weapons over the shoulders, securing weapons and equipment to withstand the wind. A safety line is attached to prevent lost weapons during SPIES operations. When the team is ready, the team leader gives a thumbs-up signal.

d. During extraction, the team RATELO maintains communications with the extraction helicopter. He gives a verbal backup to the thumbs-up signal and relays all information during the flight. His location is near or at the bottom hookup point so he can assist in giving accurate information about the extraction, clearing of obstacles, and descent.

e. Lift-off of the extraction helicopter is vertical until the SPIES rope has cleared all obstacles. With the barrel directed downward at a 45-degree angle and outward, the team members fire their individual weapons using the hip position.

f. Once the SPIES rope is clear of all obstacles, the extraction helicopter changes to horizontal flight and departs the area by the safest route. Airspeed is limited to 70 knots in moderate climates and 50 knots in cold climates while team members are attached to the SPIES rope.

g. When the extraction helicopter has reached a safe dismount area, a transaction is made into hover flight at an altitude of 250 feet AGL. A vertical descent is started with the SPIES master continuously providing information to the pilot on the distance from the ground to the lower end of the SPIES rope. The vertical descent rate of the aircraft (at touchdown) is less than 5 feet per second.

h. When the team reaches the ground, they immediately move out toward the front of the helicopter. All team members rapidly unhook themselves and teammates that need assistance. Once this is completed, they either move away from the area and set up security or assist in clearing the rope if the helicopter is going to land.

## 4-13. WATER EXTRACTION PROCEDURES

The SPIES is suitable for extracting teams from the water. For this procedure, three inflatable life vests or any type of floatation device is tied to the SPIES rope to provide buoyancy for the rope while in the water. One floatation device is tied at each end of the attachment points, and one floatation device is tied in the middle of the attachment point area just above the middle two sets of D-rings. The team members to be extracted wear their SPIES harnesses under their individual life vest. They can also wear swimming fins, masks, and snorkels (amphibious operations) to ease hooking up to the SPIES rope within the spray area beneath the hovering helicopter.

a. After the extraction helicopter has attained a stable hover above the team member's location, the SPIES master drops the SPIES rope (with floatation attached) on order from the pilot. When the team members complete the hookup to the SPIES rope, the team leader signals the SPIES master to start lift-off. Lift-off of the extraction helicopter is vertical until all team members and the bottom end of the rope have cleared the water. During the initial lift-off, team members know that they are to be dragged through the water. The team members prepare to roll on their backs until clear of the water.

b. Flight speed and altitude are the same as over land. The dismounting procedures remain the same, except when landing on a ship. Once on board, the team members take their orders from personnel in charge of the deck.

## Section IV. AFTER-OPERATIONS PROCEDURES

After-operations procedures include repairing, cleaning, and proper storage of the rappelling equipment.

### 4-14. REPAIRING AND CLEANING OF EQUIPMENT

The contaminated ropes are washed with a mild detergent, such as liquid dish soap, and cold water followed by a rinse in clean, fresh water. Ropes are dried at a temperature not to exceed 140 degrees Fahrenheit. Stubborn oil, grease, hydraulic fluid, and other petroleum stains can be removed with the cleaning agent xylene (Grade A or B, TT-X 916).

### 4-15. STORAGE OF EQUIPMENT

To avoid ultraviolet deterioration, the nylon materials should be protected from direct sunlight. The SPIES rope is stowed in an aviator's kit bag for protection when not in use. Bins or similar facilities are used for storage of SPIES equipment. Shelves used for storage should be at least 4 inches from the walls and 12 inches from the floor. Areas used for storage should be well ventilated and free of oil, acid, cleaning compounds, and other contaminants. Equipment must *not* be stowed above or near hot water pipes or heating apparatus.

CHAPTER 5

# FAST-ROPE INSERTION AND EXTRACTION SYSTEM

*The fast-rope insertion and extraction system (FRIES) is employed by small units to make rapid infiltration and exfiltration using rotary-wing aircraft in confined areas. The use of FRIES is normally restricted to situations where aircraft cannot land, and when mission requirements preclude the need for large amounts of equipment, and ammunition or heavy crew-served weapons.*

**Note:** Headquarters, Department of the Army (HQDA) policy specifies that FRIES is not approved for Army-wide use and names the Commanding General, United States Army Special Operations Command (CG USASOC) as the executive agent for FRIES doctrine. The use of FRIES is restricted to special operations forces, pathfinders, long-range surveillance units, and HQDA-approved schools with a USASOC-approved FRIES program of instruction. Requests for other units to conduct FRIES operations should be submitted to Commander, USASOC, ATTN: AOOP-TRS, Fort Bragg, NC 28307-5000.

## Section I. GENERAL

FRIES information is subject to change. Units that recognize the need to update this chapter should submit recommendations to Commander, USASOC, ATTN: AOOP-TRS, Fort Bragg, NC 28307-5000.

### 5-1. OBJECTIVES

The objectives of this chapter are to—

- Prescribe the requirements for FRIES qualification and training, including proficiency sustainment, safety, and administrative procedures.
- Establish the scope of the duties and responsibilities of key personnel who conduct FRIES operations.

### 5-2. GUIDANCE FOR COMMANDERS

Units that have HQDA approval to perform FRIES operations are authorized to conduct initial FRIES qualification and FRIES master (FRM) qualification training. Training requirements are established in this publication, USASOC Regulation 350-6, and applicable SOAR policies. Aircraft material requirements are specified by the US Army Technology Applications Program Office (TAPO) .

a. FRIES training and operations are inherently risky; therefore, safety is paramount. All training and operations require in-depth attention to detail from planning through preparation and execution.

b. The following basic guidance applies to FRIES training and operations:

(1) Unit commanders must personally approve FRIES training sites.

(2) A risk analysis and or assessment must be completed before FRIES operations.

(3) Night FRIES operations are medium risk or higher.

(4) Units authorized to conduct FRIES operations may incorporate FRIES activities in field training exercises only after personnel have become FRIES qualified.

(5) Field exercises using FRIES activities are subject to the limitations imposed by applicable directives, including mandatory support requirements.

(6) Requests for waiver of procedures or restrictions prescribed herein must be submitted in writing with endorsement of the first general officer in the requesting unit=s chain of command to HQ USASOC.

## 5-3. TRAINING PREREQUISITES

> **CAUTION**
>
> **Trainers must ensure all personnel preparing to participate in FRIES training are tested *with equipment* to ensure they have sufficient upper body strength to safely perform the full scope of roping duties.**

Before allowing soldiers to participate in FRIES training and operations, the unit commander must make sure the soldiers are physically fit. Minimum screening standards that apply to each training participant include—

- Having passed the Army physical fitness test within the past six months.
- Having passed a medical examination within the past two years.
- Being free of any injury or physical condition that could cause a potential safety hazard during FRIES training.
- Demonstration of ability to perform a controlled descent from a height of 15 feet while carrying a 40-pound load (not including the basic duty uniform and combat boots).
- Demonstration of ability to hold a static position on a FRIES rope for 20 seconds, using hands and feet to lock in while carrying a 40-pound load.
- Being qualified and current to fly FRIES activities IAW HQ USASOC and the Special Operations Aviation Regiment (SOAR) Aircrew Training Program and applicable policies (for aircrews).

**Note:** The peacetime maximum soldier load will not exceed 50 pounds. This includes helmet, weapon, vest, web gear, and rucksack. Rucksack weight will not exceed 35 pounds.

## 5-4. PERSONAL EQUIPMENT REQUIRED

The minimum personal equipment required for FRIES training or operations includes—

- Heavy leather gloves.

- A helmet with a chin strap.
- Protective goggles.
- A long-sleeve shirt or jacket, long pants, and boots.
- Hearing protection and identification tags for helicopter operations.

## Section II. FRIES QUALIFICATION TRAINING

Each major unit (commanded by an O5 or higher) is responsible for conducting its own FRIES qualification and sustainment program. FRIES trainers will be FRIES master (FRM) qualified and current. Unit training should be tailored to unit and situation needs, however proponent requisites will not be reduced without written approval of the CG USASOC.

### 5-5. INITIAL FRIES QUALIFICATION TRAINING

Before participating in fast-rope operations, personnel are briefed on the FRIES and its purpose, capabilities, and limitations. The briefing also covers the duties and responsibilities of the pilot in command (PIC), safety, FRIES master (FRM), the assistant FRIES master (AFRM), and any ground assistants. Once the FRIES briefing is conducted, the remainder of the initial training is hands-on practice of the proper FRIES operational techniques.

a. Individuals are shown the proper techniques for boarding the aircraft, moving to the door, grasping and descending, locking-in, and clearing the rope. Individuals are also shown the proper wear, hook-up, and use of the STABO extraction harness.

b. After the demonstration, all ropers participate in a practice exercise in which they properly perform the following tasks using a tower and subsequently an aircraft. All ropers must demonstrate a minimum of six properly executed FRIES descents (three without equipment and three with equipment) and at least one successful lock-in.

(1) *Perform FRIES descent and lock-in from a tower at the 34-foot level.* Training should be progressive, starting from a tower at the 34-foot level without equipment and then with equipment. All ropers must complete a successful lock-in at this level. Ropers will not progress above the 34-foot level until after demonstrating the ability to stop descent, lock-in, and hold a stationary position for 20 seconds with equipment.

(2) *Perform FRIES descent and lock-in from a tower at the 60-foot level.* The recommended succession will be no less than three successful FRIES descents on the 60-foot tower, again starting without equipment, completing a lock-in, and finishing with equipment.

(3) *Perform FRIES descent from a helicopter at 60 feet or higher.* The recommended completion should be at least two helicopter descents from 60 feet or higher (one without equipment and one with equipment).

**WARNING**

**During initial qualification, students will not perform rapid exits or have more than three soldiers on a rope at any time. During advanced training, the limit is no more than five personnel with equipment on any rope at any time. In all cases (insertion or extraction), the maximum load allowed on any fast-rope system will not exceed 1,500 pounds.**

**CAUTION**

**All ropers should be limited to no more than a total of 10 roping events in a 24-hour period; no more than 6 events should be with equipment.**

### 5-6. FRIES PROFICIENCY SUSTAINMENT TRAINING

Commanders must make sure soldiers participating in FRIES operations receive sustainment training on equipment and procedures within 24 hours before the FRIES operation. Ropers who do not attend FRIES sustainment training will not be allowed to participate in FRIES operations. As a minimum, training will include a review of the following:

- Arm-and-hand signals.
- Individual equipment riggings.
- Aircraft familiarization.
- Safety procedures.
- Any rehearsals the FRM or commander deems necessary.

## Section III. FRIES MASTER SELECTION AND QUALIFICATION TRAINING

Selection and qualification of FRIES masters (FRMs) is a unit prerogative. Units will maintain records of qualification and proficiency. These records will accompany SM to new units when a permanent change of station occurs.

### 5-7. FRIES MASTER SELECTION

Units may train and certify soldiers as FRIES-qualified FRMs. Soldiers selected by units to qualify as FRMs should be the rank of E5 or higher and should have demonstrated leadership abilities with emphasis on maturity and decisiveness.

### 5-8. FRIES MASTER PREREQUISITES

Soldiers performing FRM, AFRM, and safety duties must have completed initial FRIES qualification training, be experienced in FRIES operations, have completed FRM training, and have executed at least 10 FRIES descents and 2 extractions. An FRM candidate's experience must include no less than five descents with combat load and at least one night descent with equipment from 60 feet or higher.

### 5-9. FRIES MASTER TRAINING AND CERTIFICATION

During FRM training and certification, FRM candidates must participate in three FRIES helicopter operations (twice observing another FRM and once executing FRM duties under observation of a current FRM) and be certified by either the parent unit or an authorized school. FRMs must be proficient in FRIES operations and must demonstrate proficiency at the following:

- Inspecting, preparing, and rigging all FRIES gear.
- Inspecting and preparing aircraft for FRIES operations.
- Controlling and coordinating actions of AFRMs, safeties, and ropers.
- Preparing and conducting both insertion and extraction operations.
- Conducting troop and pilot briefings.
- Giving arm-and-hand signals.
- Giving time warnings and associated commands.
- Deploying and retrieving FRIES ropes.
- Rigging and lowering equipment.

### 5-10. FRIES MASTER REFRESHER TRAINING

Units will conduct refresher training to maintain acquired skills. FRMs who have not participated in FRIES operations during the past six months will receive refresher training by a current FRM and serve as an AFRM before performing FRM duties. Refresher training for ropers consists of an FRM briefing and participation in FRIES training.

## Section IV. KEY PERSONNEL DUTIES AND RESPONSIBILITIES

The following personnel duties and responsibilities provide baseline requirements for the safe conduct of FRIES operations. Unit standing operating procedures (SOPs) may increase (but will not reduce) training safety requirements.

### 5-11. AIR MISSION COMMANDER

The air mission commander (AMC) provides overall command and control when conducting large, complex missions. The AMC is selected or appointed by the headquarters responsible for the overall mission. He ensures that all operators, aircrew members, and support elements synchronize their actions during the conduct of the mission and the FRIES operation. An AMC is not usually needed during small-unit qualification or proficiency training events when only one helicopter is used.

### 5-12. FRIES TRAINING OFFICER

Units that conduct FRIES training appoint a training officer in charge (OIC) or a noncommissioned officer in charge (NCOIC) to manage the training. The OIC is responsible for planning and coordinating various aspects of training to include—

- Coordinating all support activities, such as procuring FRIES equipment, aircraft, training areas, medical support, and communication.

- Assigning qualified personnel to perform FRM, AFRM, ground safety, and medic duties.
- Adhering to procedures for planning, preparing, and executing the operation IAW this TC, USASOC Regulation 350-6, and the participating unit=s SOPs.
- Ensuring the FRMs, AFRMs, PICs, aircrews, safeties, and medics are briefed on the operation.
- Ensuring FRIES training events are not conducted on or over unsafe ground. (The ground must be free of loose material that could become airborne and cause an unsafe condition or obscure vision for aviators or ropers.)
- Ensuring a medic, an aid bag, a backboard, and a dedicated vehicle and driver are on site during all training.
- Ensuring positive communication with aircraft is maintained throughout training.

## 5-13. PILOT IN COMMAND

The aviation unit providing helicopter support for the FRIES training and operation appoints the PIC. The PIC is responsible for all aspects of the flight and ensures—

- Aircrew members are current and qualified to conduct FRIES operations
- Aircrew members are informed of and understand their responsibilities in fast-rope operations.
- Procedures for planning, preparation and execution are adhered to IAW this TC, USASOC Regulation 350-6, and applicable SOPs and policies.
- Aircraft is properly configured to perform the mission.
- Aircraft is at the proper altitude, airspeed, and location as briefed.
- The command to deploy ropes is not given until the aircraft is at a stabilized hover.
- The aircraft position is maintained to keep ropes in contact with the surface until all descending ropers are on the ground (or extracting ropers are securely attached for extraction).
- Ropes are fully recovered inside the aircraft (or jettisoned) before the aircraft departs the stabilized hover position at the infiltration site.
- All personnel are briefed on in-flight emergency and safety procedures.
- Ropes are never deployed with anything other than night illumination attached to the free end.
- All personnel are briefed on fire support to be provided by the aircraft including:
  - — The nature of fire support the aircraft can provide.
  - — The time the fire support starts, shifts, and stops.
  - — The primary and alternate commands and signals that start, stop, lift, or shift fires.

## 5-14. FRIES MASTER

Units conducting FRIES operations will designate one overall FRM to organize, coordinate, and supervise the activities of the day, and AFRMs as needed. AFRMs are additional FRM-qualified soldiers who assist the primary FRM. An AFRM will be designated for each roping point. AFRMs are responsible for all preparation, inspection, and command and control of all roping activities on their points. FRMs also must—

- Ensure all FRIES roping personnel, including FRMs, AFRMs, and safeties, are qualified and current and understand the proper roping procedures for FRIES operations.
- Perform safety and serviceability checks on all FRIES equipment and rigging.
- Ensure the attachment bar or points are serviceable and free of any defects or contamination and that the quick-release mechanisms and safety pins are present and serviceable and operate correctly.
- Inspect the rope to make sure it has no contamination, damage, or defects that could make it unsafe.
- Check the rope to make sure it is the correct type and length for the operation (smooth rope for infiltration or looped for exfiltration).
- Ensure the rope is properly attached with safety pin in place and back coiled.
- Ensure the extraction harnesses are serviceable and properly worn.
- Ensure the rope chemlights are correctly rigged and illuminated when needed (two at the mount, two at the end, and two 15 feet from the end).
- Confirm the proper seating arrangement, including seat belts or a safety strap for ropers.
- Coordinate with the PIC and brief the aircrew, safeties, and ropers on the following:
  - — The aircraft flight, approach, flare, and position over the target site.
  - — The height of the aircraft above the target.
  - — A description of the FRIES site (rooftop, small clearing) and size, as well as the actions of the ropers upon arrival at the site.
  - — The signal to deploy the rope, who will deploy it, and how it will be done.
  - — Time warnings and actions at each time warning.
  - — Arm-and-hand signals.
  - — The use of radios and intercoms.
- Brief the PIC and crew chief on verbal commands and final adjustments over the target.
- Confirm the correct target and inform the pilot before the deployment of ropes or ropers.

**WARNING**

**The FRM must ensure that nothing, other than rope illumination, is ever attached to the free end of the FRIES rope during deployment and that equipment is never attached and dropped.**

- Deploy the ropes when the PIC gives the command DEPLOY ROPES. (This action may be delegated to the aircrew or safety.)
- Ensure safeties and AFRMS understand the time and activity sequence and all arm-and-hand signals.
- Ensure safeties know the proper procedures to—
- Keep the aircraft on target.
  - — Clear the ropes.
  - — Recover the ropes.
  - — Jettison the ropes.
- Observe the target and pass to the PIC, through the crew chief, final aircraft position adjustments over the target to make sure conditions are safe.
- Deploy the ropes when the PIC gives the command DEPLOY ROPES. (This action may be delegated to the safety.)
- Ensure all ropers know how to hook up for extraction and know the correct arm-and-hand signals.
- Direct the aircrew or safety to inform the pilot when the aircraft is on target (the PIC replies with the command DEPLOY ROPES).

**CAUTION**

**When equipment is to be lowered, it should be lowered before the FRIES is deployed or personnel begin roping. When simultaneous, equipment and ropers will exit from opposite doors of the aircraft.**

- — The safety repeats, " DEPLOY ROPES." The FRM signals the AFRMs to deploy ropes and makes a final check, ensuring all ropes from his aircraft reach the surface and are safe for descent (at least 15 feet of rope must be on the surface).
- — When conditions are safe, the FRM commands "GO" to the ropers and points to the rope or leads the stick out.

**Note:** If the FRM leads the stick out, the AFRM automatically assumes control. He completes marshaling the remaining ropers and is the last man to exit.

- Coordinate actions with the crew chief for the positioning or storage of seat belts, safety straps, rucksacks, door bundles, and communication cords.

**Note:** During night operations, upon completing descent, the AFRM observes the the ropes of his aircraft. When he sees that all ropes are clear, he issues the prearranged ALL CLEAR hand signal to the aircrew safety. This hand signal tells the aircrew that they are free to jettison or recover the ropes, and fly away.

## 5-15. FRIES AIRCREW MEMBER AND SAFETY

FRIES safeties are aircrew members who are FRM-qualified and perform the additional duties of the safety during helicopter roping operations. Safeties assist FRMs and pilots as needed. Aircrewmen performing safety duties differ from FRMs and AFRMs in that safeties are members of the aircrew (crew chief or gunner) whose orientation is the flying portion of the operation. Conversely, AFRMs are FRM-qualified members of the ground unit who are performing FRM duties as assistants to the primary FRM and who exit with the ground unit. Safeties remain in the aircraft and operate the doors, extend and retract the rope mount bars, and wear night vision goggles (NVGs) when appropriate. Safeties must have completed FRM training IAW this TC and the aircrew training program (ATP).

a. During roping operations, the safety's four major areas of responsibility are:

(1) Observe all activities in the aircraft and assist the pilot and FRM or AFRM as needed.

(2) Serve as the communication link between the PIC and the FRM. The safety monitors aircraft communication and relays information between the PIC and the FRM, constantly keeping both informed.

(3) Assist the FRMs and AFRMs in lowering cargo and clearing the aircraft of ropers.

(4) Clear and jettison the ropes, when appropriate.

b. The safety performs the following duties:

(1) Relays time warnings, aircraft positioning commands, and other information to assist the FRM.

(2) Conducts aircraft equipment rigging and serviceability inspections, making sure mounts, doors, and ropes are ready and safe for operations both pre-flight and en route.

(3) Verifies that the aircraft is at a stabilized hover with ropes in contact with the surface and remains so (or stops stick until corrections are made).

(4) Checks the rope to make sure 15 feet of rope are on the target surface.

(5) Tells the PIC, "ROPES DEPLOYED", when appropriate.

(6) Tells the PIC, "ROPERS OUT", when the first roper has exited.

(7) Helps the AFRM in controlling and spacing each roper's exit.

(8) Monitors the rope and ropers to make sure they do not get hung up, dragged, or dropped off site.

(9) Observes the exit of the last roper until the roper reaches the surface, moves clear of the rope, and gives the ALL CLEAR signal.

(10) Tells the PIC, "ROPERS DOWN", (station name: aft, forward, left, right) when all ropers are on the surface and conditions are safe to jettison the ropes.

(11) Passes the signals as required to position the aircraft, including HOLD, MOVE LEFT, RIGHT, FORWARD, BACK, UP, or DOWN.

(12) Acts as belay man during cargo lowering when needed.

(13) Jettisons ropes upon receiving the command JETTISON ROPES from the PIC. Tells the PIC, "Ropes are jettisoned," when the ropes have been recovered or jettisoned.

(14) Tells the PIC when ALL CLEAR.

| CAUTION |
| --- |
| **This command tells the pilot the aircraft is cleared for flight.** |

(15) Deploys the rope for extraction upon receiving the PIC command DEPLOY ROPES.

(16) Observes the extracting rope and personnel and informs or guides the PIC through aircraft positioning and throughout the recovery flight.

### 5-16. FRIES ROPER

Ropers are responsible for notifying the FRM, AFRM, safety, or pilot if they observe any unsafe act or condition. Ropers may halt or call for a halt of roping for safety at any time. During FRIES insertion training, the number of ropers on the fast rope at one time is limited to three. During extraction training, the total weight per extraction bar or rope will not exceed a total of 1,500 pounds. When rucksacks are worn by ropers, the rucksacks must not exceed a weight of 35 pounds (total roper equipment weight will not exceed 50 pounds). Ropers will—

- Keep hands at head level.
- Maintain visual contact with lower ropers during their descent.
- Maintain a minimum of a 1-second interval on exit to avoid collisions.
- Keep at least two points of contact on the rope (both hands) at all times.
- Use their feet for additional breaking any time needed.
- Execute descents at a safe speed.
- Slow the rate of descent halfway down the rope to avoid landing on each other.
- Move quickly away from the rope upon arrival on the surface.
- Know correct wear of extraction (STABO) harness and extraction procedures.

| WARNING |
| --- |
| **The FRM, AFRM, and safety must make sure all personnel are on the surface and have moved clear of the ropes before aircraft fly away or ropes are jettisoned.** |

## Section V. FRIES RIGGING OF AIRCRAFT

The aviation unit is responsible for preattachment inspection and installation of FRIES anchor point mount bars and associated equipment. All aircraft used to conduct FRIES operations with US Army personnel must be rigged and equipped IAW current parent service directives--for Army aircraft, USASOC, SOAR, and TAPO directives apply. Rigging of the aircraft includes installation and removal of the seats and seat belts (or personnel restraints), and attaching and detaching the ropes.

**Figure 5-1. UH-60 helicopter**

> **CAUTION**
> **Army aircraft that have the FRIES mounts installed must have an airworthiness release for FRIES operations from the Aviation Troop Support Command (ATCOM) TAPO.**

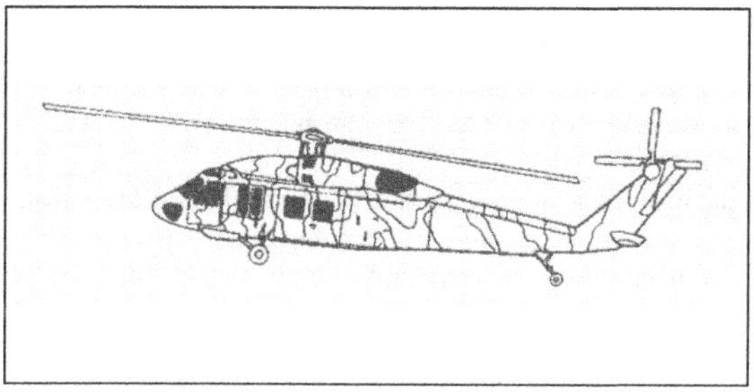

## 5-17. RIGGING OF FRIES IN UH/MH-60

The UH-60 is a twin-turbine, medium-speed, single-main-rotor helicopter (Figure 5-1).

The aviation aircrew or safety is responsible for—

- Inspecting the FRIES mount bar assembly, attachment hardware, and associated equipment to make sure they are serviceable and ready to use.
- Ensuring the bar assembly is correctly mounted to the aircraft (Figure 5-2).

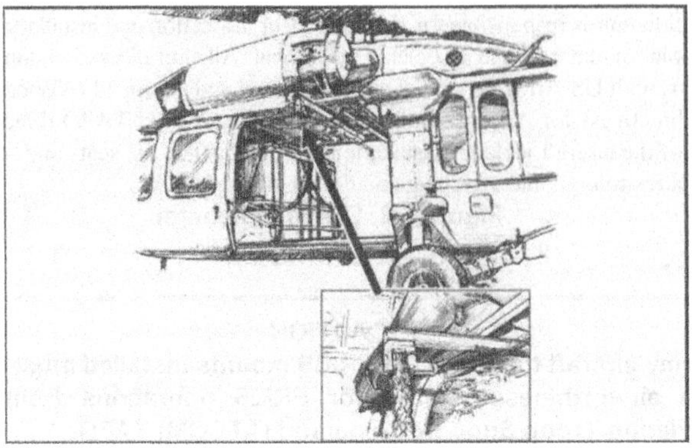

**Figure 5-2. Bar assembly correctly mounted to aircraft.**

- Ensuring the mount is complete, functional, and free of any defect.
- Ensuring the mount and the aircraft floor are clean and free of any petroleum, oils, or lubricants (POL) or other contaminants that could weaken the rope or cause personnel to slip or fall.
- Ensuring the mount and aircraft are free of sharp edges or burrs that could cut the rope.
- Ensuring no metal fatigue or structural weakness exists that could cause the system to fail.
- Extending locking and retracting mount bars as needed.
- Lowering and belaying equipment as needed.
- Opening, locking, and closing cargo doors as needed.

**Note:** For cold-weather operations or during flights of long duration, the cargo doors are closed and latched until the time specified for opening. The aircrew will operate doors or directly supervises those who do. In good weather, doors should remain locked open.

- Removing or reconfiguring troop seats.
- Providing seat belts or floor-mounted personnel restraints for ropers' use during flight.
- Attaching and detaching ropes as needed.

**Note:** Clusters of rucksacks or small door bundles (weight limited to 250 pounds) may be lowered (belayed) on rappelling ropes. Cargo lowered this way must never be free dropped (the belay must start at the time the bundle is pushed [tipped] out the door).

## 5-18. RIGGING OF FRIES IN CH/MH-47

The CH-47 is a tandem-rotor, medium-transport helicopter (Figure 5-3).

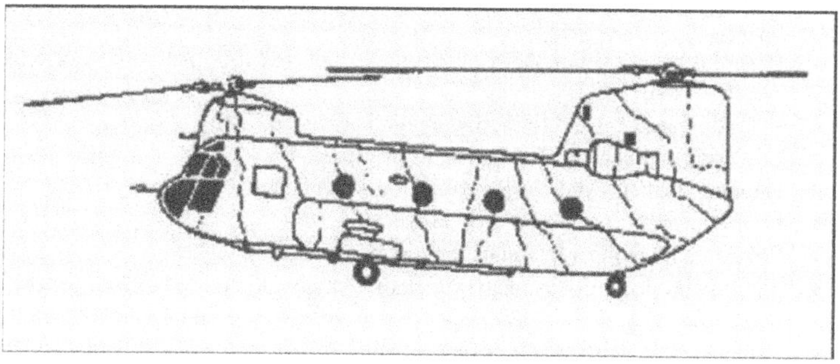

**Figure 5-3. CH-47 helicopter.**

The aviation aircrew or safety is responsible for:

- Inspecting the FRIES mount bar assembly, attachment hardware, and associated equipment to make sure they are serviceable and ready to use.
- Ensuring the bar mounts are correctly mounted to the aircraft (Figure 5-4, rear; Figure 5-5, winch).

**Figure 5-4. Bar mounts correctly mounted, rear.**

**Figure 5-5. Bar mounts correctly mounted, winch.**

- Ensuring the mount is complete, functional, and free of any defect.

- Ensuring the mount and the aircraft floor are clean and free of any POL or other contaminants that could weaken the rope or cause personnel to slip and fall.
- Ensuring the mount and aircraft are free of sharp edges or burrs that could cut the rope.
- Ensuring no metal fatigue or structural weakness exists that could cause the system to fail.
- Extending locking and retracting mount bars as needed.
- Lowering and belaying equipment as needed.
- Opening, locking, and closing cabin doors as needed.
- Removing or reconfiguring troop seats.
- Providing personnel restraints for ropers' use during flight.
- Taping and padding any sharp edges that can come in contact with ropes or personnel
- Attaching and detaching ropes as needed.

**5-19. RIGGING OF FRIES IN MH-53**

The MH-53 is a USASOC twin-engine, single-rotor, medium-transport helicopter used by the Army (Figure 5-6).

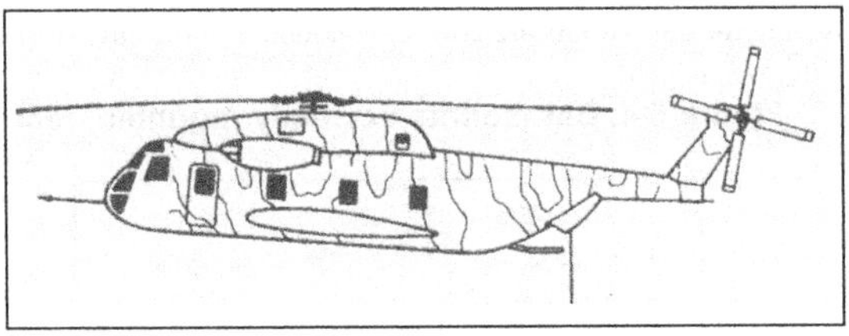

**Figure 5-6. MH-53 helicopter.**

The MH-53 primary FRIES mount bar assembly is located overhead, just aft of the rear ramp. The ramp mount bar assembly spans the cabin width and has a two-pin release mechanism on each end. The ramp position mount assembly provides two points; each will accommodate one FRIES rope or one cargo lowering rope. The ropes are carried back-coiled on the ramp.

a. The alternate position for fast-rope descent from the MH-53 is the crew entrance in the right side of the fuselage immediately behind the pilot's position. In this position, the fast-rope mount must be connected to the external hoist point. The FRIES master is responsible for inspecting the MH-53 mount system for its presence, security, and operability.

b. The aviation unit is responsible for—

- Installation and maintenance of the fast-rope mount assemblies at each position to be used.

- Configuring the aircraft to accommodate the number of soldiers and type of mission planned.
- Ensuring seat belts or restraint straps are present to secure ropers during flight.

**Note:** The MH-53 aircrew can improvise handrails to assist roper exits. When used, improvised handrails will be rigged IAW USAF special operations SOPs.

- Configuring the MH-53 for fast-roping when equipped with external 650-gallon fuel tanks.
  - — Remove the ramp door to install the cross bar.
  - — Place the bar in the brackets with the two quick-release paddles facing forward.
- Configuring aircraft armament (50-caliber machine gun and the GAU-2B ramp minigun) for FRIES operations.
  - — When using the aft right roping position, point the 50-caliber machine gun to the 6 o'clock position and secure the gun with bungee cords.
  - — When using the left aft rope, disconnect the minigun ammunition chute and stow it in the ammunition can, or disconnect it at the ammunition can and securely wrap it around the minigun. Use bungee cords to secure the ramp minigun in the firing position facing the 6 o'clock position.

**Note:** Ideally, only the right aft rope is used when the GAU-2B ramp minigun is installed. Although not recommended, the left aft rope may be used, if required.

## Section VI. EQUIPMENT MAINTENANCE AND INSPECTION

Before conducting a FRIES operation, the FRM must inspect all FRIES equipment (ropes, STABO harnesses, and mount bars) for serviceability and readiness for use.

> **WARNING**
>
> **FRIES ropes and harnesses are critical life support equipment and must be stored, cared for, and maintained as life support items.**

### 5-20. ROPES AND HARNESSES

FRIES equipment should be maintained in the same manner as a parachute. The unit rigger section can provide detailed guidance on appropriate inspection, care, and maintenance of FRIES equipment.

a. FRIES ropes and harnesses must always be stored in a clean, cool, dry space out of direct sunlight and free of chemicals or chemical vapors. Equipment that becomes wet with fresh water should be hung up to dry (indoors) on hardwood pegs. Equipment that is exposed to salt water or becomes imbedded with dirt or mud should be washed and rinsed in fresh water (within 72 hours) and hung up to dry (indoors) on hardwood pegs (out of direct sunlight).

b. Before conducting a fast-rope operation, the FRM—

- Inspects the fast rope thoroughly and carefully. Checks the rope length to ensure it is the correct rope for the operation planned.
- Checks the woven loop on the mount end for excessive wear or chemical contamination. Checks the rope along its entire length for fraying, cuts, and chemical contamination.
  - — Do not use a rope that is severely frayed. (Light fraying on the rope from normal use does not weaken the rope.)
  - — Do not use a rope when any single strand is cut halfway through or has two or more cuts that penetrate one-third or more through any strand's thickness within 1 foot of the running length of the FRIES.
- Inspects the rope for contamination of acid, alkaline compounds, saltwater, fire extinguisher solutions, or petroleum-based solvents. Changes in color caused by chemicals are usually blotchy and have an unusual odor. Although used ropes gradually change color, such changes do not indicate a decrease in strength unless the change is due to contact with strong chemicals. Changes occurring because of use are usually uniform throughout the length of the rope.

**WARNING**

**Any rope known or suspected to be chemically contaminated, cut beyond allowable limits, excessively frayed, or sun damaged must be taken out of service, marked for destruction, and turned in to the rigger section.**

- Inspects the extraction loops to the same standard as the main rope. Ensures the woven attachment loops are secure.
- Inspects the harness to ensure—
  - — Ropers are wearing the harness under all load-carrying equipment.
  - — Ropers have properly fastened all connectors.
  - — Harness material and stitching are not cut, torn, or contaminated, and all hardware is free of corrosion and is in operable condition.

**Note:** Serviceability inspection and maintenance of the extraction harness are the responsibility of the rigger-qualified personnel.

### 5-21. FRIES MOUNT BARS

The aviation unit is responsible for the installation, removal, storage, and maintenance of FRIES mount bars. The US Army Special Operations Aviation Regiment, Ft Campbell, KY, prepares and provides checklists for these procedures. Aviation units participating in FRIES operations must have and comply with current SOAR guidance.

Aviation units:

- Install and maintain the FRIES mount bar; check mount bolt torque and witness marks.
- Check the FRIES mount bar for security and operability.
- Check the condition of the mount bar; inspect for bends, dents, distortion, cracks, corrosion, and contaminants.
- Ensure the fast-rope quick-release pins are installed and operable (two each side).
- Check the security and condition of the rope-release system, as follows:
  - — The mounting bolts are installed and properly torqued.
  - — The safety pins are installed, and lanyards are present.
  - — The system locks and unlocks as designed.
- Ensure the release cables are attached at both ends and are not broken or kinked.
- Check the release handles to ensure they are securely mounted and can move freely.
- Check the latch as follows:
  - — Latch or close the release mechanism and apply a downward pressure on the release arm. Attempt to pull the release handles the release should not open.
  - — Relax pressure on the release mechanism and pull the handles the release should open.
  - — With no pressure on the release mechanism and with the safety pin installed, attempt to pull the release handles--the release should not open.

## Section VII. OPERATIONAL REQUIREMENTS AND LIMITATIONS

This section establishes medical coverage and communication support requirements for FRIES operations, including during adverse weather conditions and night operations.

### 5-22. MEDICAL SUPPORT

Dedicated medical support is required for FRIES training and demonstrations. A qualified medic with assistant, aid bags, two backboards, two stretchers, and a dedicated vehicle with a driver must be present to respond to any mishaps. The nearest medical evacuation (MEDEVAC) unit will be notified of where and when FRIES training is scheduled. MEDEVAC response will be coordinated as on call. Training or demonstrations will not be conducted without medical support present. The ground force commander may direct the FRIES aircraft to transport life or death (urgent) patients when no other MEDEVAC unit is available, providing the medical team is trained and properly equipped with the appropriate equipment for aerial transport of patients. This must be included in the air mission briefing prior to training and coordinated with the medical team providing coverage. FRIES operations that are evaluated as high risk or that occur at night must have a certified combat lifesaver on location and medics (equipped as stated above) in the immediate vicinity.

### 5-23. COMMUNICATIONS REQUIREMENTS

During peacetime training and operations, FRIES lift aircraft and ground elements will maintain continuous radio communications. Aircrew personnel clearing ropes at each FRIES station must be on the intercom system, monitoring aircraft internal communications. They must also be proficient in the use of arm-and-hand signals as precoordinated between FRIES users and aircraft crews. The designated aircrew member will relay time warnings between the PIC and FRM. Routine time warnings are 10-, 6- and 1-minute warnings to the FRM. Ground-to-air communications must be continuously maintained throughout FRIES training.

## 5-24. ADVERSE WEATHER OR TERRAIN CONDITIONS

FRIES training and operations will ***not*** be conducted under the following conditions:

- Anytime temperatures are below 32 degrees Fahrenheit or wind-chill factors that could cause cold-weather injuries exist including those caused by helicopter rotor wash or extraction cruise speed.
- Anytime water or ice on the rope prevents personnel from controlling their descent.
- Anytime the rope is exposed to freezing conditions for a sufficient length of time to freeze, thereby reducing its tensile strength.
- Anytime dust out or white out conditions restrict the aviators or the ropers vision.
- Anytime trees or vegetation on the LZ is too dense for safe training.
- Anytime the LZ is too steep or too slippery for safe training.
- Anytime winds exceed 30 knots or are too gusty for safe operations.

## 5-25. NIGHT OPERATION REQUIREMENTS

Night operations require the following procedures.

a. Attach two chemlights at the free end of the rope and two above the extraction attachment loops (15 feet from the end of the rope) to aid in determining when adequate rope is on the surface.

b. Attach two chemlights (one primary and one backup) to the loop of the rope, or to the rope mount, to guide personnel exiting the aircraft to prevent glare from obscuring vision.

c. Attach one chemlight or luminous tape to the top of each roper's helmet. If a chemlight is used, select a color different from the colors used on the ropes.

**CAUTION**

**Ropers will not wear NVGs during descent. NVGs limit depth**

**perception and cause a tunnel-vision effect. Static FRM and Safeties may wear NVGs to help confirm that fast ropes are fully deployed. The aircrew will wear NVGs as required by flight mission needs.**

## Section VIII. FRIES PROCEDURES

The risks of personal injury and equipment damage during FRIES operations can be reduced by detailed planning, thorough training, and strict adherence to safe procedures. Conducting FRIES operations with heavy loads requires personnel to be proficient in these operations. FRIES users must limit ropers to no more than 6 descents with combat load and no more than 10 total descents in any 24-hour period. The total weight of equipment soldiers may carry during training is limited to never more than 50 pounds. This section provides standardized procedures for FRMs, PICs, and ropers. These procedures are common to all units and aircraft, except as noted.

### 5-26. FRIES MASTER

The FRM performs the following series of actions.

- Briefs members of his team and aircrew.
- Inspects team members and equipment.
- Installs the FRIES rope in the aircraft and conducts safety checks.
- Relays time warnings to team members. (Time warnings are a tool to help keep aircrew and ropers' actions synchronized and can be modified according to user needs; however the 1-minute warning should always be used.)
- Breaks chemlights, if required. Two chemlights or luminous tape will be attached at the anchor point to identify the handhold point.

**WARNING**

**Equipment must never be attached to the FRIES and dropped. As little as 10 pounds of weight can exceed the load limit of the FRIES mounts when the load stops at the end of a 60-foot drop. The only items to be attached to the fast rope during deployment are the position markers.**

- Ensures the rope is properly configured for deployment (back coiled to prevent tangles).
- Ensures the team members are in order of exit no later than the 1-minute warning.
- Confirms target on final approach.
- Deploys the rope and ensures it reaches the ground. During night

operations, the rope is marked with six chemlights: two at the mount, two on the end, and two 15 feet from the end.

- Deploys personnel.
- Accounts for personnel, and signals aircrew.

> **WARNING**
>
> **Because the FRIES mount in the CH-47 helicopter is narrow, special care must be taken by ropers to avoid fouling weapons or equipment on the frame during exit.**

## 5-27. PILOT IN COMMAND

The PIC briefs the FRM, safety, and roping personnel before FRIES operations.

a. The PIC briefing includes:

- Approaching, loading, unloading, and departing the aircraft.
- Actions in the aircraft.
- Flight route and checkpoints en route.
- Altitude or height for roping.
- Time warnings to FRM as needed (recommend 10 minutes, 6 minutes, and 1 minute).

b. When the PIC is over the intended objective he calls, "DEPLOY ROPES" Fast-rope warnings and expected aircrew coordination calls are as follows:

(1) The aircrew or FRM provides any final position adjustments of the aircraft with the commands of MOVE (left, right, forward, back, up, down, or hold) as needed to place the ropes ON TARGET. (When the pilot is to hold the position, an explanation will be provided as soon as possible.)

(2) When the FRM deploys the ropes, the aircrew member calls, "ROPES DEPLOYED."

(3) When the first roper exits the aircraft, the aircrew member calls, "ROPERS OUT."

(4) The safety watches the last roper until he is on the ground and gives the prearranged signal. The safety then tells the PIC, "ALL ROPERS AWAY," and identifies his station. This alerts the PIC that the safety is going to either jettison or recover the ropes.

(5) The safety will then either recover or jettison the rope as planned, clear the aircraft all around for obstructions, and call, "ROPES CLEAR (station)" This informs the PIC that the aircraft is clear for flight.

c. The PIC also briefs on emergencies that could occur in the following situations:

- En route (when ropers are inside the aircraft).
- Before deployment of ropes.

- When fast-rope personnel are on the ropes.
- After the ropers are on the surface.

## 5-28. ROPERS

Ropers perform the following actions:

a. At the command STAND BY (given at 1-minute warning), ropers make final check of themselves and take the prepared-to-exit position.

b. At the command GO, the first roper exits the aircraft, rotates his body 90 to 180 degrees to make sure his equipment clears the aircraft, places the fast rope between the arches of his feet, and begins descent. The roper uses his hands and feet to slow his descent approximately two-thirds of the way down.

c. Subsequent ropers exit at 1-second intervals using the same procedure, but begin slow descent approximately half-way down to avoid landing on other ropers.

d. During descent, ropers keep a lookout and break as necessary to avoid landing on obstructions or on fellow ropers.

e. Ropers prepare to land just before reaching the ground by spreading their legs about shoulder-width apart with their knees slightly bent.

f. At landing, ropers quickly move clear of the ropes to avoid collisions with descending ropers.

**CAUTION**
**According to AR 95-3, anytime operations occur over water, all personnel will wear an approved personal flotation device as identified in FM 57-220 or USASOC Regulation 350-2.**

## 5-29. EQUIPMENT-LOWERING PROCEDURES

Anytime a roper's load exceeds a total of 50 pounds or his rucksack weight exceeds 35 pounds, equipment will be lowered (as clusters or bundles) on separate ropes. Lowering equipment from heights great enough to cause equipment damage necessitates belaying the equipment during descent to avert damage. Belaying devices used in mountaineering allow ropers or safeties to control equipment descent.

**Figure 5-7. Rucksacks upright on cargo deck.**

**CAUTION**
**When equipment is to be lowered, it should be lowered before the FRIES is deployed or personnel begin roping. When done**

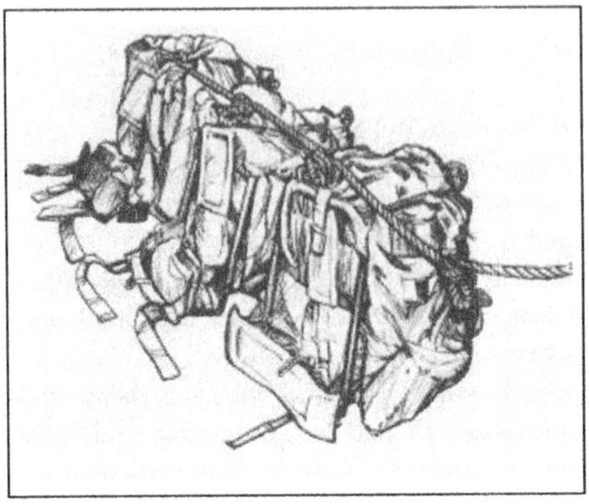

**simultaneously, equipment and ropers will exit from opposite doors of the aircraft.**

a. Rigging equipment to be lowered requires rigging rucksacks in clusters. Clusters of up to five rucksacks may be lowered (smaller clusters work better). FRMs will ensure the total weight of any cluster does not exceed 250 pounds.

(1) Place the rucksacks upright on the cargo deck of the helicopter (Figure 5-7) .

(2) Lay a sling rope across the frames between the shoulder strap attachment points.

(3) Tie a bowline knot at each end of the sling and a figure eight knot at each rucksack.

(4) Attach a snaplink to the frame of each rucksack, to the loops in the sling rope, and to the end of the lowering rope.

(5) For Lowe rucks, install the issue H-harness on the rucks and join them into clusters with a sling rope tied to the quick-release attachment buckles.

b. Rigging of the lowering rope to belay the rucksacks requires the following actions.

(1) Backcoil a rappel rope on the helicopter floor at the gunner=s position.

(2) Attach a snaplink and belay device (either a figure eight [Figure 5-8] or an SMC ladder [Figure 5-9]) to the floor tiedown (with a snaplink) inboard of the ceiling rappelling anchor point.

(3) Attach a snaplink to the ceiling rappelling ring.

(4) Route the belay rope from the backcoil, through the belay device, then up through the ceiling snaplink, and down to the cluster snaplink.

(5) Tie the end of the belay rope into a bowline with halfhitch and snap it into the cluster snaplink.

(6) The FRM identifies two ropers to push the cluster or bundle out of the aircraft. He also identifies an AFRM or a safety to belay the cluster or bundle. The FRM briefs these personnel on the time to lower the bundle, the signals to be used, and the procedure to clear the belay system before deploying the FRIES. In performing his

belaying duties, the AFRM or safety must The aviation aircrew or safety is responsible for:

- Provide sufficient slack to allow the pushers to eject the cluster from the aircraft.
- Control the descent until the cluster reaches the surface.

**Figure 5-8. Belay Device figure eight SMC** **Figure 5-9. Belay device ladder**

- When the cluster is on the surface, unsnap both the upper and lower snaplinks, drop the remainder of the rope from the aircraft, and immediately look out to make sure conditions are safe to deploy the FRIES.
- Inform the FRM and pilot when conditions are safe.

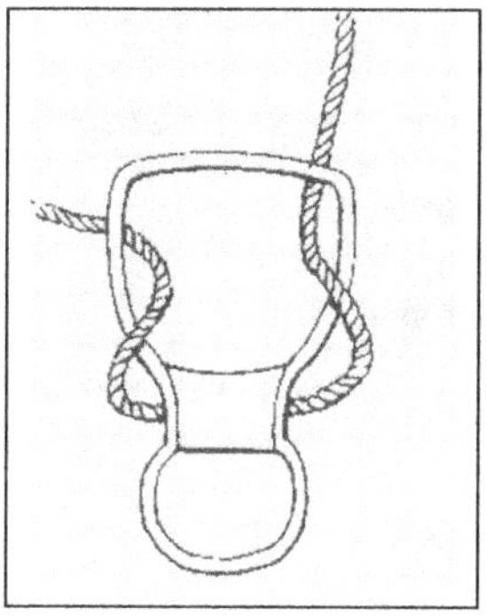

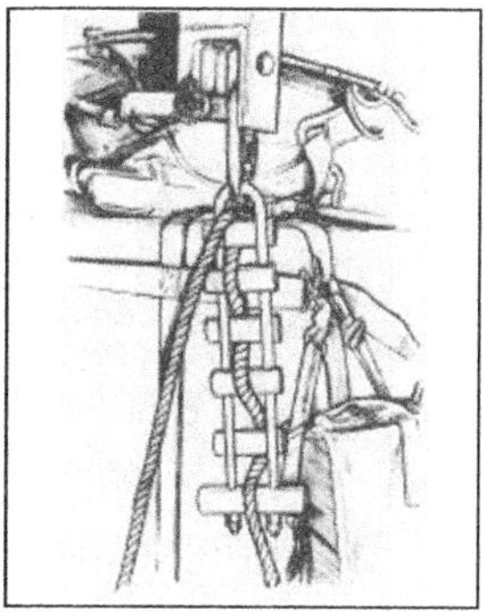

- Deploy FRIES on command

**Note:** Expect to reposition aircraft before ropers exit to avoid ropers landing on equipment.

## 5-30. HELICOPTER OPERATIONS FRM CHECKLIST

Table 5-1 shows a checklist of duties of key personnel in fast-rope operations.

| PRE-ROPING ACTIONS |
| --- |
| Receive a briefing from the OIC or the AMC. |
| Coordinate and brief all participants. |
| Rig aircraft and conduct a joint inspection. |
| Brief ropers, AFRMs, and safeties. |
| Rig and inspect ropers. |

| |
|---|
| Conduct a static rehearsal. |
| **AIRCRAFT LOADING**<br>Position equipment and personnel.<br>Make sure all personnel have straps or seat belts.<br><br>**ACTIONS IN FLIGHT**<br>Monitor the command net.<br>Monitor the aircrew net.<br>Monitor the flight route.<br><br>**ACTIONS AT 10-MINUTE WARNING** (applies to long infiltrations)<br>Issue the 10-minute time warning and GET READY.<br>Check equipment and belay system hookup.<br>Check fast-rope hookup.<br>Secure fast-rope bar in position.<br>Make sure rope is back coiled and markers are attached.<br><br>**ACTIONS AT 6-MINUTE WARNING**<br>Issue the 6-minute time warning.<br>Remove personnel restraints or seat belts.<br>Position personnel and equipment.<br>Break chemlights (night operations).<br>Open aircraft doors, if required.<br><br>**ACTIONS AT 1-MINUTE WARNING**<br>Issue the 1-minute time warning and STAND BY.<br>Position ropers in stick formation.<br><br>**ACTIONS AT FLARE**<br>Identify the target area.<br>Deploy bundles/equipment (Safeties) and clear ropes.<br>Deploy FRIES ropes (FRM and AFRM).<br>Check ropes to make sure 15 feet of rope are on the surface (FRM). |

**Table 5-1. FRM helicopter fast-rope operations checklist**

| |
|---|
| **ACTIONS FOR DESCENT**<br>~ FRM, AFRM, or safety positions the number 1 man at the rope. (The FRM may exit first or last.)<br>~ FRM issues the command GO. AFRM echoes GO command, and the ropers exit the aircraft.<br>Safety informs PIC, "ROPERS OUT." |

| |
|---|
| AFRM or safety controls the rate of exit of the ropers.<br>AFRM exits last.<br>Aircrew or safeties observe the last roper. Safety tells the PIC, ALL ROPERS AWAY, after the last roper is on the surface and signals.<br>PIC issues command JETTISON or RECOVER ROPES.<br>Aircrew or safety jettisons or recovers ropes, and issues ROPES CLEAR report to the pilot. |

**Table 5-1. FRM helicopter fast-rope operations checklist (continued).**

### 5-31. FRIES COMMANDS AND SIGNALS

The following standardized arm-and-hand signals are provided for FRIES operations. Ground-to-aircraft arm-and-hand signals are military standard as found in the Pathfinder field manual 57-38.

a. Time warnings (10-, 6-, and 1-minute) are given by holding up the appropriate number of fingers and verbally sounding off.

b. The signal for GET READY is the same as the standard Airborne signal—arms extended horizontally, fingers extended vertically and joined.

c. The signal for STAND-BY is arms extended downward at a 45-degree angle from the body with hands closed except for index fingers, which are extended, pointing to the floor

d. To execute the signal for DEPLOY ROPES, the FRM or safety forms both hands into fists at mid-thigh level and sweeps arms horizontally outward. At full extension, he opens his hands as if dropping the ropes out of the helicopter and sounds off with DEPLOY ROPES.

e. The signal for GO is arms raised to horizontal, hands closed with index fingers extended pointing to the ropes.

f. The signal for STOP or ABORT is hands closed into fists with arms raised across the forehead .

g. The signal for SIT-DOWN and DON SEAT BELTS is pointing to the floor, followed by moving both fists around waist, as if donning a belt.

h. The signal for OK is hands raised to eye level, index finger and thumb forming a circle with other fingers extended.

## Section IX. EMERGENCY ACTIONS

Procedures for correcting an emergency condition that could reasonably be encountered are described in this section. Multiple emergencies, adverse weather, or other unusual conditions may require modification of these procedures. The nature and severity of the emergency dictate the response necessary; therefore, personnel must use sound judgment in determining the correct action to be taken.

### 5-32. EMERGENCIES BEFORE ROPING STARTS

All personnel sit down, don their seat belts, and take further instructions from the pilot or crew chief.

### 5-33. EMERGENCIES AFTER ROPING STARTS

Emergency procedures to be followed after roping starts are discussed in this paragraph.

a. **Unsafe Drift or Premature Liftoff.** Anyone who detects the aircraft having drifted off the site must immediately stop training and inform the FRM, AFRM , safety, or PIC.

(1) The FRM, AFRM, safety, or roper stops stick.

(2) The ropers stop descent and lock in.

(3) The FRM or crewman informs the PIC, and guides him in moving the aircraft back on target.

(4) The unit continues operations.

b. **Rope Hung or Snagged (On Ground, Tree, Building).** Anyone who detects the rope being snagged or hung must immediately stop training and inform the FRM, AFRM, safety, crew chief, or PIC.

(1) The safety ensures ropers are off the rope and are clear.

(2) The aircraft descends or lands as needed.

(3) The safety frees the rope and informs the pilot.

(4) The unit resumes the operation.

c. **Lost Communications.** All training and operations must be planned to include the use of the intercom between the PIC or crewmen and the FRM. If the intercom fails, the following arm-and-hand signals can be used until the rope can be cleared and the intercom is restored.

- Wrists crossed diagonally above the forehead with clenched fists to signal stop stick.
- Finger pointed toward the exit and a verbal GO to signal ropers to exit.
- Open palm moved and faced in the required direction to signal aircraft movement.
- Clenched fist to stop aircraft movement.

**Note:** Training will not be conducted without radio communications between the pilots or aircrew and the FRM and the ground.

## Section X. SAFETY

Safety is everyone's responsibility. All personnel involved in FRIES operations are responsible for identifying hazardous situations and preventing injuries to personnel. Anyone who observes an unsafe condition or act is authorized to halt the operation and inform the FRM or the PIC.

### 5-34. PREFLIGHT AND INFLIGHT

Anyone who detects an inflight emergency will immediately notify the PIC, crew chief, or safety and follow aircrew instructions.

a. Before a FRIES operation, the FRM will coordinate with the PIC both routine and emergency procedures as well as any unique conditions or potential problems. He also discusses the procedures to be used during flight, during roping, and before the aircraft's departure from the roping site. The FRM then discusses the information with all the AFRMs, safeties, and crew chiefs as well. He reviews the time warnings (including arm-and-hand signals) and appropriate actions such as who will deploy ropes, verify rope status, and perform rope-clearing procedures.

b. The mission prebrief covers any aircraft equipment that, if disturbed, could cause an aircraft or fast-rope problem. The problem areas should be made known to all personnel concerned and preventive measures taken to neutralize or eliminate them where possible.

c. The FRM gives the crew chief the final position adjustment over the target, using hand signals.

d. All safety straps and communications cords that are not in use are securely stowed. All safety straps and communication cords in use should be positioned so they do not present a hazard to personnel movement in the aircraft.

e. After the PIC gives the command DEPLOY ROPES, the FRM is responsible for deploying the rope and making sure it is in contact with the surface before ropers exit.

**Note:** The safety is responsible for observing rope and ropers and for making sure the pilot maintains aircraft position until all ropers are on the surface.

**CAUTION**

**According to AR 95-3, anytime operations occur over water, all personnel will wear an approved personal flotation device as identified in FM 57-220 or USASOC Regulation 350-2.**

## 5-35. DURING ROPING

The following safety procedures apply during roping.

a. During the actual descent, fast-ropers will not use NVGs because they limit depth perception and create a tunnel vision effect.

b. During descent, ropers must watch for obstructions and lower ropers, and must break as necessary.

c. Individual ropers will lock in during emergencies by wrapping one leg around the rope once or twice and standing on the fast rope with the other foot.

d. Ropers must maintain sufficient intervals on the rope to prevent contact with lower ropers. Individual ropers exit when the head of the man in front of them disappears below the ramp.

e. Ropers execute descents at speeds commensurate with their experience and proficiency in FRIES infiltration operations. (For example, less-experienced ropers should descend slowly until they become more proficient.)

f. The number of ropers on each rope at one time must not exceed the safe limit of the rope mount point in use, as follows:

(1) Do not exceed 1,500 pounds per rope on the MH/UH-60.
(2) Do not exceed 1,500 pounds per rope on the MH/UH-47.
(3) Do not exceed 1,500 pounds per rope on the MH/UH-53.
(4) Do not exceed 1,300 pounds per rope on the MH-60G.

## 5-36. SAFETY BRIEFING

All units must present a safety briefing before conducting FRIES training or operations. The safety briefing must cover the following (as a minimum):

- The planned event (day or night, tower or helicopter, high or low hover, insertion or extraction).
- Actions on the aircraft (or tower).
- Operation (or training) location and any hazards within 500 meters.
- Identification of key personnel, their location, and their duties.
- FRIES equipment and its characteristics.
- FRIES equipment rigging or inspection.
- Roping personnel preparation (shirttails tucked in, chin strap fastened, loose straps taped, gloves and goggles on).
- Arm-and-hand signals.
- Emergencies and emergency procedures.
- Medical procedures (medic with aid bags, backboards, litters, and dedicated vehicle with driver on site).
- Communications requirements (frequencies, call signs, required reports).
- Any other information essential to safety.

## 5-37. SAFETY REMINDERS

The following safety reminders will be reviewed by all key personnel before conducting operations.

a. Ropers must wear long-sleeve shirts or jackets, full-length pants, laced boots, helmets, protective goggles, and heavy leather gloves.

b. MILES harnesses must not be worn during fast-roping.

c. Ropers will not wear NVGs.

d. Commanders must personally approve the rope site selected for night operations. The FRIES site must be large enough to permit all ropers, upon reaching the surface, to move clear of the rope and for units to consolidate or re-form.

e. All night rope operations must be assessed as medium risk or higher.

f. FRIES descent training is not authorized with a fast rope longer than 90 feet.

g. Ropers must not carry more than 50 pounds of equipment during FRIES descents. The maximum weight of rucksacks is limited to 35 pounds. When rucksacks exceed 35 pounds (or total load exceeds 50 pounds), gear must be lowered (belayed) on a separate system, preferably from the opposite door.

h. Ropers will not carry bulky equipment that could interfere with FRIES operations. Equipment that cannot be carried by the soldier during roping (due to its weight or bulk) can be lowered (belayed) during insertion. During extractions, rucksacks must be attached to the extraction loops of the FRIES rope using snaplinks.

i. Ropers must not make more than 6 combat equipment descents or a total of 10 descents in a 24-hour period.

j. FRIES training must not be conducted if—

- The aircraft is unstable.
- The fast rope is not fully deployed (15 feet of the rope must be on the surface).
- Any obstacle is present that could interfere with the descent of the ropers.
- The rope is frozen or slippery
- Anyone identifies any unsafe condition.

k. Safety personnel must—

- Position themselves so they can observe active FRIES points with out obstructing ropers.
- Conduct final rigging checks of the mount, equipment, and ropes at the 10-minute warning.
- Ensure the aircraft is over the target area and the ropes reach the ground.
- Constantly keep both the PIC and FRM informed—
  - — When the aircraft is in position for deployment.
  - — When anything is being deployed from the helicopter (ropes, bundles or personnel).
  - — When 15 feet of rope are on the surface.
  - — When the first roper exits on the fast rope.
  - — When all ropers on the surface are clear.
  - — When the ropes are recovered or jettisoned.
  - — When the aircraft is cleared for flight.
  - — When 15 feet of the extraction rope is on the surface.
  - — When extraction personnel are attached to the rope and ready for lift off.
  - — When extracted personnel are safely lowered to the ground.
  - — When extracted personnel are out from under the aircraft.
  - — Any other conditions that could impact safe operations.

**Note:** If the FRM and the safety exit, the crew chief performs rope clearing and or jettisoning duties.

**WARNING**

**Using the word CLEAR tells pilots the aircraft is ready to fly. Before any use of the word CLEAR, FRMs and safeties must ensure that all personnel are well away from the ropes and the ropes have been fully recovered or released (during insertion), or all personnel are securely attached to the ropes (during extraction).**

l. FRIES training must not be conducted in densely wooded areas or areas prone to blowing dust, sand, or snow, which can produce a hazard for ropers or the aircraft.

m. FRIES training and operations must not be conducted during severe weather such as rain, sleet, ice, or extreme cold.

n. The possibility of personnel injury during FRIES training and operations is always present. Every soldier involved in FRIES activities must maintain constant vigilance to prevent personnel injury and equipment damage.

CHAPTER 6

# EQUIPMENT

*This chapter discusses the basic equipment used for rappelling operations (Appendix B). It covers the types of ropes, the procedures for rope selection, and the care for each rope type. It also discusses the two methods of coiling and the different snaplinks used to join the equipment and rope.*

## Section I. ROPES

Ropes are the most important items of equipment for the rappeller. They provide access down the obstacle or aerial platform, while ensuring individual safety.

### 6-1. TYPES OF ROPES

Two types of ropes are available to the rappeller—*static* and *dynamic*.

a. **Static Rope.** This rope allows for minimal stretch of the rope. The static rope stretches about 5 to 15 percent at the point of failure and about 2 percent under a working load. The minimum tensile strength for an 11-mm (7/16-inch) static rope for military use is 4,500 pounds. *Kernmantle* static ropes range from 3 millimeters to 11 millimeters in diameter and are constructed similar to the 550-parachute cord. (*Kern* means core and *mantle* means sheath.) The internal core of the rope is constructed of a continuous multifilament-nylon yarn, which is spun into continuous parallel strands. This internal core is then covered with a nylon-braided outer sheath. Due to the internal parallel strand construction of the static rope, there is less spinning and kinking than with a dynamic rope. Static ropes can be used for slings, harnesses, etriers (stirrups), rappelling, rope installations, hauling lines, and where stretch of the rope is not desired.

b. **Dynamic Rope.** This rope allows for stretch within the fibers of the rope, which can be a disadvantage in rappelling, prusik climbing, and other applications. Dynamic ropes are more susceptible to abrasion and wear. They have about a 5 to 10 percent working stretch. The minimum tensile strength for an 11-mm (7/16-inch) rope for military use is 4,500 pounds. Two types of dynamic ropes are nylon-laid and kernmantle:

(1) *Nylon-laid ropes.* Synthetic fibers are the best climbing and rappelling material. Nylon has become the standard material for climbing ropes and has replaced manila, flax, hemp, and sisal. The military mountaineering critical application rope is constructed in a *laid* fashion with continuous multifilament-nylon fiber yarn twisted into strands. Three strands are twisted into a climbing rope of a specific diameter ( Figure 6-1). Nylon-laid ropes are easy to inspect for serviceability by twisting the fibers. Laid ropes tend to untwist slightly when under a load, causing kinking and spinning. They are also susceptible to abrasion. The following specifications pertain to the standard military nylon-laid climbing/rappelling rope:

- 36 1/2 meters (120 feet) long.
- 11 millimeters (7/16 inch) wide.
- At least a 4,500-pound tensile strength.
- Right-hand lay.
- One-third stretch factor (at point of failure).

- May lose as much as 15 percent of its strength when wet.
- Weighs 6 pounds when dry.

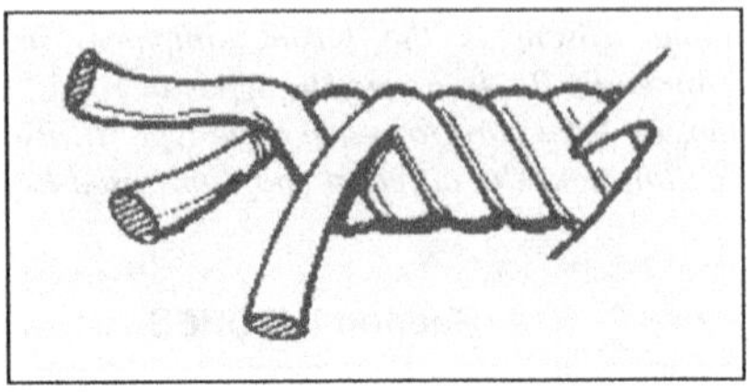

**Figure 6-1. Nylon-laid rope.**

(2) *Kernmantle ropes.* These ropes are similar to their static counterparts in that they consist of an inner core and an outer sheath. The core is constructed of continuous twisted nylon filaments, which are laid or braided together and enclosed in a tightly braided outer sheath. These ropes are well suited for climbing on rock, snow, or ice where a brief elastic stretch of the rope occurs during a fall. They have a stretch of about 40 percent at the point of failure. The breaking strength is high, and there are no exposed strands for rock crystals to work between to damage the rope ( Figure 6-2). There is less sliding friction through the snaplink and over other surfaces, since the outer sheath is smooth. Kernmantle ropes are available in many sizes, lengths, stretch factors, tensile strengths, and fall ratings.

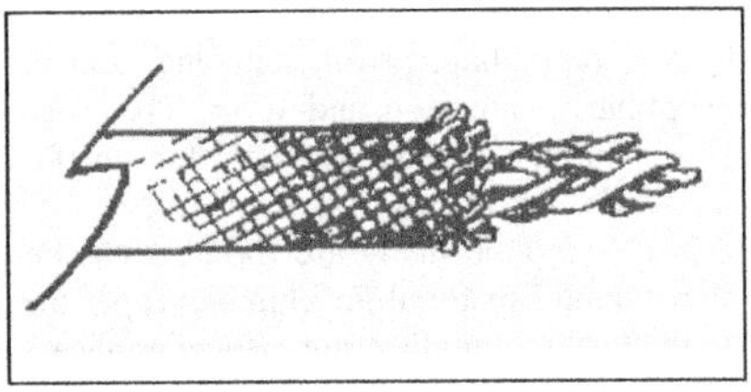

**Figure 6-2. Kernmantle (core and sheath) rope.**

## 6-2. ROPE SELECTION

Rope selection for military operations requires ropes of various types, lengths, and diameters; and depends on its intended use (static or dynamic), environment (water-repellant), climate, mission (lightweight ropes), and other factors. Static ropes should have minimal stretch; be smooth, flexible, and strong; resist abrasion and cutting; and have a low-impact force.

a. *Laid rope* is easier to inspect (by twisting the strands apart to observe the interior surfaces), has greater friction in holding knots, and is less expensive than kernmantle ropes. Its disadvantages include: much lower abrasion resistance, extremely high stretch factor for both static and dynamic loads, greater friction on rock, readiness to kink, tendency to spin a free-hanging rappeller, and greater absorption of water and dirt.

b. *Kernmantle* rope is easy to work with, has less friction, does not kink easily, and has less susceptibility to abrasion. The outer sheath protects the internal fibers from dirt, abrasion, and ultraviolet radiation. Its disadvantages are that it is hard to determine damage to the core fibers, and it is more expensive than laid rope. Kernmantle rope is not readily identifiable as dynamic or static and must be marked accordingly.

c. Rope diameter and length vary with intended use. A standard diameter for many uses is 11 millimeters (7/16 inch). For stirrups, utility ropes, and hauling lines, 7 millimeters (5/16 inch), which can also be used in double strands, is adequate. Standards to remember are that the smaller the diameter, the less the tensile strength; but the larger the diameter, the heavier the rope. The standard lengths are 36 1/2 meters (120 feet), 40 meters (135 feet), 45 meters (150 feet), and 50 meters (165 feet). Normally, standard lengths should not be cut. If a nonstandard length is needed, the rope should be precut before the mission and clearly marked.

d. A rope should be selected for a specific use. There are different types of ropes for various uses. A dynamic rope is designed for climbing, while a static rope is designed for rappelling, rescue operations, load hauling, and rope installations.

**WARNING**

**Using a rope for other than its intended use could result in personal injury or damage to equipment.**

## 6-3. CARE OF ROPES

Because the rope is the rappeller's lifeline, it requires a great amount of care and maintenance.

a. Before and after use, the rope should be carefully inspected for excessive wear, cuts, mildew, or rotten spots. If any of these defects are found, the rope is unserviceable. An unserviceable rope is destroyed by cutting it into usable parts. If no sections are usable, the rope is cut into smaller parts and discarded.

b. Ropes should never be spliced together, stepped on, or dragged. This greatly reduces the tensile strength of the rope.

c. The rope must not come in contact with sharp edges. If this is likely, the areas must be padded or taped before rappelling.

d. The rope must be kept as dry as possible. If it becomes wet, it must be dried out to prevent rotting. (Avoid excessive heat/heating during the drying process.)

## 6-4. COILING

Two methods of coiling are the *mountaineer's coil* and the *farmer's coil* ( Figure 6-3).

a. **Mountaineer's Coil.**

(1) To use the mountaineer's coil, grasp the rope about 3 feet from the end with the left hand. The right hand is then placed next to the left hand and runs along the rope until both arms are outstretched. Grasping the rope firmly, bring the hands together to form a loop, which is laid in the left hand. Repeat, forming uniform loops that run in a clockwise direction until the rope is completely coiled. If the rope tends to twist or form figure eights, give it a slight twist with the right hand when each loop is formed. Coil the rope in a clockwise direction to conform with the lay.

(2) To complete the coil, form a bight about 30 centimeters (12 inches) long with the starting end of the rope and lay it along the top of the coil. Uncoil the last loop and use it to wrap around the coil and the bight. Wrap toward the closed end of the bight, binding the final wrap across itself to lock it in place. Make six to eight wraps to secure the coil. Push the end of the rope down through the closed end of the bight. Carry the running end of the coil either in the pack (by forming a figure eight, doubling it, and placing it under the flap) or by placing it over the shoulder and under the opposite arm. If the rope to be coiled is anchored as in coiling a belay or rappel lane, start the coil near the end closest to the anchor so that the kinks work themselves out of the free end.

b. **Farmer's Coil.**

(1) The farmer's coil is easier and faster to coil than the mountaineer's coil. Establish the center of the rope either by locating its center mark or by grasping both ends of the rope and feeding them out until a bight comes up that is the center of the rope. Begin coiling with the doubled rope the same as the mountaineer's coil method. Leave about 4.6 meters (15 feet) of rope uncoiled. Then, squeeze the coils together and make four to six wraps around the middle of the coil, ensuring that the first wrap is held in place by the other wraps.

(2) Form a bight with the two running ends and place it through the bight formed by the top of the coil. Then, run the two running ends over the top of the coil and through the bight that is formed. Dress them in the center of the back of the carrier. Run the two ends over the shoulders to form shoulder straps. Then, bring them under the arms, cross them in the back over the coil, bring them around the body of the carrier, and tie them off with a square knot at the stomach.

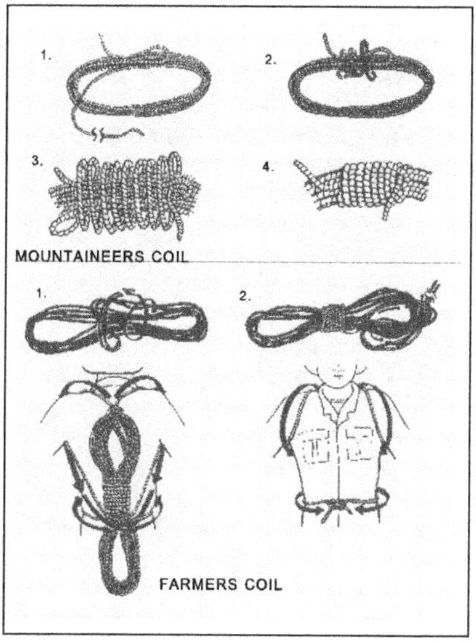


**Figure 6-3. Two methods of tying a coil.**

## Section II. SNAPLINKS

Snaplinks are used to join equipment, rope, and people into a functioning system. When used properly, they are strong, versatile items. Snaplinks are metal (steel, aluminum, or alloys) loops with a hinged, spring-loaded gate on one side. They come in many sizes, shapes, and strengths (with and without locking gates) and in many types of metal. Steel is the strongest snaplink, but it is heavier.

### 6-5. DESCRIPTION

Snaplinks are available in many sizes and shapes to suit different needs. A locking snaplink is best for rappelling. Nonlocking snaplinks are easy to operate, but should be used only where they cannot be accidentally opened. Hollow snaplinks should be avoided since their use is limited. The following information applies to all snaplinks.

a. The weakest part of a snaplink is the gate. The gate must be closed before applying a load. When the gate is open, snaplinks should have little or no lateral movement of the gate.

b. Locking pins should be checked to ensure that they are not loose, worn, or corroded.

c. The metal should be checked for any cracks, grooves, burrs, flaws, or rust.

d. The spring-loaded gate should automatically close securely from an open to a closed position with no gap between the locking pin and notch.

e. If an engraver is used to mark snaplinks, it should be applied only to the gate, never to the load-bearing side.

f. A snaplink should never be side-loaded (across the gate) since this reduces the overall strength to the point of gate failure.

## 6-6. TYPES OF SNAPLINKS

The two types of snaplinks used during military operations are the standard snaplink (Figure 6-4) and the locking gate snaplink ( Figure 6-5). Both snaplinks are available in a D-shaped variation or an oval variation.

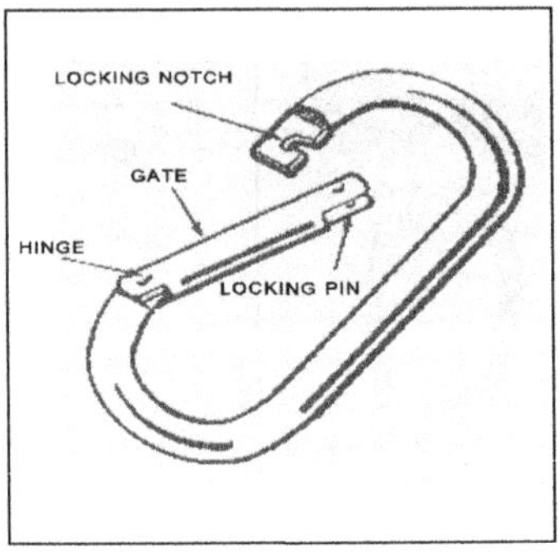


**Figure 6-4. Standard snaplink (oval).**

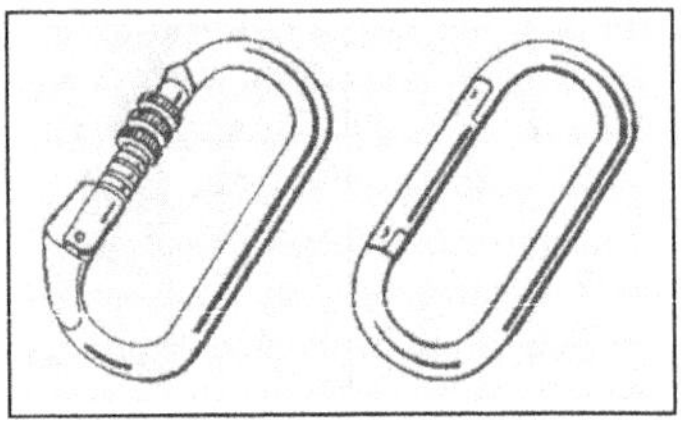

**Figure 6-5. D-shaped snaplinks.**

a. **D-Shaped Snaplinks.** D-shaped snaplinks are stronger than the oval type because the shape directs the largest part of the load to be applied to the longer, stronger side opposite the gate (Figure 6-5). D-shaped snaplinks are made of steel or aluminum alloys, and are available in many sizes and thicknesses, all of which can be with or without locking gates. Their strength and durability vary, which must be considered before use.

b. **Oval Snaplinks.** Oval snaplinks are versatile and have different applications. They are made of steel or aluminum alloys and are available in many sizes and thicknesses. Both sides of an oval snaplink bear the strain equally under load weight. Many modified ovals are available and most can be with or without the locking gate. All must meet military standards or UIAA specifications.

c. **Locking Snaplinks.** Locking snaplinks have a locking mechanism with a threaded sleeve on the gate. The sleeve screws tightly over the gate opening end or hinge end to hold the gate closed (Figure 6-6). A reverse locking gate is needed to prevent a moving rope from unscrewing the sleeve. The locking sleeve and threads should be kept free of dirt and grit. If the sleeve is forced to close, it may strip the threads. The locking mechanism ages and weakens after repeated use and should be routinely inspected for wear. The characteristics of locking snaplinks are as follows:

- Material: steel.
- Approximate strength: 2,000 pounds with the gate closed.
- Gate: spring-loaded without locking sleeve.
- Weight: 4.3 ounces.

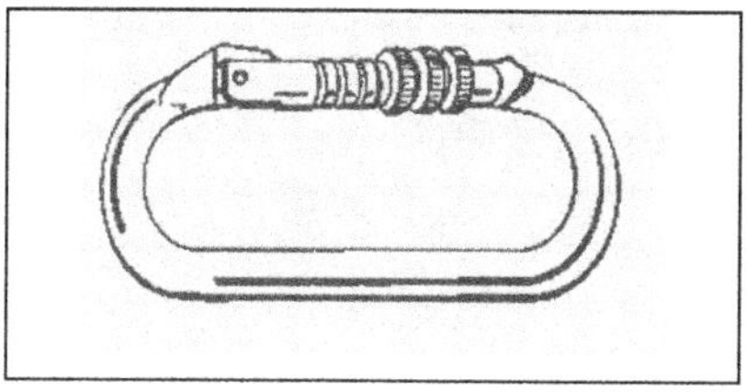

**Figure 6-6. Locking snaplink.**

## 6-7. INSPECTION

All rappelling material and individual rappelling equipment must be inspected before conducting a rappelling operation. The rappel master is responsible for ensuring the serviceability of rappelling material and individual rappelling equipment.

a. **Snaplinks.** Snaplinks should be inspected daily before, during, and after use.

(1) The metal should be checked for cracks, grooves, burrs, rust, and flaws. The gate should open and close freely without binding. There should be no lateral movement when the gate is open. The gate spring action should snap shut when released. The locking notch should have a slant or slot so that the gate remains shut under the impact of a rappeller's fall. The gate pins should not work their way out of their holes and should not be shorter than their holes. If there is a locking mechanism, it should be inspected to ensure that threads are not stripped and that the sleeve tightly locks the gate.

(2) If burrs, grooves, or rough areas are identified, the snaplink should not be used. Rust should be removed with steel wool and the spot rubbed with oil or solvent. The spring should be lubricated as needed. The snaplink must be boiled in water for 20 to 30 seconds to remove the cleaning agents since solvents and oils cause dirt to cling to the snaplink and to rub off on the ropes. It is better to use a dry graphite-based lubricant on snaplinks since such lubricants do not attract dirt.

b. **Ropes.** The ropes are inspected for serviceability. A rappel rope is unserviceable if it is saturated with petroleum products, is mildewed, shows excessive wear, has soft spots (internal damage), or if one strand is cut more than one-half of its diameter. Ropes should be inspected daily—before, during, and after use. An applicable annotation should be made in a corresponding rope log (DA Form 5752-R) ( Figure 6-7).

| ROPE LOG (USAGE AND HISTORY)<br>For use of this form, see TC 90-6-1, the proponent agency is TRADOC | | | UNIT ID MARKING<br>D-11-86-5 |
|---|---|---|---|
| NSN<br>8455-73560366 | DOCUMENT NUMBER<br>W330SV 61890007 | SERIAL NUMBER<br>6-6-6933 | MFR LOT NUMBER<br>G620135115103 |
| DATE OF MFR<br>08 FEB 87 | ISSUE DATE<br>01 OCT 87 | DATE IN SERVICE<br>17 NOV 87 | LENGTH<br>45M (150 FT) |
| DIAMETER<br>11mm (7/16 inch) | FIBER<br>TYPE 6.6 NYLON | COLOR<br>OD Green | CONSTRUCTION<br>DYNAMIC KERMANTLE |

INSPECT ROBE FOR DAMAGE OR EXCESSIVE WEAR EACH TIME IT IS DEPLOYED AND AGAIN AFTER EACH USE. IMMEDIATELY RETIRE ALL SUSPECT ROPES.

| DATE USED | LOCATION | TYPE OF USE | ROPE EXPOSURE | INSPECTOR'S INITIAL/DATE | ROPE CONDITION AND COMMENTS |
|---|---|---|---|---|---|
| 20 NOV 87 | Yonah Mtn GA | Direct Aid Climb 5.7 A2 | 65°F clear & dry Smoothe Granite | DCE 20 NOV 87 | |
| 21 NOV 87 | Yonah Mtn GA | Top Rope Belayed Balance Climb | 68°F Clear & dry Smoothe Granite | DCH 21 NOV 87 | |
| 15-18 Jan 88 | Gulcana Glacier ALASKA | ROPED Glacier Travel | 2°F snow storm | JEH 15-18 Jan 88 | |
| 27 Jan 88 | Smuggler's Notch VT | Front Point Ice Climb | 35°F Cloudy No precip wet rope Due to ice than | AJS 27 Jan 88 | SUSTAINED ONE 15 FT Leader FALL |
| 28 Jan 88 | Smuggler's Notch VT | Front Point Ice Climb | 20°F Cloudy & Snowy | JML 28 Jan 88 | |
| 8 Feb 88 | Mt Shasta CAL | Flat Foot Ice Climb | 8°F Lightly Snowing | DCH 8 Feb 88 | Approx ONE inch of Sheath Slippage on End of rope |
| 22-24 Feb 88 | DeCos Mtn NM | Two man Party Rock Climb 5.7+ | 45°F Clear & DRY GRANITE | LLG 22 Feb 88 | |
| 10-12 Jun 88 | Mt McKinley ALASKA | ROPED PARTY Climb CLASS VI | 20-53°F ONE DAY of RAIN | JMK 10-12 Jun 88 | Abrasion to Rope at Points of Tie-in (Ends & Middle) |

DA FORM 5752-R, MAY 89

**Figure 6-7. Example of completed DA Form 5752-R.**

c. **Gloves.** The gloves are inspected for serviceability. They are unserviceable if holes are found in the friction-bearing surfaces or along the seams.

d. **Helmet.** The helmet is inspected for correct assembly and serviceability. It is unserviceable if it cannot be properly assembled or if any straps are cut or torn one-quarter of its width.

(1) Ensure that the chin strap attaching tabs are secured to the helmet by the locking nuts.

(2) Ensure that the pull-the-dot tab has four plies of nylon and that the chin strap is routed from the inside of the buckle around the horizontal bar and back toward the inside of the buckle.

## Section III. ALTERNATE METHODS OF DESCENT

Snaplinks or carabiners that are used for rappelling are placed through the sewn loops of the seat harness or the sewn loops of the seat-chest combination.

### 6-8. FIGURE-EIGHT DESCENDER

The rappeller attaches a locking carabiner to the harness (Figure 6-8). He routes the rappel rope up through the large hole and places the bight over the collar. He inserts the figure eight with the rappel rope attached into the locking carabiner, and locks it down. The rappeller brakes to the rear and descends as in a seat hip rappel. If heavy loads are to be descended, he uses a double wrap around the collar.

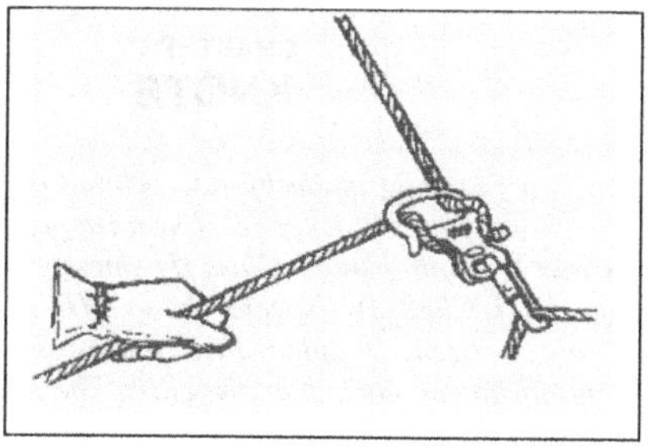

Figure 6-8. Figure-eight descender.

## 6-9. MUNTER HITCH

The rappeller attaches a large radius snaplink to the harness or rappel seat. He ties a Munter hitch and clip into the snaplink (Figure 6-9, page 6-9). He brakes to the front and descends with an L-shape body position.

**Notes:** 1. The guide hand should remain on the standing end of the rope to keep the rappeller's head and face away from the hardware.

2. The Munter hitch creates significant rope-to-rope friction and may cause premature wear on nylon ropes.

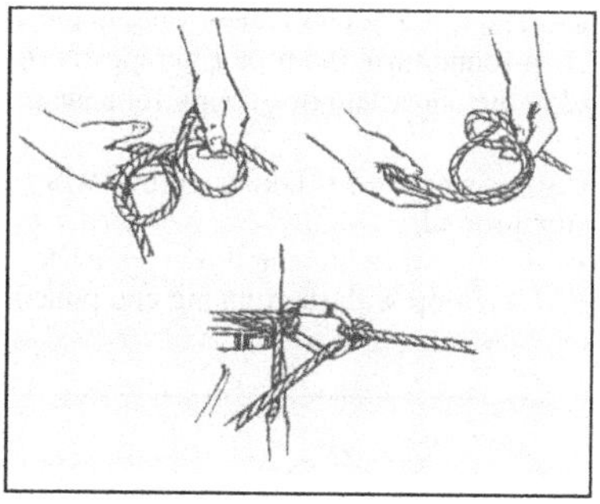

Figure 6-9. Munter hitch.

CHAPTER 7

# KNOTS

*A knot is a fastening made by intertwining or tying pieces of string, cord, rope, or webbing. All knots used by a rappeller are divided into four classes: Class I—joining knots, Class II—anchor knots, Class III—middle rope knots, and Class IV—special knots. These classes of knots are intended only as a general guide since some of the knots discussed herein may be appropriate in more than one class. The skill of knot tying may be lost if it is not used and practiced. With experience and practice, knot tying becomes instinctive and helps the soldier in many situations.*

## 7-1. MILITARY MOUNTAINEERING TERMINOLOGY

The terms explained below are the most commonly used in military mountaineering (Figure 7-1).

a. A *bight of rope* is a simple bend of rope in which the rope does not cross itself.

b. A *loop* is a bend of rope in which the rope does cross itself.

c. A *half hitch* is a loop that runs around an object and locks itself.

d. The *running end* (working end) of the rope is the free end of the rope that can be used.

e. The *standing part* of the rope is the part that is static (anchored, coiled); the remaining part of the rope not being used (also called static end).

f. The *lay* of the rope is the same as the twist of the rope.

g. A *round turn* is a single complete wrap of the rope around an object providing 360-degree contact. The running end leaves the completed circle in the same direction as the standing part. In a round turn, the rope is wrapped around an object 1 1/2 times.

h. The *pigtail* is the short length of rope remaining at the end after tying a knot or coiling a rope.

i. *Back feeding* (or stacking) is taking off one wrap at a time from a coil, and letting it fall naturally to the ground.

j. *Dress down the knot* is tightening down a knot to its functioning form.

k. An *overhand* is a loop with the running end pulled through the loop.

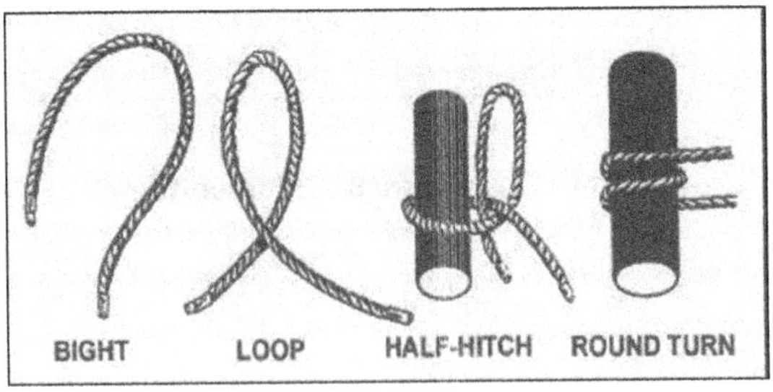


**Figure 7-1. Examples of rope terminology.**

## Section I. JOINING KNOTS

### 7-2. SQUARE KNOT

The square knot is used to tie the ends of two ropes of equal diameter (Figure 7-2).

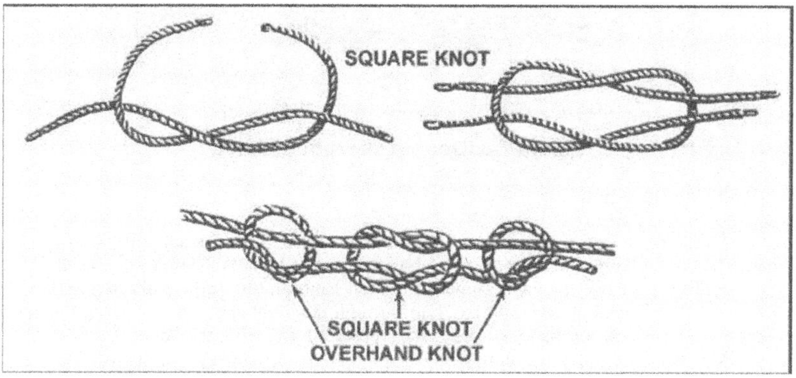


**Figure 7-2. Square knot.**

a. **Tying the Knot.**

STEP 1. Holding one working end in each hand, place the working end in the right hand over the one in the left hand.
STEP 2. Pull it under and back over the top of the rope in the left hand.
STEP 3. Place the working end in the left hand over the one in the right hand and repeat STEP 2.
STEP 4. Dress the knot down and secure it with an overhand knot on each side of the square knot.

b. **Checkpoints.**

(1) There are two interlocking bights.

(2) The standing parts are on the same side and properly secured with overhand knots.

### 7-3. DOUBLE SHEET BEND KNOT

The double sheet bend knot is used to tie the ends of two or more ropes of equal or unequal diameter (Figure 7-3). When a single rope is tied to a multiple of ropes, the bight is formed with the multiple of ropes.

Figure 7-3. Double sheet bend knot.

a. **Tying the Knot.**

STEP 1. Form a 30-centimeter (12-inch) bight in the left hand with the rope or ropes. Ensure that the short end of the bight is facing to the inside.

STEP 2. Place the index finger of the left hand on top of the bight. Bring the rope in the right hand up through the bight and over the index finger of the left hand, so the working end is away from the body.

STEP 3. Reach through the loop and grasp the working end of the rope and bring it back towards the body. Place it between the index finger and the bight, forming a round turn. Repeat this one more time, wrapping toward the tip of the index finger.

STEP 4. Remove the finger and dress the knot down.

b. **Checkpoints.**

(1) The two wraps around the bight are held in place by a locking bar.

(2) The two standing parts of the ropes exit parallel and from opposite ends of the knot. The two working ends form an L and exit the knot at a 90-degree angle and on opposite sides of the knot.

## 7-4. FISHERMAN'S KNOT

The fisherman's knot is used to tie two ropes of similar or dissimilar materials (Figure 7-4).

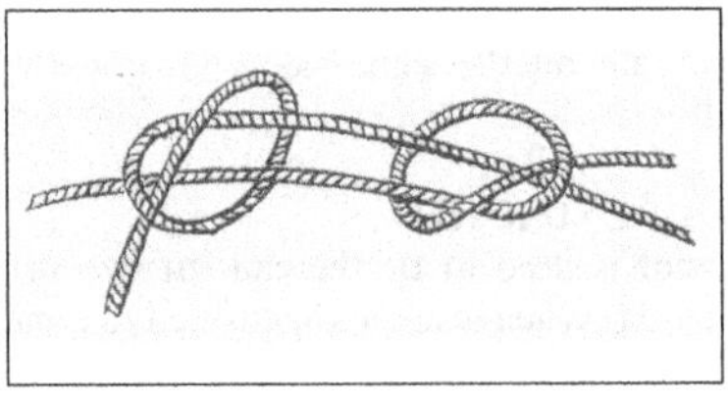

Figure 7-4. Fisherman's knot.

a. **Tying the Knot.**

STEP 1. Tie an overhand knot in one end of the rope.
STEP 2. Pass the working end of the other rope through the first overhand knot. Tie an overhand knot around the standing part of the first rope with the working end of the second rope.
STEP 3. Tightly dress down each overhand knot and tightly draw the knots together.

b. **Checkpoints.**

(1) The two separate overhand knots are tied tightly around the long standing part of the opposing rope.

(2) The two overhand knots are drawn snug.

## 7-5. DOUBLE FISHERMAN'S KNOT

The double fisherman's knot (also called double English or grapevine) is used to tie two ropes of similar or dissimilar materials (Figure 7-5).

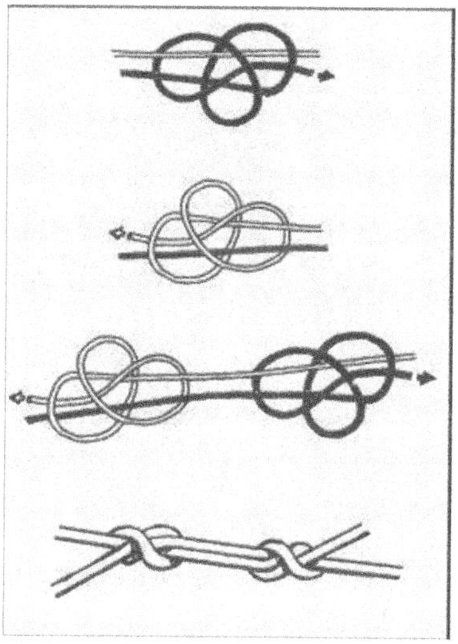

**Figure 7-5. Double fisherman's knot.**

a. **Tying the Knot.**

STEP 1. With the working end of one rope, tie two wraps around the standing part of another rope.
STEP 2. Insert the working end (STEP 1) back through the two wraps and draw it tight.

STEP 3. With the working end of the other rope, which contains the standing part (STEPS 1 and 2), tie two wraps around the standing part of the other rope (the working end in STEP 1). Insert the working end back through the two wraps and draw tight.

STEP 4. Pull on the opposing ends to bring the two knots together.

b. **Checkpoints.**

(1) Two wraps with the working end running through and drawn tight.

(2) Both wraps are drawn together with all the slack out of the knot.

### 7-6. WATER KNOT

The water knot is used to attach two webbing ends (Figure 7-6). It is also called a ring bend, overhand retrace, or tape knot. It is used in runners and harnesses.

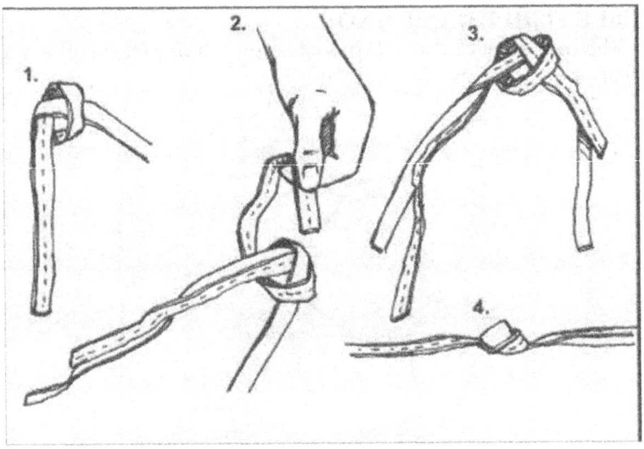


**Figure 7-6. Water knot**

a. **Tying the Knot.**

STEP 1. Tie an overhand knot in one of the ends.

STEP 2. Feed the other end back through the knot, following the path of the first rope in reverse.

b. **Checkpoints.**

(1) There are two overhand knots, one retracing the other.

(2) There is no slack in the knot, with the working ends coming out of the knot in opposite directions (at least 5 centimeters [2 inches] long).

## Section II. ANCHOR KNOTS

### 7-7. BOWLINE KNOT

The bowline knot is used to tie a single fixed loop in the end of a rope (Figure 7-7). It does not slip under strain and is easily untied. This knot is always used when there is alternating tension. It can also be used to tie the end of a rope to an anchor.

**Figure 7-7. Bowline knot.**

a. **Tying the Knot.**

STEP 1. Bring the working end of the rope around the anchor, from right to left (as the rappeller faces the anchor).
STEP 2. Form an overhand loop in the standing part of the rope (on the rappeller's right) toward the anchor.
STEP 3. Reach through the loop and pull up a bight.
STEP 4. Place the working end of the rope (on the rappeller's left) through the bight, and bring it back onto itself. Dress the knot down.
STEP 5. Form an overhand knot with the tail from the bight.

b. **Checkpoints.**

(1) The loop is locked into place by a bight.

(2) The short portion of the bight is on the inside and on the loop around the anchor (or inside the fixed loop).

**7-8. ROUND TURN AND TWO HALF HITCHES KNOT**The round turn and two half hitches knot is used to tie the end of a rope to an anchor and must have constant tension (Figure 7-8).

**Figure 7-8. Round turn and two half hitches knot.**

a. **Tying the Knot.**

STEP 1. Route the rope around the anchor from right to left and wrap down (must have two wraps in the rear of the anchor and one in the front). Run the loop around the object to provide 360-degree contact, distributing the load over the anchor.

STEP 2. Bring the working end of the rope left to right and over the standing part, forming a half hitch (first half hitch).

STEP 3. Repeat STEP 2 (last half hitch has a 15-centimeter [6-inch] pigtail).

STEP 4. Dress the knot down.

b. **Checkpoints.**

(1) A complete round turn should exist around the anchor with no crosses.

(2) Two half hitches should be held in place by a locking bar with no less than a 15-centimeter (6-inch) tail remaining.

(3) When the two half hitches are twisted to the left or the right, they will appear to for a figure eight.

(4) Looking at the top of the half hitches, the rope should appear to be one continuous rope.

## 7-9. FIGURE-EIGHT RETRACE KNOT

The figure-eight retrace knot is also called a rerouted figure eight. Although it produces the same result as a figure-eight loop, by tying the knot in a retrace it can be used to fasten the rope to trees or to places where the loop cannot be used (Figure 7-9).

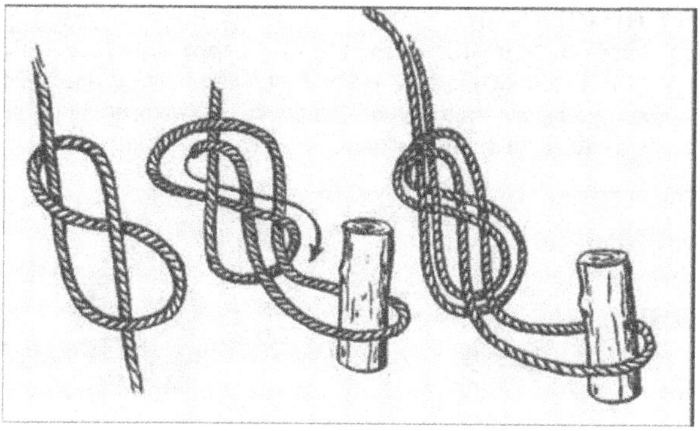

**Figure 7-9. Figure-eight retrace.**

a. **Tying the Knot.**

STEP 1. Use a length of rope long enough to go around the anchor, leaving enough rope to work with.

STEP 2. Tie a figure-eight knot in the standing part of the rope, leaving enough rope to go around the anchor. To tie a figure-eight knot:

- Form a loop in the rope.
- Wrap the working end around the standing part.
- Route the working end through the loop.
- The finished knot is dressed loosely.

STEP 3. Place the working end around the anchor point.

STEP 4. With the working end, insert the rope back through the loop of the knot in reverse.

STEP 5. Keep the original figure eight as the outside rope and retrace the knot around the wrap and back to the long sttanding part.

STEP 6. Safety off with an overhand knot.

b. **Checkpoints.**

(1) At least 15 centimeters (6 inches) of rope extends past the figure-eight knot.

(2) The original figure eight, tied with the standing end, remains as the outermost rope throughout the knot.

(3) The knot is dressed tightly.

## 7-10. CLOVE HITCH KNOT

The clove hitch knot can be used in the middle of the rope as well as at the end of the rope (Figure 7-10). The knot must have constant tension on it once tied to prevent slipping. It can be used as either an anchor or middle knot, depending on how it is tied. (Do not use a clove hitch knot on metal except on a picket-hold-fast.)

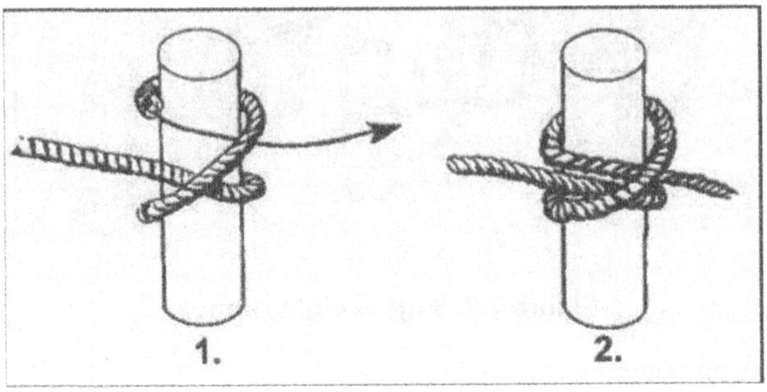


**Figure 7-10. Clove hitch knot.**

a. **Tying the Knot.**

(1) If there is access over the top of the anchor, the knot is tied as follows:

STEP 1. Hold rope in both hands, palms down with hands together. Slide the left hand to the left from 20 to 25 centimeters (8 to 10 inches).

STEP 2. Form a loop away from and back toward the right.

STEP 3. Slide the right hand from 20 to 25 centimeters (8 to 10 inches) to the right. Form a loop inward and back to the left hand.

STEP 4. Place the left loop on top of the right loop. Place both loops over the anchor and pull both ends of the rope in opposite directions to complete the knot.

(2) If the clove hitch knot must go around the anchor, tie the knot as follows (for instructional purposes, assume that the anchor is horizontal):

STEP 1. Place 76 centimeters (30 inches) of rope over the top of the anchor. Hold the standing end in the left hand. With the right hand, reach under the horizontal anchor, grasp the working end, and bring it inward.

STEP 2. Place the working end of the rope over the standing end (to form a loop). Hold the loop in the left hand. Place the working end over the anchor from 20 to 25 centimeters (8 to 10 inches) to the left of the loop.

STEP 3. With the right hand, reach down to the left-hand side of the loop under the anchor. Grasp the working end of the rope and bring it up and outward.

STEP 4. Dress down the knot.

b. **Checkpoints.**
(1) The knot has two round turns around the anchor with locking bar.
(2) The locking bar is facing 90 degrees from the direction of pull.
(3) The ends exit 180 degrees from each other.
(4) The knot has more than a 15-centimeter (6-inch) pigtail remaining.

## SECTION III. MIDDLE ROPE KNOTS

### 7-11. WIREMAN'S KNOT
The wireman's knot forms a single, fixed loop in the middle of the rope (Figure 7-11).

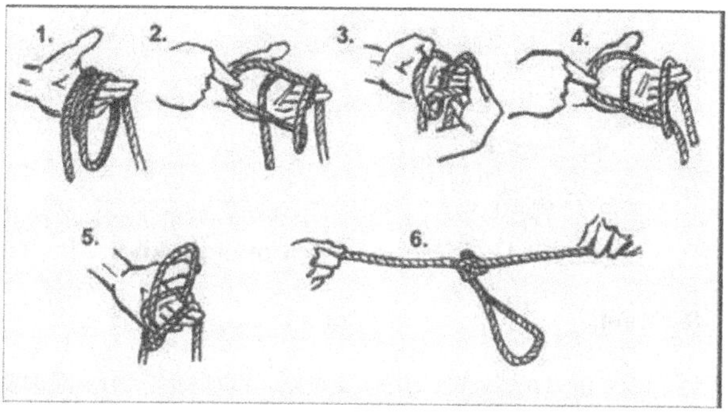


**Figure 7-11. Wireman's knot.**

a. **Tying the Knot.**

STEP 1. When tying this knot, face the anchor that the tie-off system will be tied to. Take up the slack from the anchor, and wrap two turns around the left hand (palm up) from left to right.
STEP 2. Take up a loop of 30 centimeters (12 inches) in the second round turn to create the fixed loop of the knot.
STEP 3. Name the wraps from the palm to the fingertips: heel, palm, and fingertip.
STEP 4. Grasp the palm wrap with the right thumb and forefinger, and place it over the heel wrap.
STEP 5. Grasp the heel wrap and place it over the fingertip wrap.
STEP 6. Grasp the fingertip wrap and place it over the palm wrap.
STEP 7. Grasp the palm wrap and pull up to form a fixed loop.
STEP 8. Dress the knot down by pulling on the fixed loop and the two working ends.
STEP 9. Pull the working ends apart to finish the knot.

b. **Checkpoints.**

(1) The completed knot should have four separate bights locking down on themselves with the fixed loop exiting from the top of the knot and laying toward the near side anchor point.

(2) Both ends should exit opposite each other without any bends.

## 7-12. DIRECTIONAL FIGURE-EIGHT KNOT

The directional figure-eight knot forms a single, fixed loop in the middle of the rope that lays back along the standing part of the rope (Figure 7-12).

**Figure 7-12. Directional figure-eight knot.**

a. **Tying the Knot.**

STEP 1. Face the far side anchor so that when the knot is tied, it lays inward.

STEP 2. Lay the rope from the far side anchor over the left palm. Make one wrap around the palm.

STEP 3. With the wrap thus formed, tie a figure-eight knot around the standing part that leads to the far side anchor.

STEP 4. Dress the knot down with the tail and the bight together.

b. **Checkpoints.**

(1) The loop should be large enough to accept a snaplink but no larger than a helmet-size loop.

(2) The tail and bight must be together.

(3) The figure eight is tied tightly.

(4) The bight in the knot faces back toward the near side.

## 7-13. BOWLINE-ON-A-BIGHT KNOT

The bowline-on-a-bight knot is used to form two fixed loops in the middle of a rope (Figure 7-13).

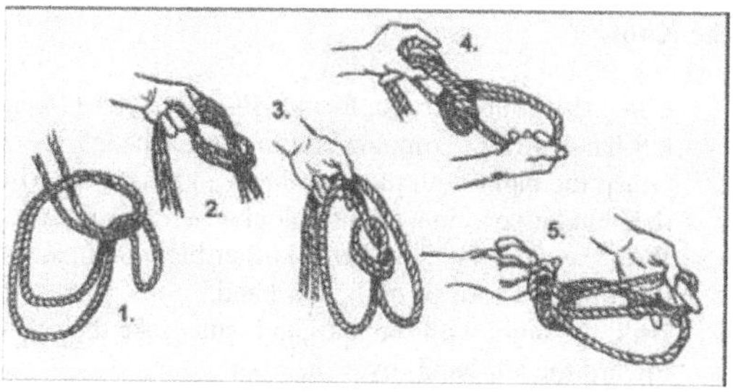


Figure 7-13. Bowline-on-a-bight knot.

a. **Tying the Knot.**

STEP 1. Form a bight in the rope about twice as long as the finished loops will be.
STEP 2. Tie an overhand knot on a bight.
STEP 3. Hold the overhand knot in the left hand so that the bight is running down and outward.
STEP 4. Grasp the bight with the right hand, fold it back over the overhand knot so that the overhand knot goes through the bight.
STEP 5. From the end (apex) of the bight, follow the bight back to where it forms the cross in the overhand knot. Grasp the two ropes that run down and outward and pull up, forming two loops.
STEP 6. Pull the two ropes out of the overhand knot and dress the knot down.
STEP 7. A final dress is required: grasp the ends of the two fixed loops and pull, spreading them apart to ensure the loops do not slip.

b. **Checkpoints.**

(1) There are two fixed loops that will not slip.
(2) There are no twists in the knot.
(3) A double loop is held in place by a bight.

## 7-14. FIGURE-EIGHT-ON-A-BIGHT KNOT

The figure-eight-on-a-bight knot is used to form two fixed loops in the middle of a rope . (Figure 7-14).

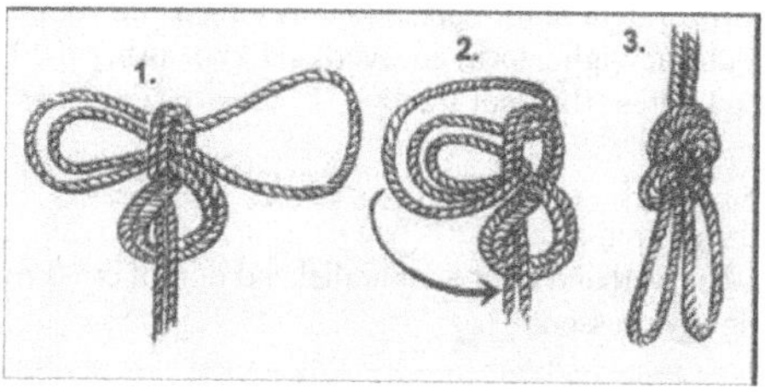


Figure 7-14. Figure-eight-on-a-bight knot.

a. **Tying the Knot.**

STEP 1. Using the doubled rope, form a 46-centimeter (18-inch) bight in the left hand with the running end facing to the left.

STEP 2. Grasp the bight with the right hand and make a 360-degree turn around the standing end in a counterclockwise direction.

STEP 3. With the working end, form another bight and place that bight through the loop just formed in the left hand.

STEP 4. Hold the bight with the left hand, and place the original bight (moving toward the left hand) over the knot.

STEP 5. Dress the knot down.

b. **Checkpoints.**

(1) There is a double figure eight with two loops.

(2) The knot will not slip.

## 7-15. OVERHAND LOOP KNOT

The overhand loop knot forms a single loop in the middle of a rope. It should not be used in a transport tightening system (Figure 7-15).

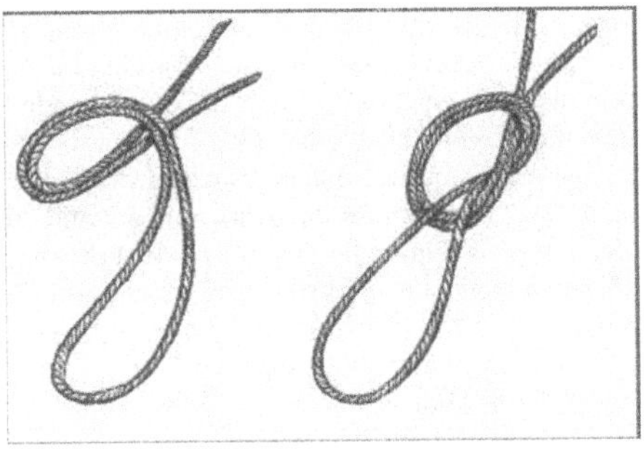

**Figure 7-15. Overhand loop knot**

a. **Tying the Knot.**

STEP 1. Form a bight in the rope about as long as the loop needed.

STEP 2. With the bight, form an overhand knot using the bight as the working end. Dress the knot tightly.

b. **Checkpoints.**

(1) The loop is the desired size.

(2) The ropes in the overhand knot are parallel and do not cross over each other.

(3) The knot is tightly dressed.

## 7-16. FIGURE-EIGHT LOOP KNOT

The figure-eight loop knot is a variation of the overhand loop and is easy to untie after it has held weight (Figure 7-16).

**Figure 7-16. Figure eight loop knot.**

a. **Tying the Knot.**

STEP 1. Form a bight in the rope about as large as the diameter of the desired loop.
STEP 2. With the bight as the working end, form a loop in rope (standing part).
STEP 3. Wrap the working end around the standing part 360 degrees and feed the working end through the loop. Dress the knot tightly.

b. **Checkpoints.**

(1) The loop is the desired size.
(2) The ropes in the loop are parallel and do not cross over each other.
(3) The knot is tightly dressed.

### SECTION IV. SPECIAL KNOTS

## 7-17. SINGLE BUTTERFLY KNOT

The single butterfly knot is used to form a single, fixed loop in the middle of the rope without using the ends (Figure 7-17). The butterfly can be used for the middle man in a rope party as well as in a transport tightening system. The knot can be hard to untie when heavy weight has been placed on it for extended periods.

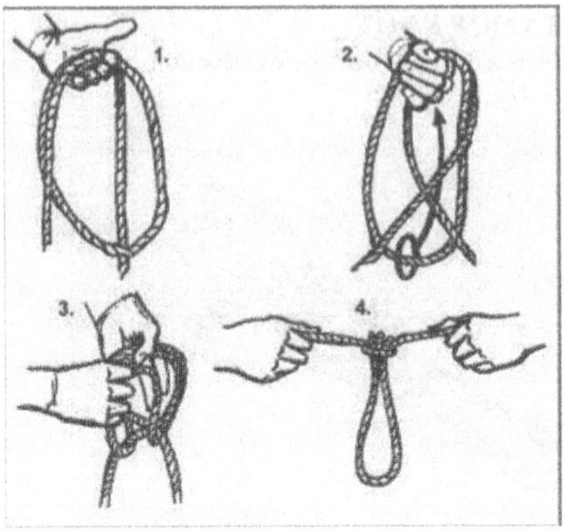


**Figure 7-17. Single butterfly knot.**

a. **Tying the Knot.**

STEP 1. Form a loop over the palm of the left hand.
STEP 2. With the right hand, cross the two long-standing ends; reach under and grasp the bottom of the loop formed in STEP 1.
STEP 3. Route the bottom of the loop over the crossed standing ends and back through the top of the loop.
STEP 4. Hold the newly formed loop and dress the knot down, removing all of the slack out of the knot.

b. **Checkpoints.**

(1) The wings of the knot are dressed down tightly and as close together as possible.
(2) The ropes between the wings are parallel with no crossovers.
(3) The loop is no larger than a helmet but large enough to accept a snaplink.
(4) All ropes in the knot are tightly dressed.

**7-18. PRUSIK KNOT**

The prusik knot is used to put a moveable rope on a fixed rope such as a prusik ascent or a tightening system. This knot can be tied as a middle, finger, or end knot.

a. **Tying the Knot.**

(1) *First method.* The prusik knot is tied at the middle or end of a rope; there are many ways to do this (Figure 7-18). The middle-of-the-rope prusik knot can be tied with a short rope to a long rope as follows:

STEP 1. Double the short rope, forming a bight, with the working ends even. Lay it over the long rope so that the closed end of the bight is 30 centimeters (12 inches) below the long rope and the remaining part of the rope (working ends) is the closest to the rappeller; spread the working end apart.

STEP 2. Reach down through the 30-centimeter (12-inch) bight. Pull up both of the working ends and lay them over the long rope. Repeat this process making sure that the working ends pass in the middle of the first two wraps. Now there are four wraps and a locking bar working across them on the long rope.

STEP 3. Dress the wraps and locking bar down to ensure they are tight and not twisted. Tie an overhand knot with both ropes to prevent the knot from slipping during periods of variable tension.

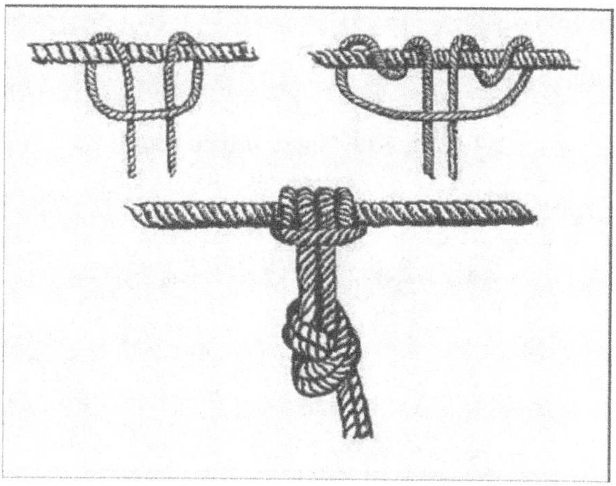

**Figure 7-18. Middle-of-the-rope prusik knot.**

(2) *Second method.* Another way of tying the prusik knot is the finger prusik (Figure 7-19).

STEP 1. Lay the rope across the thumb and index finger of the left hand.

STEP 2. Wrap the rope around the index finger twice (wrapping away from the rappeller). Wrap the rope around the thumb twice (wrapping towards the rappeller).

STEP 3. Place the thumb and index finger tip to tip. Push the wraps together so that the prusik is on the left thumb. Four wraps and a locking bar should be across them.

STEP 4. Pass the running end of the rope through the wraps of the prusik knot.

STEP 5. Dress the wraps and locking bar down to ensure they are tight and not twisted.

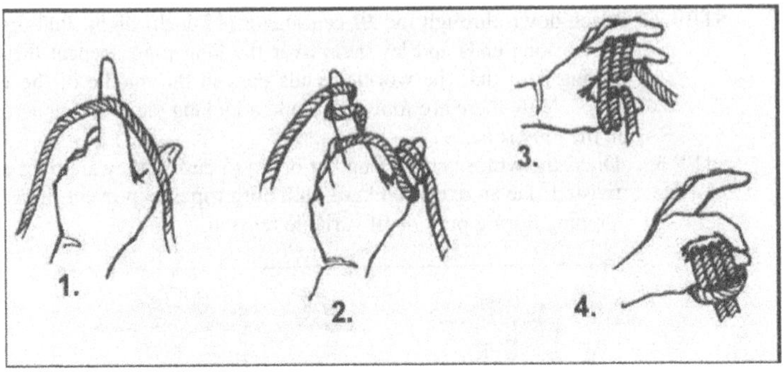


**Figure 7-19. Finger prusik knot.**

(3) *Third method.* Another way of tying a prusik knot is the end-of-the-rope knot (Figure 7-20).

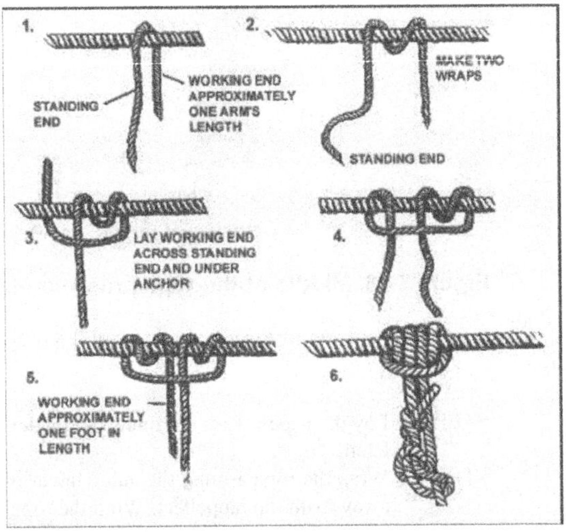


**Figure 7-20. End-of-the-rope prusik knot.**

STEP 1. Using an arm's length of rope, place it over the long rope.
STEP 2. Form a complete round turn in the rope.
STEP 3. Cross over the standing part of the short rope with the working end of the short rope.
STEP 4. Lay the working end under the long rope.
STEP 5. Form a complete round turn in the rope, working back toward the middle of the knot.

STEP 6. Four wraps and a locking bar should run across them on the long rope. Dress the wraps and locking bar down. Ensure they are tight, parallel, and not twisted.

STEP 7. Finish the knot with a bowline to ensure that the prusik knot will not slip out during periods of varying tension.

b. **Checkpoints.**

(1) Four wraps with a locking bar.

(2) The locking bar faces the rappeller.

(3) The knot is tight and dressed down with no ropes twisted or crossed.

(4) Other than a finger prusik, the knot should contain an overhand or bowline to prevent slipping.

## 7-19. BACHMAN KNOT

The bachman knot provides a means of using a makeshift mechanized ascender (Figure 7-21).

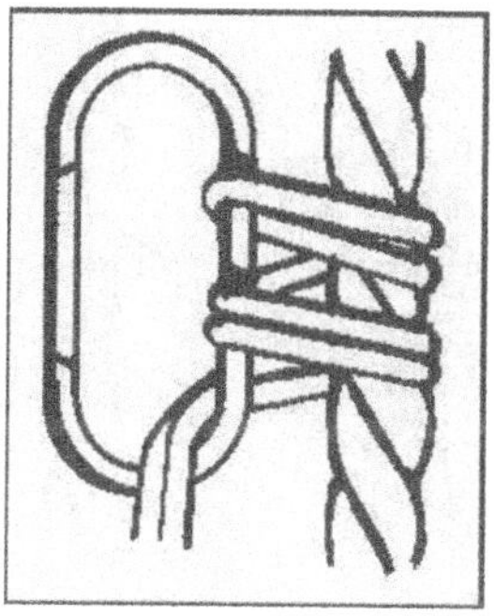

**Figure 7-21. Bachman knot.**

a. **Tying the Knot.**

STEP 1. Find the middle of a utility rope and insert it into a snaplink.

STEP 2. Place the snaplink and utility rope next to a long climbing rope.

STEP 3. With the two ropes parallel fromthe snaplink, make two or more wraps wraps around the climbing rope and through the inside portion of the snaplink.

**Note:** The rope can be tied into an etrier (stirrup) and used as a prusik-friction principal ascender.

b. **Checkpoints.**

(1) The bight of the climbing rope is at the top of the snaplink.

(2) The two ropes run parallel without twisting or crossing.

(3) Two or more wraps are made around the long climbing rope and through the inside portion of the snaplink.

## 7-20. BOWLINE-ON-A-COIL KNOT

The bowline-on-a-coil knot is an expedient tie-in used by rappellers in two- and three-man party climbs when a climbing harness is not available (Figure 7-22).

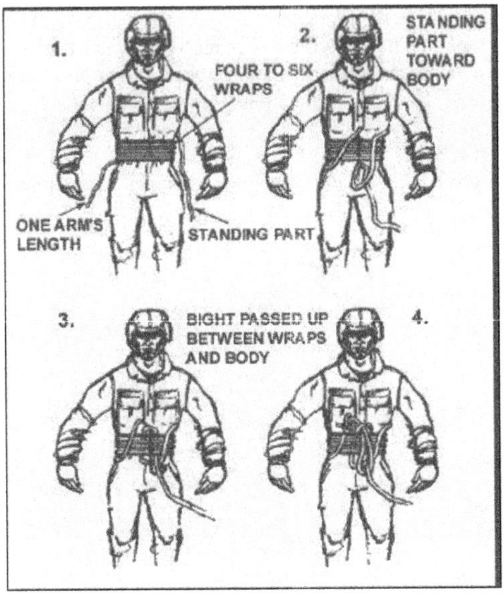


**Figure 7-22. Bowline-on-a-coil knot.**

**Note:** A safer tie in, when possible, is a swiss seat and figure eight around the shoulder, to avoid damaging the ribs if the climber falls.

a. **Tying the Knot.**

STEP 1. To tie a bowline-on-a-coil knot, start the same as tying a bowline around the waist. (The waist of the rappeller is the anchor point. The rappeller has about 40 centimeters [16 inches] of rope in the right hand. The standing end is coming from the remainder of the rope in his left hand.)

STEP 2. Draw slack from the standing end of the rope in the left hand, wrapping enough coils around the body to complete a minimum of four (but no more than six) coils.

STEP 3. With the rope in the left hand, make a loop inward with the standing end.

STEP 4. Slide the loop up between the horizontal wraps and the body, ensuring that the loop does not come uncrossed.

STEP 5. Bring a bight up through the loop. Bring the working end of the rope in the right hand through the bight and back onto itself.

***OR***

Insert the short working end to the right through the loop (just passed through the wraps). Pass the working end around the long standing end and back on itself, forming a bight.

STEP 6. Dress the knot down.

STEP 7. Safety off with an overhand knot against the knot formed on the top single coil. From 10 to 15 centimeters (4 to 6 inches) of rope should be left (standing part).

b. **Checkpoints.**

(1) There are a minimum of four (maximum of six) parallel wraps; the top and bottom ropes cross forward of the hips.

(2) The loop must be underneath all wraps.

(3) Other checkpoints are the same as the bowline.

## 7-21. THREE-LOOP BOWLINE KNOT

The three-loop bowline knot is used to form three fixed loops in the middle of a rope (Figure 7-23). It is used in a self-equalizing anchor system.

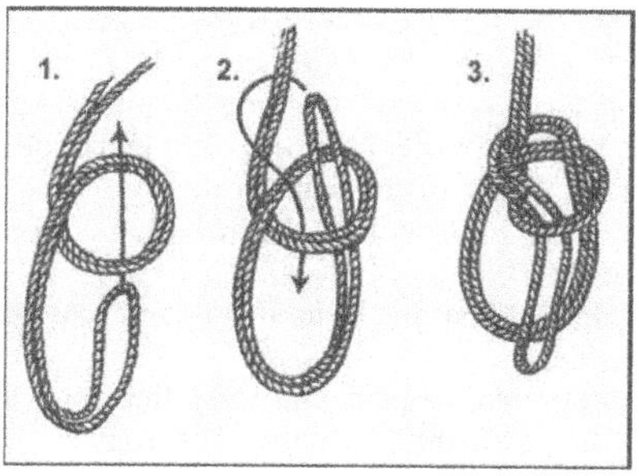


**Figure 7-23. Three-loop bowline knot.**

a. **Tying the Knot.**

STEP 1. Form about a 60 centimeter (24-inch) bight in the rope, laying it across the left hand, palm up, with the bight to the front.

STEP 2. With the right hand, form a loop with both ropes. Hold it in the left hand.

STEP 3. With the right hand, bring the long bight through the loop(formed in STEP 2) and around and around the two standing ends.

STEP 4. Follow the bight around the standing ends and back down into the loop.

STEP 5. Hold all three loops in the left hand. With the right hand, grasp both standing ropes and pull while dressing up the bowline.

b. **Checkpoints.**

(1) Checkpoints are the same as a regular bowline, except that it is tied with two ropes.

(2) An overhand knot must be tied if the third loop is not to be used as a secondary anchor.

## 7-22. FIGURE-EIGHT SLIP KNOT

The figure-eight slip knot forms an adjustable bight in a rope (Figure 7-24).

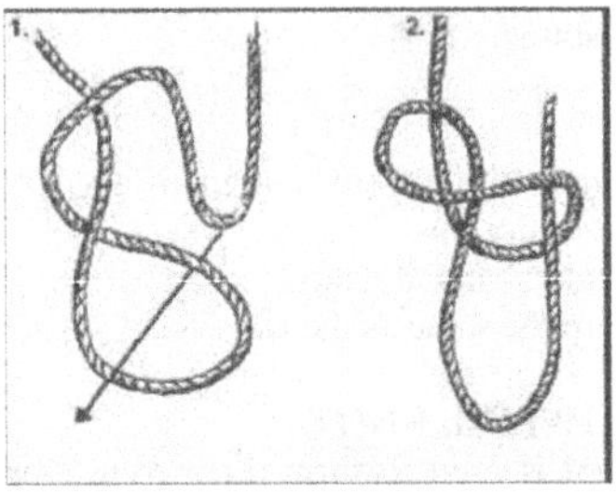


**Figure 7-24. Figure-eight slip knot.**

a. **Tying the Knot.**

STEP 1. Form a 30-centimeter (12-inch) bight in the end of the rope.

STEP 2. Hold the center of the bight in the right hand. Hold the two parallel ropes from the bight in the left hand about 30 centimeters (12 inches) up the rope.

STEP 3. With the center of the bight in the right hand, twist two complete turns clockwise.

STEP 4. Reach through the bight and grasp the long standing end of the rope. Pull another bight (from the long standing end) back through the original bight.

STEP 5. Pull down on the short working end of the rope and dress the knot down.

STEP 6. If the knot is to be used in a transport tightening system, take the working end of the rope and form a half hitch around the loop of the figure-eight knot.

b. **Checkpoints.**

(1) The knot is in the shape of a figure eight.

(2) The sliding portion of the rope is the long working end of the rope.

## Section V. SPECIAL-PURPOSE KNOTS

### 7-23. KLEIMHIEST KNOT

The kleimhiest knot provides a moveable, easily adjustable, high-tension knot capable of holding extremely heavy loads while being pulled tight (Figure 7-25). This system requires a six-man pulling team when used in a transport tightening system.

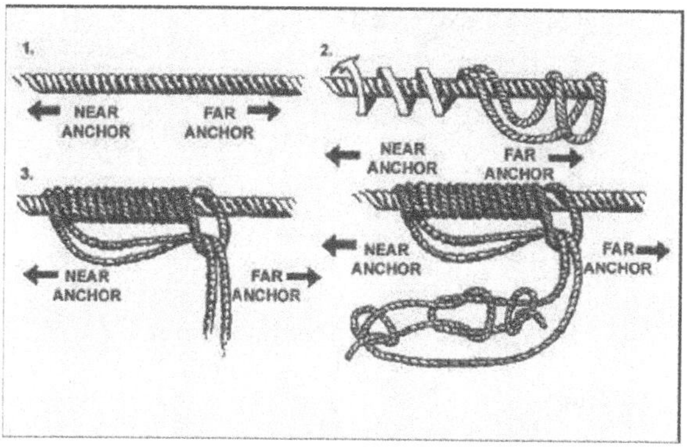


**Figure 7-25. Kleimhiest knot.**

a. **Tying the Knot.**

STEP 1. Using a utility rope, offset the ends by 30 centimeters (12inches). With the ends offset, find the center of the rope and form a bight. Lay the bight over a horizontal rope near the far side anchor.

STEP 2. Wrap the tails of the utility rope around the horizontal rope back toward the near side anchor. Wrap at least four complete turns.

STEP 3. Pass the remaining tails of the utility rope through the bight (see STEP 1).

STEP 4. Join the two ends of the tail with a joining knot (square knot with overhand safety knots).

STEP 5. Dress the knot down tightly so that all wraps are touching. Insert two snaplinks (opposing gates) into the large loop tied with the tails.

b. **Checkpoints.**

(1) The bight is closest to the far side anchor.
(2) All wraps are tight and touching.
(3) The ends of the utility rope are properly secured with the adjoining knot.

## 7-24. OVERHAND KNOT

The overhand knot (Figure 7-26) is used to make a knotted rope for a handline, to secure the ends of other knots, and to make stirrups in direct-aid climbing. This knot can also be used to temporarily whip the end of a rope. It is used with other knots.

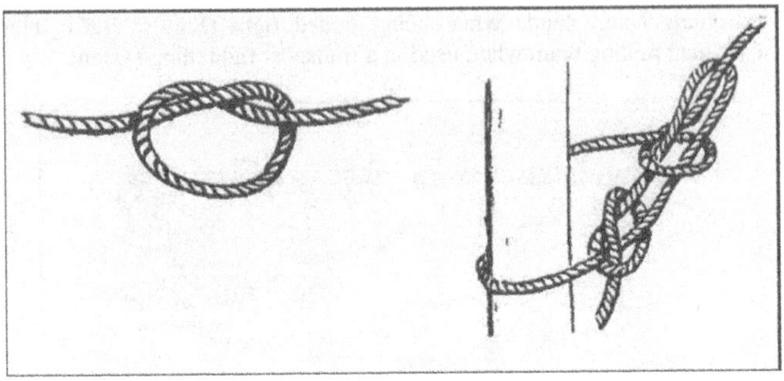

**Figure 7-26. Overhand knot.**

a. **Tying the Knot.**

STEP 1. Make a loop in the rope.
STEP 2. Pass the working end of the rope through the loop.
STEP 3. Dress down the knot.

b. **Checkpoints.**
(1) Running ends exit the loop in opposite directions.
(2) The knot is dressed down tightly.

## 7-25. FROST KNOT

The frost knot is used when working with flat webbing (Figure 7-27). It holds better than a standard overhand knot or square knot.

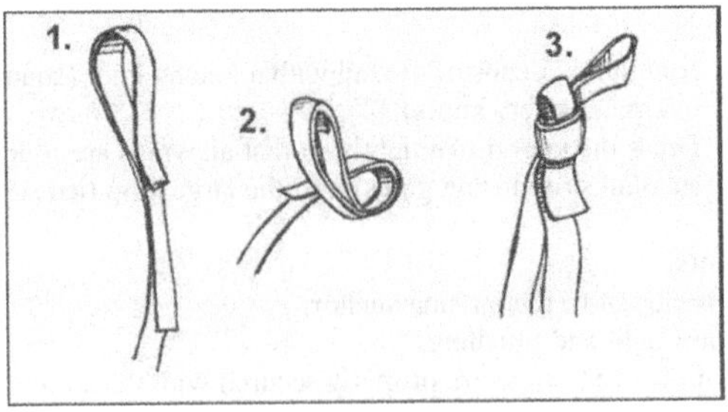


**Figure 7-27. Frost knot.**

a. **Tying the Knot.**

STEP 1. Lap one end (a bight) of webbing over the other (about 25 to 30 centimeters [10 to 12 inches]).
STEP 2. Tie an overhand knot with the newly formed triple-strand webbing; dress tightly.

b. **Checkpoints.**
(1) The tails of the webbing run in opposite directions.
(2) Three strands of webbing are formed into a tight overhand knot.
(3) A bight and tail exit the top of the overhand knot.

## 7-26. GIRTH HITCH KNOT

The girth hitch knot is used in tying a runner to a partly driven piton (Figure 7-28).

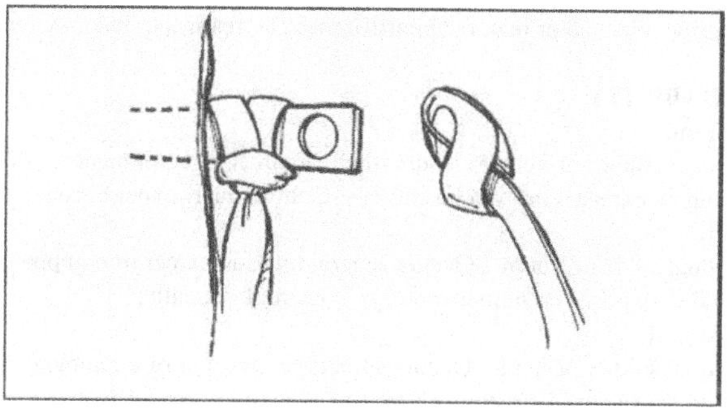

**Figure 7-28. Girth hitch knot.**

a. **Tying the Knot.**

STEP 1. Bring the standing ends back through the bight of a length of rope or webbing.
STEP 2. Dress the knot tightly.

b. **Checkpoints.**
(1) Two wraps exist with a locking bar running across the wraps.
(2) The knot is dressed tightly.

APPENDIX A

# RISK ASSESSMENT

*This appendix explains risk assessment for rappel operations. A risk assessement matrix is shown in Table A-1.*

## A-1. EFFECT

a. **Catastrophic.** Death or permanent total disability, system loss, major property damage.

b. **Critical.** Permanent partial disability, temporary total disability in excess of three months, major system damage, significant property damage.

c. **Moderate.** Minor injury, lost workday accident, compensable injury or illness, minor system damage, minor property damage.

d. **Negligible.** First aid or minor supportive medical treatment, minor system impairment.

## A-2. PROBABILITY

a. **Frequent.**

(1) Individual soldier or item—Occurs often in career or equipment service life.

(2) All soldiers exposed or item inventory—Continuously experienced.

b. **Likely.**

(1) Individual soldier or item—Occurs several times in career or equipment service life.

(2) All soldier exposed or item inventory—Occurs frequently.

c. **Occasional.**

(1) Individual soldier or item—Occurs sometime in career or equipment service life.

(2) All soldiers exposed or item inventory—Occurs sporadically, or several times in inventory service life.

d. **Seldom.**

(1) Individual soldier or item—Possible to occur in career or equipment service life.

(2) All soldiers exposed or item inventory—Remote chance of occurrence; expected to occur sometime in inventory service life.

e. **Unlikely.**

(1) Individual soldier or item—Can assume will not occur in career or equipment service life.

(2) All soldiers exposed or item inventory—Possible, but improbable; occurs only very rarely.

## A-3. RISK LEVELS

a. **Extremely High.** Loss of ability to accomplish mission.

b. **High.** Significantly degrades mission capabilities in terms of required mission standards.

c. **Medium.** Degrades mission capabilities in terms of required mission standards.

d. **Low.** Little or no impact on mission accomplishment.

| RISK ASSESSMENT | | | HAZARD PROBABILITY | | | | |
|---|---|---|---|---|---|---|---|
| | | | Frequent | Likely | Occasional | Seldom | Unlikely |
| | | | A | B | C | D | E |
| EFFECT | Catastrophic | I | ***Extremely High*** | | | | |
| | Critical | II | | | ***High*** | | |
| | Moderate | III | | ***Medium*** | | | |
| | Negligible | IV | | | ***Low*** | | |

**Table A-1. Risk assessment matrix.**

APPENDIX B

# BASIC EQUIPMENT FOR RAPPEL OPERATIONS

*The equipment listed in this appendix is the standard basic items required to conduct ground and air rappelling operations.*

## B-1. GROUND OPERATIONS

For ground operations, the following equipment is needed:

a. Nylon rope, 120 feet (NSN 4020-00-931-8793).
b. Snaplinks (NSN 8465-00-360-0228).
c. Gloves (NSN 8415-00-268-7868).
d. Rope coiling log or standard deployment bag.
e. Rappel seat.
f. Rappel rings (used as expedient anchor points).

## B-2. AIR OPERATIONS

For air operations, the following equipment is needed:

a. All items in paragraph B-1.
b. Donut ring (UH-1H operations only) (locally produced TSC item).
c. Safety floating ring.
d. Two 6-foot safety ropes (with snaplinks).
e. Eight 120-foot nylon ropes.
f. Monkey harness for static rappel master.
g. Sixteen snaplinks (UH-60 operations).
h. Twenty-two snaplinks (UH-1H operations).
i. FM communications.
j. Aircraft communications headset.

# GLOSSARY

## ACRONYMS AND ABBREVIATIONS

| | |
|---|---|
| AFRM | assistant fast-rope master |
| AGL | above ground level |
| AMC | air mission commander |
| AR | Army regulation |
| ATP | aircrew training program |
| attn | attention |
| BDU | battle dress uniform |
| CG | commanding general |
| CGU | modified helicopter safety strap |
| chemlight | chemical light |
| cm | centimeter |
| DA | Department of the Army |
| D-bag | deployment bag |
| F | Fahrenheit |
| FM | frequency modulated; field manual |
| FRIES | fast-rope insertion/extraction system |
| FRM | fast-rope master |
| GSO | ground safety officer |
| HQ | headquarters |
| IAW | in accordance with |
| ICS | interagency communications system |
| intercomm | intercommunication |
| LBE | load-bearing equipment |
| LZ | landing zone |
| MEDEVAC | medical evacuation |
| mm | millimeter |
| mph | miles per hour |
| MSO | mountaineering safety officer |
| NCO | noncommissioned officer |
| NCOIC | noncommissioned officer in charge |
| NSN | national stock number |
| NVG | night vision goggles |

| | |
|---|---|
| NWTC | Northern Warfare Training Center |
| OIC | officer in charge |
| PIC | pilot in command |
| POI | program of instruction |
| POL | petroleum, oils, and lubricants |
| RATELO | radiotelephone operator |
| RSO | rappel safety officer |
| SFC | sergeant first class |
| SOAR | Special Operations Aviation Regiment |
| SOP | standing operating procedures |
| SPIES | special patrol infiltration/exfiltration system |
| STABO | a system for extracting personnel by helicopter (the combined first letters of the surnames of the five persons who designed the system) |
| SSO | STABO system operator |
| TAPO | Technology Applications Program Office |
| TC | training circular |
| TM | technical manual |
| TRADOC | Training Doctrine Command |
| TSC | training support center |
| UIAA | Union of International Alpine Association |
| UH | utility helicopter |
| US | United States |
| USAF | United States Air Force |
| USAIS | United States Army Infantry School |
| USASOC | United States Army Special Operations Command |
| VIP | very important person |

# REFERENCES

## DOCUMENTS NEEDED

These documents must be available to the intended users of this publication.


| | |
|---|---|
| DA Form 2028 | Recommended Changes to Publications and Blank Forms. 1 February 1974. |
| DA Form 5752-R | Rope History and Usage. May 1989. |
| FM 57-38* | Pathfinder Operations. 9 April 1993. |
| FM 7-93* | Long Range Surveillance Unit Operations. 3 October 1995. |
| TC 90-6-1* | Military Mountaineering. 26 April 1989. |


## READINGS RECOMMENDED

These sources contain relevant supplemental information.


| | |
|---|---|
| AR 95-3 | Aviation: General Provisions, Training, and Standardization, A Resource Management. 27 September 1990. |
| FM 8-35 | Evacuation of the Sick and Wounded. 22 December 1983 |
| FM 57-220 | Basic Parachuting Techniques and Training. 1 June 1990. |
| TM 10-1670-262-12 | Operator's and Organizational Maintenance Manual (Including Repair Parts and Special Tools List) for Personnel STABO Extraction System and Cargo Personnel Lowering Anchor Device. 7 April 1988. |
| USASOC Reg 350-2 | Training Airborne Operations. 1 June 1995. |
| USASOC Reg 350-6 | Training Special Operations Forces Infiltration/Exfiltration Techniques. 12 May 1995. |
| | The Air Assault School Handbook. |


*This source was also used to develop this publication.

# INDEX

**TC 21-24**
**10 SEPTEMBER 1997**

By Order of the Secretary of the Army:

DENNIS J. REIMER
*General, United States Army*
*Chief of Staff*

Official:

Joel B. Hudson

JOEL B. HUDSON
*Administrative Assistant to the*
*Secretary of the Army*
03807

**DISTRIBUTION:**

*Active Army, Army National Guard, and U.S. Army Reserve*: To be distributed in accordance with the initial distribution number 115059, requirements for TC 21-24.